Regional Geology Reviews

Series Editors

Roland Oberhänsli, Institute of Earth and Environmental Sciences, University of Potsdam, Potsdam, Brandenburg, Germany

Francois Roure, Direction Geologie-Geochimie-Geophysique, Institut Francais du Petrole, Rueil Malmaison Cedex, France

Dirk Frei, Department of Earth Sciences, University of the Western Cape, Bellville, South Africa

The Geology of series seeks to systematically present the geology of each country, region and continent on Earth. Each book aims to provide the reader with the state-of-the-art understanding of a regions geology with subsequent updated editions appearing every 5 to 10 years and accompanied by an online "must read" reference list, which will be updated each year. The books should form the basis of understanding that students, researchers and professional geologists require when beginning investigations in a particular area and are encouraged to include as much information as possible such as: Maps and Cross-sections, Past and current models, Geophysical investigations, Geochemical Datasets, Economic Geology, Geotourism (Geoparks etc), Geo-environmental/ecological concerns, etc.

Abd el-aziz Khairy Abd el-aal •
Jasem Mohammed Al-Awadhi •
Ali Al-Dousari

Editors

The Geology of Kuwait

Editors
Abd el-aziz Khairy Abd el-aal
Environment and Life Sciences Research Center
Kuwait Institute for Scientific Research
Kuwait City, Kuwait

Jasem Mohammed Al-Awadhi
Department of Earth and Environmental Science
Kuwait University
Kuwait City, Kuwait

Ali Al-Dousari
Environment and Life Sciences Research Center
Kuwait Institute for Scientific Research
Kuwait City, Kuwait

ISSN 2364-6438   ISSN 2364-6446   (electronic)
Regional Geology Reviews
ISBN 978-3-031-16726-3   ISBN 978-3-031-16727-0   (eBook)
https://doi.org/10.1007/978-3-031-16727-0

This Springer imprint is published by the registered company Springer Nature Switzerland AG
The registered company address is: Gewerbestrasse 11, 6330 Cham, Switzerland

Preface

Kuwait's geographical and regional location in the Arab region and its natural resources, which include oil and natural gas, makes the need for geological and geophysical manuscripts very significant and of great value to the State of Kuwait. From a purely geological point of view, the State of Kuwait needs more of these geological and geophysical studies because of their great impact as Kuwait is a virgin country with promising natural resources, not only oil and gas but also groundwater, mineral resources, and various types of sand. Definitely, the book not only sheds light on the geological structure, natural resources, and underground reservoirs but also sheds light on the natural risks facing Kuwait, such as the risks of earthquakes, floods, and sand dune movements. This book is considered as one of the most important resources for those working in the educational field, professors, and students as well as those working in the field of environmental protection research, and also a very important source for workers in the field of hazard mitigation and petroleum exploration. This book includes a set of chapters covering all aspects of geosciences. Generally, this book is not only important for those interested in geological sciences in Kuwait but also for those interested in geological sciences also in the region, such as Iran, Saudi Arabia, and the region of the Arab Gulf states. One of the most important reasons that led us to publish this book is the lack of a publication of books, researches, and references in the geosciences of Kuwait and its surrounding areas. This book fundamentally includes basic information in geological sciences in addition to practical applications, researches, examples, and cases from the State of Kuwait, which will be complementary to all that were published in the Kingdom of Saudi Arabia, Iraq, and The Arabian Peninsula previously. The authors of this book are among the great scholars in the field of geosciences who are known for their impacts and international publishing in this field.

The book provides up-to-date research outcomes on the geological sciences of Kuwait. Furthermore, the book updates the knowledge on the tectonic, structure, environmental topics, and natural resources of Kuwait. This book is the first of its kind in the State of Kuwait, which deals with various geological topics. The book contains 10 chapters including surface geology of Kuwait, subsurface stratigraphy of Kuwait, sand dunes, marine geology of Kuwait, structure of Kuwait, petroleum geology of Kuwait, seismicity of Kuwait, Geo- and Environmental Hazard in Kuwait, groundwater in Kuwait, and application of remote sensing science in Kuwait.

Kuwait City, Kuwait

Abd el-aziz Khairy Abd el-aal
Jasem Mohammed Al-Awadhi
Ali Al-Dousari

Acknowledgments The authors deeply thank the Kuwait Institute for Scientific Research (KISR) for continuous technical support and provision of all material and logistical capabilities during the preparation of this work. All thanks and gratitude to **Dr. Abdullah Al-Enezi** for his effective contribution to the production of this book. This work was funded and published by the Kuwait Institute for Scientific Research.

Contents

About the Editors

Prof. Abd el-aziz Khairy Abd el-aal is a geologist and geophysicist professor. He is now working at Kuwait Institute for Scientific Research, Kuwait, since 2019. In 1999 he joined the National Research Institute of Astronomy and Geophysics (NRIAG) Helwan, Egypt. Formerly, he was the director of the Egyptian National Seismic Network (ENSN). He obtained a Bachelor's degree in Geological and Geophysical Sciences with a very good grade with honors from the Faculty of Science, Sohag University, Egypt, in 1992. In 2003 he received his M.Sc. in applied geophysics. In 2006, he was awarded Ph.D. degree in applied geophysics. His research interests are mainly focused on engineering seismology, tectonics, hazard mitigation, seismic exploration, and environmental sciences. He was working in a shallow and deep seismic group in NRIAG.

He conducted many shallow seismic profiles and site investigations in different parts of Egypt. The subject of his Ph.D. is certainly in seismic risk reduction. He has worked in many field-related sub-disciplines of Earth Sciences including engineering geology, geotechnical engineering, tectonics, crustal deformation, installing seismic networks, and seismic hazard mitigation. He has a good experience in teaching seismic courses in many universities in Egypt. He has participated as first and co-author in more than 85 published papers and book chapters in international indexed and refereed scientific Journals and conferences. He is working as associate editor of many international scientific journals such as the Arabian Journal of Geosciences (AJGS), Journal of Petroleum Exploration and Production Technology (JPEPT) etc. He is now acting as a referee in most high-standing international journals publish on seismology and hazard mitigation subjects. Professor Abd el-aziz Khairy Abd el-aal is a member of many local, regional, and international scientific committees. He has received several awards from the Egyptian Academy of Scientific Research, Egypt, in applied geophysical and Geological Sciences.

Prof. Dr. Jasem Mohammed Al-Awadhi is now a Professor at the Department of Earth and Environmental Sciences, Faculty of Science Kuwait University. He got his B.Sc. in Civil Engineering from Kuwait University, Kuwait, in January 1986. His Master in Civil Engineering from Colorado State University, U.S.A., January 1992. The Ph.D. in Civil Engineering: from University of Aberdeen, May 1996. The Professional interests are Environmental Science—Desert Science—Sediment Transport, Erosion and Deposition—Sand Control Measures and Modeling—Air Pollution Modeling and Seismology—Environmental Impact Assessment Studies. He has more than 60 published papers in refereed scientific Journals and conferences in the English language. He also has published more than 15 books, book chapters, reports, and scientific papers in the Arabic language. He was a supervisor for many Master and Ph.D. thesis. He was awarded many scientific prizes and awards. He has also led many major projects and leads. He has conducted many workshops, training courses, and conferences.

He is a member of many local and international organizations like Kuwait Environmental Protection Society, Kuwait. Committee member for rehabilitation of gravel quarry areas, Environment Public Authority, Kuwait. Member, National Committee for Compacting Desertification, Environment Public Authority, Kuwait. Member, Scottish International Resource Project, Scotland. Member, Middle East Seismology Fume, USA. Founder, GCC Dry Land Studies Network.

He touches many academic courses: Earth and Planets, Basic Geology, Land Degradation, Desertification, Pollution and Desert, Process of Aeolian Sediment Transport, Environmental Geology, Desert Soil, Rangeland Management, Sand Movement and Dust fallout, and Special Studies.

Dr. Ali Al-Dousari is one of the leading scientists in the area of land degradation (desertification) and rehabilitation methods in the Arab region, especially in the Arabian Peninsula and has significant experience in addressing land degradation linked to socioeconomic effects. He has won academic prizes locally and regionally (The Wildlife Award from the Arabian Gulf Cooperation Council (GCC)); the Abdul Hameed Shoman Award for Best Arab Researchers; the State Incentive Award; the Best Arab Book Award in the field of the environment; and the best editor in AJGS Springer Award) and acts as the Kuwait representation active member of the Committee of Science and Technology (CST) on the United Nations Committee for Compacting Desertification (UNCCD) (2016–2019), as well as being a member of many local, regional, and global scientific committees. Dr. Ali is on the editorial boards of many local and international journals (Springer). He is the author of more than 80 books, chapters, and published papers in refereed scientific journals and conferences.

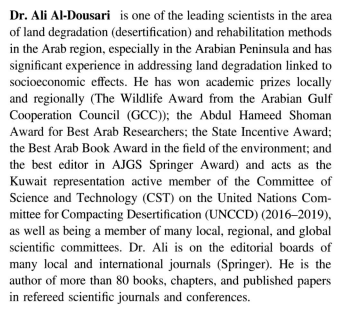

Surface Geology of Kuwait

Yaqoub AlRefaei, Ali Najem, Aimen Amer, and Faisal Al-Qattan

Abstract

This chapter represents a comprehensive review of Kuwait's surface geology and stratigraphy from previous works accomplished by numerous geoscience researchers in the past decades. The surface of Kuwait is characterized by nearly flat topography, featureless to gently undulating, apart from a few tens of meters of escarpments in the north and south, and flat low to moderately elevated hills and ridges. It predominantly consists of siliciclastic sediments and sedimentary rock units ranging in age from Middle Eocene to Holocene. The main stratigraphic exposed successions are located in Jal Az-Zor escarpment, Al-Subyiah (Bahrah) area, Ahmadi Quarry, the Khiran Ridges, and the Enjefa Beach. The oldest exposed rock units are represented by the Middle Eocene Dammam Formation, which is exposed at the Ahmadi Quarry, whereas the youngest recent deposits cover most of Kuwait's surficial area and lie on top of the Kuwait Group's deposits. This chapter will illustrate the geology and stratigraphy of Kuwait's surface sediments and sedimentary rock strata. Recommendations and future insights were also documented as part of the way forward to improve the presently available work for the surface geology of Kuwait.

Y. AlRefaei (✉)
Earth and Environmental Sciences Department, Kuwait University, P.O. Box 5969 13060 Safat, Kuwait
e-mail: yaqoub.alsayed@ku.edu.kw

A. Najem · A. Amer
Schlumberger Oilfield Eastern LTD., Block No.6, Building 193 and 194, P.O.Box 9056 61001 Eastern Ahmadi, Kuwait
e-mail: AGadalla@slb.com

A. Amer
e-mail: aamer@slb.com

F. Al-Qattan
Kuwait Oil Company, Exploration Group-Exploration Operations, P.O Box 9758 61008 Ahmadi, Kuwait
e-mail: FQattan@kockw.com

1.1 Introduction

The surface geology of Kuwait is dominated by the presence of flat to near-flat, featureless to gently undulating, sands and gravels covered deserts with few low to moderately elevated hills and ridges (Al-Awadi et al., 1997; Khalaf et al., 1984). These elevated hills and ridges are in both northern (e.g., Jal Az-Zor Escarpment and Al-Subyiah Ridges) and southern (e.g., the Ahmadi Quarry and the Khiran Ridges) parts of Kuwait (Al-Awadi et al., 1997; Khalaf et al., 1984). Kuwait's surface and near-surface deposits are dominated by siliciclastic sediments and sedimentary rock units, with some carbonate deposits in the southern regions. The age of these deposits ranges from Middle Paleogene (Middle Eocene) to Quaternary (recent/Holocene) (Milton, 1967). The oldest exposed rock units are represented by the Middle Eocene Dammam Formation, which is exposed at the Ahmadi Quarry, whereas the youngest recent deposits cover most of Kuwait's surficial area and lie on top of the Kuwait Group's deposits (Milton, 1967).

This chapter will illustrate the geology and stratigraphy of Kuwait's surface sediments and sedimentary rock strata. The description of the sedimentary rock units within outcrops of Jal Az-Zor escarpment, Al-Subyiah (Bahrah) area, Ahmadi Quarry, the Khiran Ridges, the Enjefa Beach, along with the surface sediment cover of Kuwait will be demonstrated in the subsequent sections (Fig. 1.1).

1.1.1 Regional Geological Setting and Surface Stratigraphy of Kuwait

Kuwait is located in the northeastern part of the Arabian Plate, particularly on the eastern flank of the Arabian Platform. The platform consists of a thick sequence of siliciclastic and carbonate sedimentary rock strata deposited on top of the Arabian Shield basement rocks. These rock units within the platform are gently dipping 1–2 degrees toward

© The Author(s) 2023
A. el-aziz K. Abd el-aal et al. (eds.), *The Geology of Kuwait*, Regional Geology Reviews,
https://doi.org/10.1007/978-3-031-16727-0_1

Fig. 1.1 Surface geologic map of Kuwait showing the age and location of outcrops and sediment cover (Modified after Khalaf et al., 1984 and Amer & Al-Hajeri, 2020)

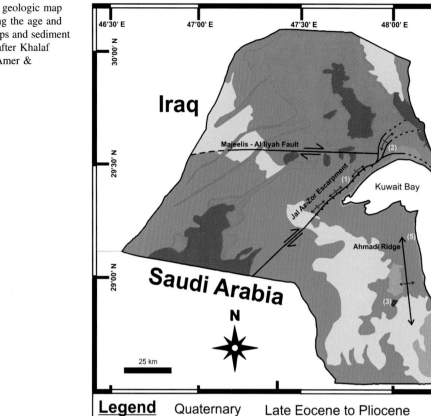

the northeast and thicken further away from the Arabian Shield towards the Zagros Mountains to the East (Owen & Nasr, 1958). The exposed units of this sequence in Kuwait are represented by the carbonate Middle Eocene Dammam Formation and the clastic Oligo-Pleistocene to the recent Kuwait Group (Owen & Nasr, 1958; Amer & Al-Hajeri, 2019). The Dammam Formation is mainly composed of carbonates with evaporites, shale deposits, and minor chert (Al-Awadi et al., 1997). The uppermost portion of the formation is exposed at the Ahmadi Quarry. The Kuwait Group overlies the Dammam Formation unconformably. This Group is dominated by clastic deposits that range in age from the Early Oligocene to Pleistocene (Amer & Al-Hajeri, 2019). The Kuwait Group is divided into three major formations—from oldest to youngest—the Ghar Formation, the Lower Fars Formation, and the Dibdibba Formation (Owen & Nasr, 1958). The Ghar Formation is believed to be exposed et al.-Subyiah (Bahrah) area, whereas the Lower

Fars and Dibdibba Formations are exposed at Jal Az-Zor and as surface sediment cover.

One of the earliest geological reports about Kuwait was by Gregory (1929), where he briefly discussed the Fars Group and the Euphrates limestone in neighboring Iraq. At this time, as the Arabian Gulf and the Persian region became an active area for oil exploration, several attempts by surface geologists were increasingly reported to be the first to discover oil in the area. Following Gregory, several research and studies were conducted by Cox and Rhodes (1935), Own and Nasr (1958), Milton (1967), Fuchs et al. (1968), and Salman (1979). Another phase of research and studies was started in the 1980s by Al-Sarawi (1982) and Khalaf et al. (1984). From the 1980s, we can notice a gap in the research related to Jal Az-Zor in particular, while regional studies emerged such as Ziegler (2001) and Sharland et al. (2004) for the whole Arabian Gulf regional geology. These studies were important in the context of understanding the tectonostratigraphic evolution. During the last 5 years, several attempts by geologists from oil companies to study the escarpment resulted in a few important publications such as Tanoli et al. (2015), Amer et al. (2017, 2019a), Benham et al. (2018), Amer and Al-Hajeri (2019), and Tanoli et al. (2019).

Gregory (1929) mentioned that within Kuwait, the presence of horizontal beds that have been described as the Kuwait series with little to be known about the age of such beds. It was concluded that nothing was known about the age of these rock units. The first stratigraphic work conducted over the area was by Owen and Nasr (1958). In this work, the authors introduced for the first time the term "Kuwait Group," and they divided the group into Ghar, Lower Fars, and Dibdibba formations. Owen and Nasr (1958) described the Ghar Formation as sands associated with subordinate gravels and occasional clays. The Lower Fars Formation was described as an evaporitic sequence characterized by anhydrite, gypsum, marls, and shallow-water limestones. The Dibdibba Formation represents the third Kuwait Group unit, and it is composed of sands, gravels, and subordinate marls. The age range given to the Kuwait Group by these authors was Miocene to Pleistocene (Owen & Nasr, 1958). The name "Dibdibba beds" was first used by Macfadyen (1938) in Iraq. The terms of Kuwait Group and it is subunits are informally used by Kuwait Oil Company in the subsurface. Tanoli et al. (2015) described the Ghar and Lower Fars formations of the Kuwait Group to have identical lithologies.

The Kuwait Group formations of Ghar, Lower Fars, and Dibdibba, as later adopted by Bergstrom and Aten (1965), focused mainly on groundwater, but no attempt to date the formations were made. Khalaf and El-Sayed (1989) studied the western part of the Jal Az-Zor escarpment with two cores from boreholes drilled, one at the top and the other at its foot focused on the calcrete development and types within the Dibdibba units. Salman (1979), in his Kuwait University Master thesis, suggested significant changes to the Kuwait Group nomenclature: it was the first time the Oligocene was introduced to Kuwait, where Ghar Formation was deposited. The Lower Fars Formation was divided into Mutla and Jal Az-Zor, extending from the Miocene to Pliocene. The Dibdibba Formation was of Pleistocene in age. This nomenclature was not popular, except for the partial utilization by Al-Awadi et al. (1997).

Fuchs et al. (1968) identified the boundary between Lower Fars and Dibdibba formations based on heavy minerals content and a lithostratigraphic correlation between the Lower Fars Formation in Kuwait with the Fars Group in Iran. Fuchs et al. (1968) described a 1–2 m thick fossiliferous layer at the base of Jal Az-Zor escarpment that contains *Ostrea Lamimaginula* and *Clausinella.* They suggested

that this unit belongs to the Late Miocene based on this biostratigraphic contents.

A radiometric isotope age dating work carried out by Amer and Al-Hajeri (2019) of oyster shells from the same bed indicated that they are Burdigalian in age (Early Miocene). Moreover, recent research by Amer et al. (2019a) found that the heavy minerals in this unit are not sufficient to differentiate between Lower Fars and Dibdibba formations. The chemostratigraphic sedimentary provenance study work done by Amer et al. (2017) compared sediments of Jal Az-Zor to the Arabian Shield in Saudi Arabia and concluded that the source of sediments was identified to come from the ophiolite basaltic and andesitic outcrops along an area known as Darb Zubaydah in the Arabian Shield.

In 1996, a newly adopted stratigraphic chart was published by Mukhopadhyay et al. (1996), suggesting an overall Miocene to Pleistocene age for the Kuwait Group. However, the Lower Fars Formation was restricted to the Middle Miocene, and the Dibdibba Formation was extended Late Miocene. The area's stratigraphy was yet again on an appointment to change with the work of Al-Ameri et al. (2011), Duane et al. (2015). following the work of (Sharland et al., 2004). These two authors suggested no erosion at the top and base of the Lower Fars Formation, and they presented a stratigraphic scenario similar to Mukhopadhyay et al. (1996). The only difference was slight variations in age suggested for Dibdibba Formation with no apparent justification.

Prior to this time, all attempts have been primarily based on lithostratigraphic correlation to neighboring Saudi Arabia, Iraq, and Iran. Tanoli et al. (2019) attempted to refine the age and nomenclature of the Lower Fars Formation based on subsurface core analyses. They suggested replacing this formation name with Jal Az-Zor Formation, and based on biostratigraphic taxonomic groups, the age was determined to be late Burdigalian–early Langhian (Early to Middle Miocene). However, though the Lower Fars Formation naming has been suggested to change to Jal Az-Zor Formation, neither correlation nor analysis was performed on the Jal Az-Zor exposure, and all the analysis was mainly focused on five wells drilled in the subsurface north Kuwait. Furthermore, some of the fauna presented had a long taxon ranging, making the ages define uncertain, and the upper and lower boundaries of the proposed Jal Az-Zor Formation were left undefined, adding additional perplexity to an already equivocal nomenclature.

Amer and Al-Hajeri (2019) proposed utilizing numerical radiometric dating of strontium isotope ratio ($^{87}Sr/^{86}Sr$) over subsurface and surface exposures of Kuwait. The researchers collected samples along the Jal Az-Zor escarpment near Mutla and Sabriyah roadcut, and from the subsurface, by sampling core data from a well drilled in north Kuwait. They used the global strontium isotope seawater curve of Veizer et al. (1999) to determine the exact age of the Kuwait Group. They found that the age ranged from Priabonian to Burdigalian (Late Eocene to Early Miocene). These workers did not exclude the middle and late Miocene from the Kuwait Group, as the upper section of the drilled well was not possible to core, and the upper section of the Jal Az-Zor escarpment was not fully covered in their novel study. This work suggests that the Oligocene in Kuwait is not eroded or non-deposited as previously thought (Mukhopadhyay et al., 1996; Sharland et al., 2001, 2004; Ziegler, 2001; Al-Ameri et al., 2011; Duane et al., 2015; Tanoli et al., 2019). Consequently, the tectonostratigraphic evolution of the area will need to be reassessed to accommodate such findings (Fig. 1.2).

Furthermore, Amer and Al-Hajeri (2019) presented a 2D seismic section acquired along the drilled well and the Jal Az-Zor escarpment that was interpreted after a well-to-seismic tie. This work included the incorporation of the outcrop exposed rocks at the Jal Az-Zor, and the analysis showed a unique perspective on the relationship between surface and the subsurface age equivalent units (Fig. 1.3). Based on this work, Amer and Al-Hajeri (2019) suggested renaming the undifferentiated Kuwait Group of Owen and Nasr (1958) to the Kuwait Formation, eliminating the equivocal tripartite subdivision believed to be inadequate to Kuwait.

1.2 Surface Recent (Holocene) Sediment Cover of Kuwait

These surficial deposits of Kuwait have two major depositional environment settings: desert-related depositions or coast-related depositions (Khalaf et al., 1984). Khalaf et al. (1984) have divided Kuwait's deserts into four physiographic regions: Al-Dibdibba gravel (Northern Kuwait), the sand flat area (Sothern Kuwait), the coastal flat (southern Kuwait), and the coastal hills at Jal Az-Zor at North and the Ahmadi Ridge in the South of Kuwait.

In the North of Kuwait, the surface sediments are mainly siliciclastics that vary in size from gravel size to very coarse to pebbly sand size (Khalaf et al., 1984). These sediments are derived and transported from the Hejaz mountains in the western parts to the eastern parts of the Arabian Plate by flooding events at Wadi Al-Batin (Khalaf et al., 1982). In the South of Kuwait, sediments are mainly sand to mud-sized and rich in carbonate deposits. The finer sizes in the southern

Age Dating Method	Lithostratigraphic Correlation and Radiometric Isotopes	Lithostratigraphic Correlation and Biostratigraphy	Lithostratigraphic Correlation	Lithostratigraphic Correlation	Lithostratigraphic Correlation	Lithostratigraphic Correlation	Lithostratigraphic Correlation	Lithostratigraphic Correlation	
Suggested Formation Name									
Geological Time	Amer and Al-Hajeri (2019)	Tanoli et al.(2019)	Duane et al.(2015)	Al-Ameri et al.(2011)	Sharland et al.(2004)	Mukhopadhyay et al.(1996)	Salman(1979)	Owen & Nasr (1958) Kuwiat	Owen & Nasr (1958) Iraq
Holocene	Unnamed Recent Deposits	Dibdibba		Dibdibba	Dibdibba		Dibdibba	Undifferentiated Kuwait Group	Dibdibba
Pleistocene	Unnamed Alluvium and Shoreface Deposits					Dibdibba			
Pliocene	?		Dibdibba						
Miocene Late	Kuwait Formation			Lower Fars	Lower Fars	Lower Fars	Lower Fars / Jal-Az-Zor (Member A / Member B)		Lower Fars
Miocene Middle		Jal Az-Zor	Lower Fars						
Miocene Early		Ghar	Ghar	Ghar	Ghar	Ghar	Ghar / Mutia		Ghar
Oligocene Late								Ghar	
Oligocene Middle									
Oligocene Early									
Eocene Late		?		Dammam	Dammam			Dammam	Dammam
Eocene Middle	Dammam	Dammam	?			Dammam	Dammam		Rus
Eocene Early	Rus	Rus / Radhuma		Rus	Rus / Umm Er Radhuma	Rus			Radhuma

Fig. 1.2 Stratigraphic chart compilation over Paleogene-Neogene section in Kuwait. The color scheme is as follows; (light, dark yellow, orange, light gold) clastic dominated; (purple) evaporites; (light and dark blue) carbonate. Red diamond symbols indicate zones of age control by the respective authors

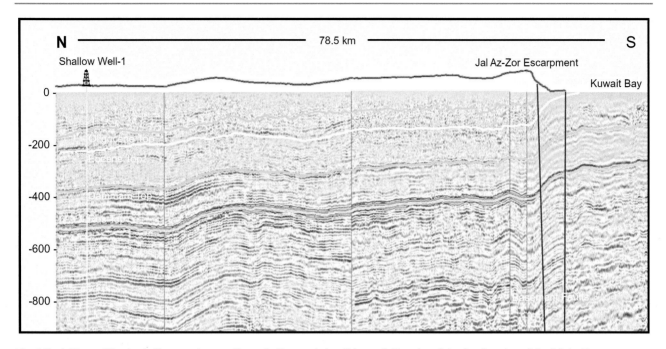

Fig. 1.3 Arbitrary 2D seismic line crossing a radiometrically sampled well in north Kuwait and the dated section of the Jal Az-Zor escarpment. The interpretation shows that the Oligocene is outcropping at the foot of the escarpment. Displayed scale in (ms). (After Amer & Al-Hajeri, 2019)

parts of Kuwait are believed to result from the flood's energy decreasing southward (AlShuaibi & Khalaf, 2011). The sediments at the southern coasts of Kuwait are also characterized by modern carbonate deposits of oolite, dolomite, and microbial mats (e.g., Al-Khiran and Azzor areas) (Gischler & Lomando, 2005). Khalaf et al. (1984) subdivided these surface deposits into six classes based on their characteristics and mode of occurrence, which are aeolian and sand deposits, desert plain deposits, residual gravel deposits, playa deposits, coastal deposits, and slope and alluvial fan deposits. The aeolian and sand deposits are sand sheets, sand dunes and drifts, and wadi fill. They cover more than 50% of Kuwait's surficial area (Al-Sulaimi & Mukhopadhyay, 2000; Khalaf et al., 1984). The desert plain deposits are Holocene aged of various grain sizes laid down in shallow and broad depressions by running rainwater. They can be found in scattered fields in the northern and western areas of Kuwait. The residual gravel deposits consist of various grain sizes, from gravel to clay. They were formed due to wind deflation of the Dibdibba deposits where fine-grained sediments were removed, leaving behind the larger and heavier grains as a desert pavement layer. These deposits are mainly found in the northern areas of Kuwait.

The playa deposits are mainly associated with drainage systems in the northwestern deserts of Kuwait (Al-Sulaimi & Mukhopadhyay, 2000). They can be found in low areas and depressions seasonally filled with rainwater forming ephemeral lakes. Sediments that fill those playas are believed

to be transported and deposited by running water from the drainage systems within the area. The coastal deposits are found along Kuwait's shoreline, and they are represented by the coastal dunes, coastal plains, beaches, tidal flats, and sabkha deposits. They mainly consist of siliciclastics, carbonates, and evaporitic deposits. The slope and alluvial fan deposits can be found at the foothill of the relatively steep slope hills and escarpments of Jal Az-Zor, Al-Subyiah, Al-Ahmadi Ridge, and Wadi Al-Batin (Khalaf et al., 1984). They consist of a mixture of coarse- and fine-grained sediments with an abundance of calcareous boulder-sized blocks that fell down-cliff due to gravitational force.

1.3   Jal Az-Zor Escarpments

The Jal Az-Zor escarpment represents the most pronounced and studied exposure over the surface geology of Kuwait (Fig. 1.4). The stratigraphic successions of the Jal Az-Zor escarpment are predominantly comprised of siliciclastic deposits. It extends for almost 60 km with a cliff-like face of about 36 m from the escarpment base (Amer et al., 2019a) and reaches up to 135 m from the mean sea level (Al-Sarawi, 1982). It exhibits a predominant NE-SW trend as a result of deep-seated basement faults (Own & Naser, 1958, Fuchs et al., 1968; Bou-Rabee & Kleinkopf, 1994; Al-Anzi, 1995; Singh et al., 2011; Amer et al., 2017; Amer & Al-Hajeri, 2020). The importance of this fault is discussed in the structural geology chapter.

5 m

5 m

Fig. 1.4 A section of the Jal Az-Zor Escarpment showing a few exposed lithofacies

1.3.1 Stratigraphic Successions of Jal Az-Zor Escarpment

The stratigraphic successions of Jal Az-Zor escarpment were recently studied intensively by researchers in the last five years, from sedimentology to stratigraphy up to conceptual depositional models. Amer et al. (2019a) introduced a detailed sedimentological description, analysis, facies, facies association, and conceptual depositional setting of the Jal Az-Zor stratigraphic successions. The study used a descriptive approach to the main units, both laterally and vertically, as several traverses were conducted in different locations. Though, the limitation of the exposed rock as well as the accessibility led to combining the stratigraphic successions in one composite stratigraphic column of the area. The study was described and interpreted at the main location (764,249.00 m E, 3,260,452.00 m N) and concluded the facies into 16 genetically related facies and six facies associations (Table 1.1). The six facies associations are an estuarine complex, tidal channel complex, shoreface complex, terrestrial facies, barrier island complex, and back-barrier complex.

The paleocurrent analysis of these genetic-related facies and facies associations measured by Amer et al. (2019a) indicates that the axis of the estuary, tidal, and fluvial channel systems is towards the NE direction suggesting an NW–SW paleoshoreline at the time of deposition. The produced facies model demonstrates a complex architecture that can be expressed in the lateral and vertical facies changes (Fig. 1.8).

The main facies, facies association, and depositional setting of the stratigraphic units described at the Jal Az-Zor escarpment presented herein are adopted after Amer et al. (2019a). The briefed description of these facies recognized are:

1. *Bioturbated cross-bedded facies*:

It is observed as the basal unit of most sequences over the Jal Az-Zor escarpment. It occurs in 1–3 m thick units characterized by large-scale low-angle cross-bedding, poorly sorted calcareous sands with the presence of granules and pebble grain sizes. It contains a high level of bioturbation, which is considered the main characteristic of this unit, dominated by Thalassinoides, Skolithos, Ophiomorpha, and less common Gyrolithes. XRD analysis of samples taken from these units suggests that the rocks are composed of 46% quartz, 20% illite, 27% muscovite, and 7% calcite.

2. *Cross-bedded conglomeratic facies*:

These facies form large concave lenses, trough cross-bedding, with the absence of bioturbation activity; those are generally sandwiched between the highly bioturbated cross-bedded facies. The presence of these facies is very localized within the studied area. It is characterized by an erosional base with a relatively flat top.

3. *Red claystone facies*:

Bedded, reddish-to-pinkish color with absence of bioturbation. The thickness of this unit can reach a maximum of 1.5 m, and dramatic vertical and lateral changes in facies do occur. In some locations, they tend to change into yellowish-green claystone facies. Tracing these facies over a short distance reveals that it changes laterally over a short distance. The XRD analysis resulted in quartz that reaches 48%, followed by muscovite representing 28.5%, illite 23%, and calcite 0.5%.

Table 1.1 Facies, facies associations, and proposed depositional setting observed in the Jal Az-Zor escarpments (Amer et al., 2019a)

Facies association	No	Facies	Abbreviation	Depositional environment
FA-1 Estuarine complex	1	Bioturbated cross-bedded facies	(F1)	Estuarine central bay (channel bar)
	2	Cross-bedded conglomeratic facies	(F2)	Estuarine channel fill
	3	Red claystone facies	(F3a)	Distal Estuarine intertidal flat
	4	Reddish-green claystone facies	(F3b)	Proximal Estuarine intertidal flat
	5	Calcareous sandstone facies	(F4)	Estuary mouth bar (shoreface/inlet shoal)
FA-2 Tidal channel complex	6	Cross-bedded sandstone facies	(F5a)	Tidal channel fill
	7	Lateral accretion red sandstone facies	(F5b)	Tidal channel point bar
	8	Bedded red muddy sandstone facies	(F5c)	Tidal channel overbank
FA-3 Shoreface complex	9	Fine-grained bedded sandstone facies	(F6b)	Foreshore
	10	Microbialites facies	(F6c)	Intertidal to supratidal
FA-4 Terrestrial facies	11	Yellowish-green siltstone facies	(F8)	Fluvial overbank
	12	Trough cross-bedded sandstone facies	(F7)	Fluvial channel fill
FA-5 Barrier island complex	13	Coarsening-upward calcareous sandstone	(F9)	Barrier island
	14	Fossiliferous sandstone facies	(F6a)	Fossiliferous shoal
	15	Microbialites facies	(F6c)	Intertidal to supratidal
FA-6 Back-barrier complex	16	Unconsolidated silty-sand facies	(F6d)	Backshore/berm
	17	Calcareous sandstone lenses	(F10)	Shoal inlet
	18	Yellowish-green siltstone facies	(F8)	Back-barrier lagoon
	19	Microbialites facies	(F6c)	Intertidal to supratidal

4. *Reddish-green claystone facies*:

These facies are characterized by the yellowish-green color of clay grain size sediments with the likelihood of silts towards the top. The thickness of this unit can reach 1.5 m. They are laterally changing into the Red claystone facies. Water escape structures were observed towards the top of the section. It has an erosional surface at the top where Ophiomorpha is observed to burrow from the overlaying Calcareous sandstone facies. The XRD analysis of one sample taken shows that these facies exhibit a similar mineralogical composition to red claystone facies with the addition of nimite that belongs to the chlorite group and represents 25%.

5. *Calcareous sandstone facies*:

These facies are the major occurrence and distinctive in the exposed successions; hence, it is the cliff-forming units in the Jal Az-Zor area. It is characterized by an amalgamated blocky pattern of poorly sorted calcareous sandstones to gravelly sandstones. It is associated with faint cross-bedding, scouring bases, and channeling features. It has various thicknesses across the escarpment with an average of 4 m. Grain size is dominated by medium-grained sands that vary from fine to very coarse, and in some cases, granules form thick layers. Reworked lithoclasts and bioclasts are abundant towards the base; those degrade and fade upwards. Faint Bioturbation is presently represented by Skolithos burrows towards the base and midsection, whereas Ophiomorpha is common towards the top. The XRD analysis of seven samples revealed an average of 38% calcite, 37% quartz, 20% muscovite, and 5% illite. It is believed that most of the calcite is related to the cement between the quartz grains, which might be the reason for extreme hardness.

6. *Cross-bedded sandstone facies*:

These facies are intermittently present over the Jal Az-Zor escarpment, mainly in the lower and middle sections of the escarpment successions. They are characterized by the fine-to-medium gain size of red and yellow sandstones representing bundles, trough, and planner cross-stratified, poorly to loosely cement sands. The thickness of these units is averaging tens of centimeters with an average of 70 cm. Bioturbation is dominated by Skolithos and Thalassinoides with the highest concentration towards the base of the facies. Uncommon fossil fragments such as echinoderms can be observed.

7. *Lateral accretion red sandstone facies*:

These facies are located at the lowermost part of the studied location of the Jal Az-Zor escarpment. It is characterized by large-scale inclined clinoformal-like beds, and in some cases, sigmoidal bedding is observed of fine-to-coarse red and yellowish sandstone. The thickness is averaging 2 m in general. The bioturbation is dominated by Skolithos, which is more intense towards the base. The lower contact is erosional and is marked by lag deposits.

8. *Bedded red muddy sandstone facies*:

These facies are thin units that occurred with the lateral accretion of red sandstone facies at the top and are considered the top cover for these facies. It is characterized by bedded red-to-yellow fine-grained sandstones, with occasional thin yellowish-green claystone beds lacking bioturbation. The thickness of these units is considered in tens of centimeters.

9. *Fine-grained bedded sandstone facies*:

These facies are composed of fine- to medium-grained calcareous sandstone. It generally exhibits a coarsening-upward succession, red-stained laminated sands, and in some localities,

it laterally transitions to trough cross-bedded sandstone facies. The thickness of these facies ranges from 1 to 2 m.

10. *Microbialites facies*:

These facies are characterized by white grayish colored, carbonate-dominated sediments with a common mixture of sand, silt, and green muds. The main characteristic of this unit is that commonly found at the top of the sequences overlaying several facies such as calcareous sandstone facies, fine-grained-bedded sandstone facies, and coarsening-upward calcareous sandstone facies. It has been suggested that diagenesis may have played a role in partially altering the surface exposures of these biogenic layers forming a caliche (calcrete) crust in some localities (Khalaf & El-Sayed, 1989). The upper part of the facies is characterized by clotted and small domal structures. The XRD analysis results are 71.4% of calcite, 28.6% of quartz, and a high amount of phosphorus contents. The thickness of this unit varies between 5 and 50 cm.

11. *Trough cross-bedded sandstone facies*:

These facies are comprised of light brown-colored sandstones that vary from fine to very coarse grain sizes. It is thin in thickness, with a maximum measured thickness of 1 m, with a small scale trough cross-bedding with the absence of bioturbation. It is observed mainly at the upper part of the Jal Az-Zor escarpment. The base is marked by lag deposits, while the upper contact is marked by soft-sediment deformation and water escape structures in places.

12. *Yellowish-green siltstone facies*:

These facies are mainly comprised of loose cemented yellowish-green-colored sandy siltstone. It ranges in thickness between a few centimeters to 30 cm with a localized extend. Overall, they are underlain by the calcareous sandstone facies or microbialites facies and overlain by the bioturbated cross-bedded facies. The unit is overall massive with no visible internal structures. The XRD analysis of these facies shows 47% quartz, 28% muscovite, and 25% nimite. The latter is responsible for the yellowish-green color facies and infers a terrestrial origin of such deposits.

13. *Coarsening-upward calcareous sandstone facies*:

These facies are characterized by white-colored, fine- to medium-grained sandstones with poorly sorted grains at the top of calcareous sandstone lithology. Towards to top, granules, pebble-sized-grained, and reworked lithoclasts are observed. The base of this unit is generally covered, and the maximum measured thickness is around 1.5 m. These lithofacies are only found towards the foot of the Jal Az-Zor escarpment and are generally capped by calcrete-like units of microbialites facies and occasionally fine-grained bedded. sandstone facies. The XRD analysis shows the content of 61% quartz, 35% calcite, and 4% illite.

14. *Fossiliferous sandstone facies*:

These facies are only observed in one unit in the area towards the foot of the Jal Az-Zor escarpment. They are dominated by fossiliferous calcareous sandstones with abundant gastropods, bivalves, and occasional oyster shells. The thickness of these facies is around 0.2–1 m, and it fits the description of the fossiliferous sandstone bed described by Fuchs et al. (1968). The XRD analysis shows 52% quartz, 44% calcite, and 5% illite.

15. *Unconsolidated silty-sand facies*:

These facies are represented by structureless, loosely cemented, unconsolidated, gray-to-white silt and sands with white calcrete lithoclasts. The sands are fine- to very fine-grained, and faint red silty laminae and mud can be observed. These facies have a thickness that varies from 2 to 4 m observed in the middle of the escarpment succession.

16. *Calcareous sandstone lenses*:

These facies are characterized by lenses of white calcareous sandstones that exhibit an erosional base and flat upper boundaries. The lower part of the lens is dominated by calcite, and towards to top, it transitions into microbialites facies. Lithoclasts of microbialites facies are commonly found in these facies. Internal grading is evident with inclined beds that exhibit an NW dip direction and exhibit possible lateral accretion surface. These lenses have a limited lateral extent and can only be found at some localities. The thickness of this unit can reach 1.5 m, and it can be partially associated with yellowish-green siltstones of F8a facies (Figs. 1.5, 1.6, 1.7 and 1.8).

1.4 Al-Subiyah Area

The outcrops of the Al-Subiyah area are located at the North and northwestern coastal plain of Kuwait Bay, trending northeast to southwest and covering an approximate area of 95 square kilometers (Fig. 1.1). This area is also known as Bahrah as the Bahrah oil field is the biggest part of the area. These outcrops are bounded by Jal Az-Zor escarpments from the North and the northwest and by Kuwait Bay's northern coast from the South and the southeast. Outcrops at the Al-Subiyah area are characterized by relatively low relief, flat-top, terrace-like, elongated, and scattered hills geomorphologically formed by aeolian, fluvial, and sea-level fluctuation erosional processes (Dalongeville & Sanlaville, 1987; Al-Asfour, 1982). These outcrops are surrounded by coastal sabkha and tidal flat environments in recent times and are covered with aeolian sand dunes and sand sheets. Beds within these hills are generally dipping 2–5 degrees towards the northeast (Khalaf et al., 2019; Al-Hajeri et al., 2020). These outcrops consist mainly of siliciclastic rock units deposited in terrestrial to shallow marine environments during the Oligo-Miocene times. They are believed to be the uppermost part of the Oligocene–Lower Miocene Ghar Formation based on the surface and subsurface correlation of the Ghar Formation in southern Iraq (Al-Juboury et al., 2010; Khalaf et al., 2019). Milton (1967) was the first to describe these outcrops as calcareous cross-bedded sandstone interbedded with green-colored claystone and tied them to the deposits above the Dammam Formation unconformity in a subsurface drilled well within the same area (Khalaf et al., 2019). The stratigraphy of the Ghar Formation and the Kuwait Group were illustrated in previous sections. It is assumed from the stratigraphic superposition. aspect that the stratigraphic units of the Al-Subiyah area are the basal units of Jal Az-Zor successions.

1.4.1 Al-Subiyah (Bahrah) Outcrop Stratigraphy and Lithofacies

Khalaf et al. (2019) conducted a sedimentological and petrographical study on several Ghar Formation outcrops at the Bahrah area. They stated that these outcrops are dominated by calcareous clastic lithofacies and subdivide these rock units into four major lithofacies:

1. *The Fawn Sandstone Facies*:

This lithofacies is the most dominant within Bahrah Area. Khalaf et al. (2019) further subdivided these lithofacies into three subfacies which are:

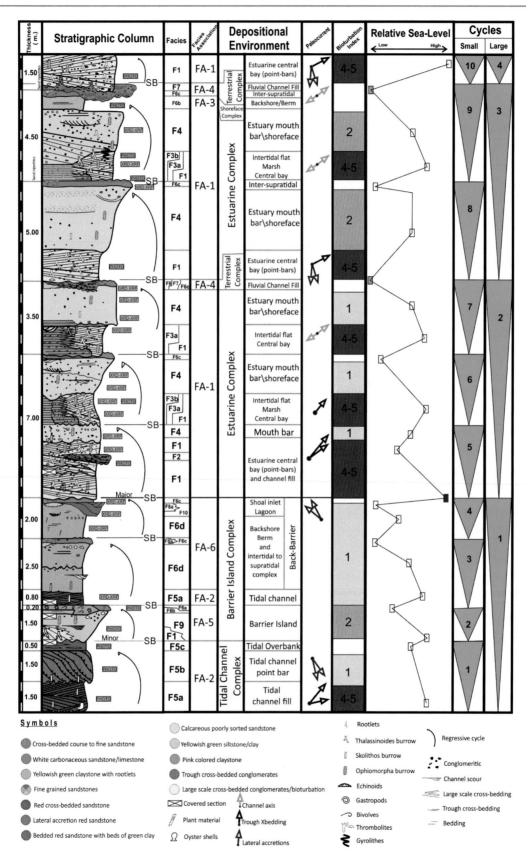

Fig. 1.5 A representative stratigraphic sequence of Jal Az-Zor Escarpment (after Amer et al., 2019a)

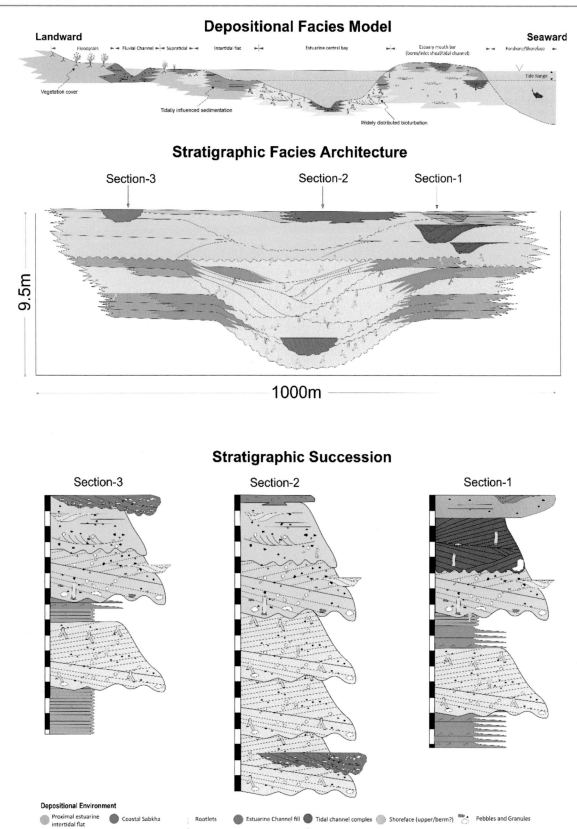

Fig. 1.6 Depositional model and facies architecture of the Jal Az-Zor escarpment (after Amer et al., 2019a)

Fig. 1.7 Examples of the facies described at Jal Az-Zor Escarpment (after Amer et al., 2019a)

a. The Cross-Bedded Sandstone Subfacies: these subfacies are most dominant within the area. It is composed of buff to reddish colored, medium-grained, hard, clean, cross-bedded, well cemented by calcite, sandstone containing some re-worked fossils (gastropods and bivalves), and borrows (Al-Hajeri et al., 2020).

b. The Planar Thin-Bedded Subfacies: These subfacies consist of red-colored, hard, clean, planar bedded, rippled, arenite cemented with calcite that interbedded with the cross-bedded sandstone and the palustrine dolomicrite lithofacies.

c. The Massive Bioturbated Sandstone Subfacies: This unit consists of extensively bioturbated sandstone with large-sized borrows. It usually overlies the palustrine sandy dolomicrite lithofacies.

2. *The Palestine Sandy Dolomicrite Lithofacies:*

This lithofacies is the second most dominant in the Al-Subiyah area. It is usually interbedded with the fawn thin-bedded and cross-bedded sandstone lithofacies. It occurs as discontinuous lenses of whitish marly sediments

Fig. 1.8 Examples of the facies described at Jal Az-Zor escarpment (after Amer et al., 2019a)

that are generally heavily bioturbated. This lithofacies is characterized by various sizes of cracks and exhibits large karstic hollows.

3. *The Mudstone Lithofacies*:

This lithofacies consists of grayish to reddish colored, thinly laminated to flaky mudstone enclosing thin bands and lenses of hard sandy mud. These lithofacies can be found mainly in the middle section of the Ghar Formation.

4. *The Carbonate Lithofacies*:

This lithofacies is limited in occurrence, and it is represented with coquina, quartzitic limestone, oolitic limestone, and dolostone. The coquina quartzitic limestone is found at the top of the Ghar Formation and superimposed by the Lower Fars clastic deposits. This lithofacies comprises of white-colored, chalky, and very porous limestone, commonly with herringbone trough cross-bedding. The oolitic limestone and dolostone are generally found as discrete thin beds within the middle and the lower section of the sequence (Fig. 1.9).

Fig. 1.9 Examples of the facies described et al.-Subyiah area (After Khalaf et al., 2019)

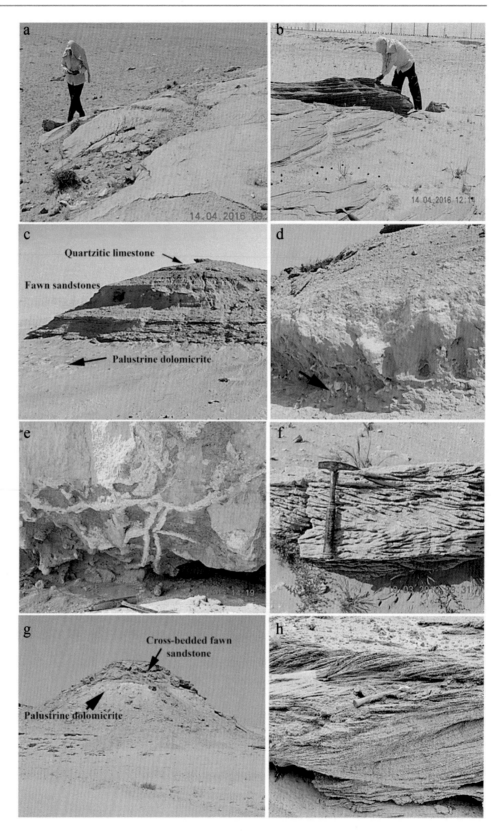

1.4.2 The Clastic Intrusion Zones at Al-Subyiah (Bahrah) Area

Clastic intrusions are distinctive tabular sedimentary structures highly associated with soft-sediment deformation of unconsolidated sediments (Jolly & Lonergan, 2002). Clastic intrusions are represented by many terminologies in literature such as sand volcanoes, sand blows, sand dykes, sand pipes, injectites, mud volcanoes, etc. They vary in length and diameter between a few centimeters to around several meters (Jolly & Lonergan, 2002). They can be found in various depositional settings in both shallow and deep depositional environments (e.g., shallow marine, glacial, lacustrine, fluvial, tidal flat, and deep marine environments) (Jolly & Lonergan, 2002; Duane et al., 2015; Al-Hajeri et al., 2020). They are often formed within tectonically active areas with high sedimentation rates; however, they can also be formed from near-surface sources within low tectonic activity regions (Jolly & Lonergan, 2002; Al-Hajeri et al., 2020).

Numerous sand intrusions can be found in northern Kuwait Bay et al.-Subyiah (Bahrah) area. They can be found within the basal portion of the Ghar Formation within the cross-bedded sandstone unit (more frequent in the fawn sandstone lithofacies) covering the coastal plain of the Bahrah area (Duane et al., 2015; Al-Hajeri et al., 2020). Duane et al. (2015) documented such clastic intrusions with various shapes and diameters within three zones in Bahrah coastal/tidal flat area. They identified these intrusions as mud volcanoes and seismites that were ejected from the feeder system (possibly Dammam Aquifer) due to seismic activities within the area and filled with mud and brecciated sandstone clasts along with dolomite, calcite, and partial marine bivalves. Duane et al. (2015) suggested that these mud volcanoes formed due to seismic induced liquefaction processes within the area (Fig. 1.10).

Recently, Al-Hajeri et al. (2020) thoroughly studied the clastic intrusions within the Bahrah area. They analyzed these intrusions quantitatively and qualitatively and proposed a conceptual model for the formation mechanism of these sand intrusions, calling them sand injectites. Al-Hajeri et al. (2020) identified two types of sand injectites with diverse forms and patterns in two different locations within the Bahrah area: the sand injection pipes and the polygonal to semi-circular sinuous sand injectites (Table 1.2). The sand injection pipes are associated with the cross-bedded sandstone lithofacies (the fawn sandstone lithofacies) within Al-Subyiah Ridges. They are characterized by their cylindrical shape and well calcite-cemented outer boundary. They ranged in color from red to yellow with very fine to coarse sand grain size. Around 200 pipes were identified with an average of 5-m height within the area and with diameters

ranging from a few centimeters to meters. The larger intrusions have a relative trend of northeast–southwest and northwest–southeast. The polygonal to semi-circular sinuous sand injectites are found within a separate reddish to brownish lithified terrane within Bahrah tidal flat/Sabkha area. Features within this area vary in size, shape, patterns, relief, and distribution. They can be divided into three groups: 1. Breccia-filled, 2. Yellow sand-filled, and 3. suture line.

Al-Hajeri et al. (2020) proposed a non-tectonic-related conceptual model for the formation of these sand intrusions, which is called: "the near-surface complex focused fluid injection conceptual model." This model suggests that the sediments and the injected fluids are both from the same source. A near-surface geobody that is dominated by loose, very porous, and unconsolidated sand is formed due to the deposition of the paleoshoreline. Water is trapped within this geobody due to its high porosity. This process increases the pore pressure within these near-surface geobodies. Biological activities—mainly bacterial activities—within the near-surface hydrocarbon reservoir facilitate further pressure build-up. Then loose unconsolidated sediments within the geobody are injected due to the fluids' overpressure forming sand injectites in more lithified sandstone host rock.

1.5 The Ahmadi Quarry

The Ahmadi Quarry is located in the southeastern part of Kuwait near the Ahmadi City (29°02′23.2″N, 48°04′36.0′E) (Fig. 1.1). Structurally, it is located on the eastern edge of the Ahmadi Structural Ridge. The quarry is administrated by the National Industries Company and covers approximately 10 square kilometers. The Ahmadi Ridge is an asymmetrical north–northwest-trending anticline structure with an approximate topographic height of 100 m (Carman, 1996) (Fig. 1.1).

In Kuwait's subsurface, the Dammam Formation and the Kuwait Group are primary aquifers covering useable brackish water in Kuwait. The Dammam Formation is a limestone–dolomite sequence of Middle Eocene age with approximately 600–700 feet thickness (183–213 m) (Bergstrom & Aten, 1965; Burdon & Al-Sharhan, 1968). These deposits were precipitated in shallow marine/marginal environments forming an extensive carbonate platform that extends over the Arabian Plate's eastern regions during the Eocene. The thickness of the Dammam deposits slightly increases towards the East (Schlumberger, 1972; Lababidi & Hamdan, 1985). The Dammam Formation is unconformably overlain by Kuwait Group, represented by Ghar, and underlain by Rus Formation. A karstification zone was

Table 1.2 Types of sand injectites with various forms and patterns at the Al-Subyiah (Bahrah) area described by Al-Hajeri et al. (2020).

Pipe type	Description	Map view photo
Circular with annular rings	Geometry:isolated circular to semi-circular standalone vertical to sub-vertical pipe forming a cylindrical geometry Size:from few centimeters to decimeter in diameter Grain size:pipe fill is fine to medium and the host rock is medium to coarse sand. The fill material is well cemented with $CaCO_3$ Rings:decrease in frequency inward and their range in thickness in few millimeters	
Hollow circular pipe	Geometry:isolated circular to semi-circular standalone vertical to sub-vertical pipe forming a cylindrical geometry Size: from few centimeters to few decimeters in diameter were recorded in the field Grain size: pipe fill is loose fine to medium sand (verv easy to remove with hand) and the host rock is medium to coarse sand. The boundary is very well cemented with $CaCO_3$	
Pipes associated with cemented veins	Geometry: isolated circular to semi-circular standalone and others are cross cuttin g two or more pipes with vertical to sub-vertical pipe forming a cylindrical geometry Size: from few centimeters to decimeter in diameter Grain size: pipe fill is fine to medium and the host rock is medium to coarse sand. The fill material (if present) is well cemented with $CaCO_3$ Rings: some of them are with ring and their frequency decrease inward and their range in thickness in few millimeters Veins: most of the veins stop along the boundary of the pipe and rarely cross-cut the pipe structure. Most of the veins are linear and few are circular to sub-circular. The veins are composed of fine sand cemented with $CaCO_3$ making them very resistive to erosion	
Hypotrochoid pipe	Geometry: circular to sub-circular, single or group of pipes in the middle surrounded with a group of distinct arcs and circles of cemented veins Size: decimeter up to few meters in diameter Grain size:pipe fill is fine to medium and the host rock is medium to coarse sand. The fill material (if present) is well cemented with $CaCO_3$ Veins: their geometry form arcs that cross-cut each other and most of the veins stop along the boundary of the pipe. The veins are composed of fine sand cemented with $CaCO3$ making them very resistive to erosion. Some of them show striation marks on them indicating vertical movement along them	

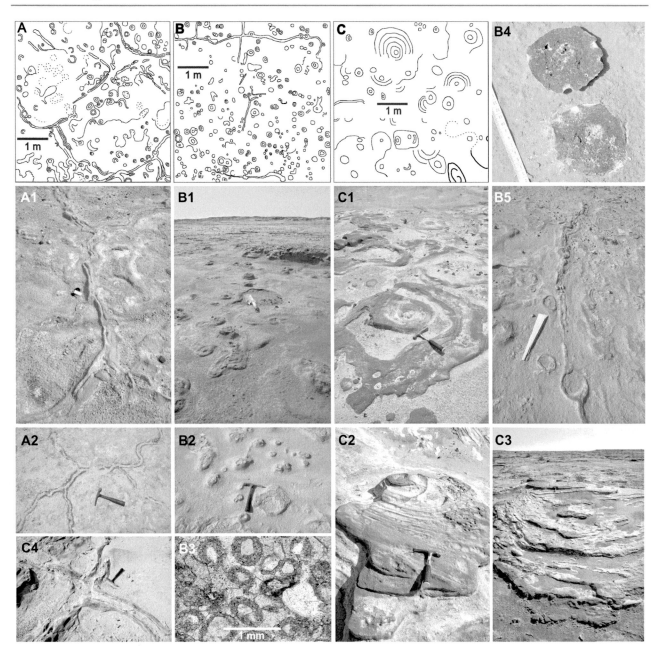

Fig. 1.10 Various types and shapes of mud volcanoes (sand intrusions) et al.-Subyiah (Bahrah) area (After Duane et al., 2015)

identified within the topmost part of the Dammam Formation that marks a major disconformity between the Dammam Formation and the overlying Mio-Pleistocene Kuwait Group clastic deposits (Owen & Nasr, 1958; Milton, 1967; Burdon & Al-Sharhan, 1968; Khalaf et al., 1989; Khalaf & Abdullah, 2014). The Ahmadi Quarry has nearly 10–25 m exposure of the topmost part of the Dammam Formation and the lower part of the Kuwait Group clastic deposits (Al-Awadi et al., 1997) (Fig. 1.11).

Fig. 1.11 A section of the Ahmadi Quarry showing the uppermost part of the Dammam Formation and the lowermost of the Kuwait Group

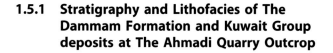

1.5.1 Stratigraphy and Lithofacies of The Dammam Formation and Kuwait Group deposits at The Ahmadi Quarry Outcrop

The Ahmadi Quarry outcrops consist of two major units: the carbonate-dominated uppermost part of the Dammam Formation and the clastic-dominated Kuwait Group. These units were studied by Al-Awadi et al. (1997), Khalaf and Abdullah (2013), Khalaf and Abdullah (2014), and recently by Khalaf et al. (2017). These studies mainly focused on the Dammam Formation lithofacies, whereas the Kuwait Group deposits at the quarry gain less attention and their lithofacies are not well identified. The lithofacies of Ahmadi Quarry can be summarized from oldest to youngest as follows (Al-Awadi et al., 1997; Khalaf et al., 2017) (Fig. 1.12):

1.5.1.1 The Dammam Formation

1. Chalky dolostone Unit: This unit is composed of white, porous, thick, friable, and granular chalky dolostone with thin chert bands and lenses (Al-Awadi et al., 1997; Khalaf et al., 2017). This unit is around 10 m thick in the Ahmadi Quarry (Al-Awadi et al., 1997; Khalaf et al., 2017). The chert bands are thin and discontinue, with sharp straight tops and irregular bottoms that formed along the bedding planes. These chert bands vary in thickness through this unit from a few centimeters to up to 50 cm (Khalaf et al., 2017). Chert was observed as lens-shaped within this unit. The upper contact of the chalky dolostone unit is gradational. There are a few fossil-rich zones within this unit in the Ahmadi Quarry. Fossils found in the Ahmadi Quarry include shell fragments of Gastropods and Pelecypods (Al-Awadi et al., 1997).

2. Karstic Dolostone Unit: This unit was first identified by Burdon and Al-Sharhan (1968) at the Ahmadi Quarry. It overlies the chalky dolostone unit and underlies the karst carapace unit. Both upper and lower contacts of this unit are gradational. The unit is composed of hard, karstic-rich dolostone with chert bands and lenses (Khalaf & Abdullah, 2014; Khalaf et al., 2017). This unit is approximately around 9 m thick in the northern parts of the quarry, whereas it thins out to less than 1 m and completely eroded towards the southern parts of the quarry (Khalaf et al., 2017).

3. Karst carapace Unit: This unit overlies the karstic dolostone unit with gradual contact. The upper contact of the unit is unconformable with the Kuwait Group deposits. This unit consists of white to yellowish-orange color, hard, dense calcitized, and silcretized dolocretic duricrust with pseudobreccia filling the cracks of the dolostone (Khalaf, 2011; Khalaf & Abdullah, 2014; Khalaf et al., 2017). The karst carapace unit has a thickness of around 3 m in the northern parts of the quarry and thins out towards the south (Fig. 1.13).

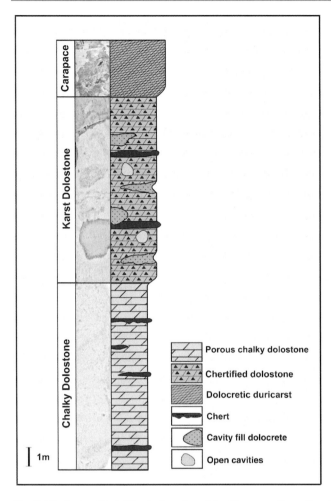

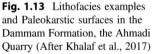

Fig. 1.12 A simplified lithostratigraphy of the Dammam Formation, the Ahmadi Quarry (after Khalaf et al., 2017)

Diagenesis and Diagenetic Stages of the Dammam Formation:

All the Dammam Formation units were subjected to various stages of diagenesis. These units were initially deposited as calcareous wackestone and mudstone within tidal flat and backshore lagoons settings and then later were altered completely to dolostone through dolomitization diagenetic processes (Khalaf et al., 2017). However, the sedimentary textures and structures of the original rock type were retained, allowing the identification of their facies and their depositional environment (Al-Awadi et al., 1997; Khalaf et al., 2017). Other forms of diagenesis within the formation include chertification, silcretization, dissolution, compaction, fracturing and karstification, stage cavity fill dolocretes, and karst carapace (Khalaf, 2011; Khalaf & Abdullah, 2014; Khalaf et al., 2017).

1.5.1.2 Kuwait Group Deposits

The Kuwait Group stratigraphy and lithofacies were discussed in previous sections. In the Ahmadi Quarry, only a few meters of the lower portion of the Kuwait Group (i.e., Ghar Formation) is exposed, whereas the topmost portion of the group is completely eroded. A karstification zone in the top part of the Dammam Formation marks the lower boundary of the Mio-Pleistocene Kuwait Group clastic deposits with a major erosional surface (disconformity) (Khalaf & Abdullah, 2014). A half-meter calcareous zone is about a meter above this unconformity (Al-Awadi et al., 1997). The Ghar Formation, which is exposed at the Ahmadi Quarry, consists of loose red to white–yellow sands without any conspicuous layering and cementation (Al-Awadi et al., 1997).

Fig. 1.13 Lithofacies examples and Paleokarstic surfaces in the Dammam Formation, the Ahmadi Quarry (After Khalaf et al., 2017)

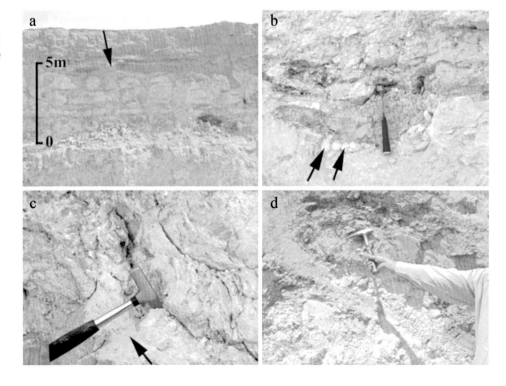

1.6   The Khiran Ridges

The Khiran Ridges are located in the southern part of Kuwait et al.-Khairan coastal area (Fig. 1.1). They are a series of lithified parallel coastal ridges trending North–Northwest to South–Southeast parallel to the Arabian Gulf Coast. These ridges vary in height and length, ranging from 5 to 15 m high and 450 to 4000 m in length, covering an area of around 100 square meters (Picha & Saleh, 1977; Picha, 1978; Al-Hurban & Al-Sulaimi, 2009). These ridges were studied by several researchers (e.g., Saleh, 1975; Picha & Saleh, 1977; Picha, 1978; Al-Sarawi et al., 1993; Al-Hurban & Al-Sulaimi, 2009). The age of these ridges is not precisely defined and varies from Pleistocene to Holocene. The formation of these ridges was linked to the Quaternary eustatic sea-level fluctuation during different phases of sea level that occurred due to the glacial and interglacial episodes (Saleh, 1975; Al-Sarawi et al., 1993). The older ridges are located further inland, whereas the younger ridges are closer to the present-day shoreline (Picha & Saleh, 1977; Picha, 1978). The older sediments at the base of the Khiran Ridges are believed to be of fluvial terrestrial origin, which were eroded and transported from the source lands west of the Arabian Plate during the Pleistocene Epoch (Al-Sarawi et al., 1993). However, the younger sediments at the top of these ridges are mainly Holocene in age (Picha & Saleh, 1977; Picha, 1978). Gunatilaka (1986) estimated that the formation of the Khiran Ridges is from 8,450 to 7,650 ± 70 (yBP) to 2,190 ± 70 yBP, whereas Al-Sarawi et al. (1993) believed that these ridges formed in the past 8,000 years during the Holocene eustatic sea-level fluctuations (Fig. 1.14).

1.6.1   Stratigraphy and Lithofacies of The Khiran Ridges and Outcrops

The Khiran Ridges and outcrops are mainly composed of siliciclastic deposits deposited in shallow marine barriers, ancient and recent beaches, and coastal dunes environments (Picha, 1978; Al-Hurban & Al-Sulaimi, 2009). Their lithology varies from oolitic marine sandstone to calcareous coastal aeolian sandstone (Picha, 1978; Al-Hurban & Al-Sulaimi, 2009). Picha and Saleh (1977) and Picha (1978) identified and described five lithostratigraphic complexes within the Khiran Ridges based on aerial photographs as follows:

(1) The Oolitic-Quartzose Sandstone Complex: This complex is considered the oldest and further inland. Its age is estimated to be the Late Pleistocene. This complex is exposed along a coastal cliff and can also be found in the subsurface by drilling at the base of the oolitic ridge complex. Most of the oolitic-quartzose sandstones lie below the present sealevel and are saturated with saline groundwater. This complex represents transgressive marine facies that are composed of coarse, slightly cemented, cross-bedded sandstone with well-rounded quartz grains, skeletal fragments, and oolites. The base of this complex shows a herringbone-type of cross-stratification with broken shell fragments that indicate a more tidal-influenced depositional setting. The uppermost part of this complex is marked by a 50 cm thick fossiliferous zone abundant with skeletal debris that mainly consists of gastropod shells. The

Fig. 1.14 Panoramic view of one of Al-Khiran Ridges, South of Kuwait

bedding of the upper part of this bridge complex is laminar and dipping towards the shoreline (i.e., East). Saleh (1975) described traces of root structure (e.g., rhizocretion) at the top of the southern complexes, which indicates a more coastal aeolian depositional setting.

(2) The Quartz-Oolitic Sandstone Complex: This complex is also estimated to be a late Pleistocene in age. It is located farthermost inland with a height of 15 m. This complex consists of several sets of cross-bedded marine and aeolian sandstones.

(3) The Older Oolitic Limestone Complex: This Pleistocene-aged complex is composed of cross-stratified aeolian-formed parallel ridges. They are 4–10 m high. They are composed almost entirely of chalky white ooids with only a small admixture of skeletal fragments, pellets, and quartz grains that are subaerially cemented by granular sparry calcite. The oolites lost their original luster due to partial dissolution and recrystallization of the cortex under atmospheric conditions. The nuclei of oolites are mostly quartz grains and structureless carbonate pellets, less frequently skeletal fragments, feldspars, and other detrital minerals.

(4) The Younger Oolitic Limestone Complex: This complex is Pleistocene in age. It forms ridges along the present coastline. These ridges are 2–8 m in height. They can be easily distinguished from the older oolitic ridges by their creamy-colored and highly lustrous hard surface of oolites that are little affected by subaerial processes. They are almost entirely aragonitic and display a very low degree of lithification.

(5) The Holocene oolitic Sediments: Represent the recent (Holocene) unconsolidated sediments on beaches, tidal flats, khors, and coastal dunes.

Al-Hurban and Al-Sulaimi (2009) revisited the Khiran Ridge complexes and furtherly studied their morphological and lithological characteristics. They described and identified two different settings of parallel ridges within the Khiran area, which are:

(1) **The Oolitic Limestone Composition Sets:** Two outcrops of the oolitic limestone outcrops were identified by Al-Hurban and Al-Sulaimi (2009) within the Khiran area: the creamy-colored outcrops and the chalky-white colored outcrops. These color differentiation are referred to as the differences in the lithification and diagenetic processes. These ridges range between 2 and 8 m high, 100 and 4000 m long, and around 100 and 300 m wide (Picha, 1978; Al-Hurban & Al-Sulaimi, 2009). These ridges are characterized by a thin fossil-rich unit

(bivalves and gastropods) at the base of the section indicating a marine depositional setting. Aeolian cross-bedding can be found within some of these ridges indicating the South to North flow direction. Fluvial influence is also noted at the base of these ridges, where well-rounded gravel-size oolitic clasts can be found. Rock units within the creamy ridge are composed of low cemented and very well sorted to well-sorted oolites. The nuclei of these ooids are mainly carbonate grains. The chalky-white colored ridge is composed primarily of larger-sized concentric oolites with carbonates, quartz, less common shell fragments, feldspar, amphibolite, and other detrital nuclei. These nuclei are relatively larger than the ones at the creamy-colored ridge. It is believed that the greater size of ooids is linked to the secondary oolitization process during the Holocene.

(2) **The Sandstone-Composition Sets:**
The sandstone-rich ridge sets at the Khiran area are approximately 19–20 m high, 100–200 m wide, and 2.5–3 km long. They are steeper on the western side and obliquely oriented to the present-day shoreline. They consist of thick, fine-sized cross-bedded sandstone deposited in terrestrial to marginal-transitional marine depositional settings followed by laminated to graded bedded sandstone with 10–30 cm fossiliferous zone at the top. The upper portion of the sandstone ridge comprises medium-sized, planar-cross bedded, calcareous sandstone with ooid grains. The amount of ooid grains diminishes very rapidly towards the top of the unit.

1.7 The Enjefa Beach Outcrop

The Enjefa Beach represents a modest exposure located along the coastline of the greater Kuwait City district of Salwa and extends for approximately 900 m in length (Fig. 1.1). The composite height of the exposed stratigraphic sequence has a total thickness of 3 m (Amer et al., 2018). The age of the exposed lithofacies was largely assumed to be Quaternary in age (Khalaf, 1988). A detailed radiometric dating study was not done until Tanoli (2015) performed ^{14}C radiocarbon dating on two shell fragments extracted from the upper portion of the exposed units. The radiometric age of these two samples revealed late Holocene (2260 ± 30 and 3160 ± 30 years BP). Though Tanoli et al. (2012) and Tanoli (2015) attempted to study the sedimentary structures and lithofacies overview of the exposed unit, it was not until Amer et al. (2018) studied the detailed facies architecture and tectonostratigraphic significance of this exposure and how it documented the final stages of Kuwait's uplift and emergence. Amer et al. (2018) suggested that the Ahmadi Ridge has played a significant role in shaping the depositional patterns of the lateral accretion facies of the tidal channel facies association. This work was followed by a detailed 3D geological model reconstruction for outcrop by Amer et al. (2019b) for educational purposes. The following section will briefly discuss the consist of facies associations, tectonostratigraphic significance, and 3D geological model of the Enjefa Beach outcrop.

1.7.1 Facies and Facies Associations of Enjefa Beach Outcrop

The Enjefa Beach exposure is composed of marginal marine deposits that consist of two genetically related facies associations. These are shoreface facies association and tidal channel facies association (Amer et al., 2018) (Fig. 1.1).

1. Shoreface facies association:

Based on the work performed by Amer et al., (2018, 2019b), the shoreface facies association can be divided into three genetically depositional facies. These are middle shoreface, upper shoreface, and foreshore facies.

The middle shoreface facies have been characterized by a thin bed of 30 cm at the base of the sequence (low tide waterline), fine-grained sands, trough cross-bedded, and extensive bioturbation (Thalassinoides). The upper shoreface facies are relatively thin ($\sim$50 cm), fine-to-coarse grain size sands that are characterized by large-scale trough cross-bedding (trough width $\geq$ 2.5 m), and ophiomorpha burrows. Amer et al. (2018) indicated that the paleocurrent direction of these facies is predominantly to the NW. The third facies over this facies association is the foreshore facies that rarely exceed 20 cm in thickness. These facies are represented by planar-laminated fine to medium grain-sized sands and gently dip towards the present-day shoreline. In places, these facies can be represented by bioclastic facies (Fig. 1.15).

Fig. 1.15 Outcrop photographs of the Enjefa Beach outcrop show; **b** foreshore facies that are bounded by erosional surfaces; **c** middle shoreface highly bioturbated interval; **d** upper shoreface large-scale trough cross-bedding top view; **e** top view of middle shoreface horizontal Thalassinoides burrows; **f** upper shoreface Ophiomorpha burrowing) Tidal channel-fill/abandonment facies with lag deposits (Modified after Amer et al., 2018)

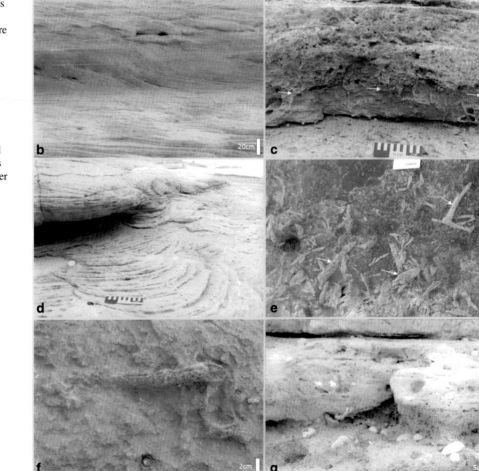

2. Tidal facies association:

The tidal channel facies association is dominated by lateral accretion facies, trough cross-bedded channel facies, and channel fill/abandonment facies. The lateral accretion surfaces are characterized by inclined bedding surfaces that record the point bar's migration associated with ophiomorpha burrows. The paleocurrent measurements performed by Amer et al. (2018) indicate a unidirectional NNW migration path. Amer et al. (2018) suggested that the Ahmadi Ridge has played a significant role in shaping the depositional patterns of the lateral accretion facies of the tidal channel facies association. The unimodal paleocurrent direction of the lateral accretion surfaces shows a predominant NNW direction. It was observed that this direction is parallel to the Late Holocene and the present-day shoreline, and no southern trend is found. In a normal point bar system, and because of the migration nature within a meandering tidal channel system, it would be expected to have multiple lateral accretion directions. However, over the Enjefa Beach outcrop, only a unidirectional paleocurrent is observed. The second facies is represented by cross-bedded channel facies. These facies are composed of fine-to-coarse, relatively small-scale trough cross-bedding (trough width ±0.5 m) and a bimodal paleocurrent direction to the WNW and WSW.

The third and last facies are Channel fill/abandonment facies. These facies are represented by fine-to-medium-grained sands, abundant fossil fragments such as gastropods, bivalves, and coral fragments (Fig. 1.16).

1.7.2 Enjefa Beach Outcrop 3D Geological Model

In general, modeling can be classified into two techniques: analogue and numerical modeling (Nieuwland, 2003). Over the Enjefa Beach outcrop, the analogues approach was used. Though using this method can raise questions on continuity and scale (McClay, 1990; Brun et al., 1994), the Enjefa Beach 3D geological model can be used to better understand the sedimentary processes and for educational purposes.

The Enjefa 3D static model developed by Amer et al. (2019b) was based on the geostatistical modeling approach introduced by Amer (2017) (Fig. 1.17). The model developed expressed the present-day digital elevation as the top boundary of the model and the lowest point of the section as the base. Three measured sections were used as input to the model and used to distribute the shoreface and tidal channel facies associations. The unique aspect of this model is that it captures the lateral accretion migration towards the NNW direction, supporting the understanding of sedimentary

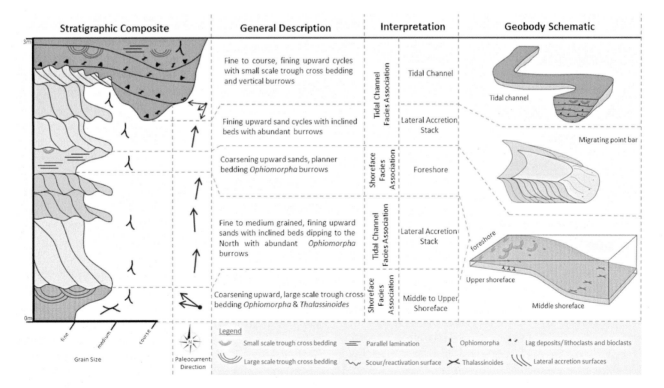

Fig. 1.16 Stratigraphic composite plot of the Enjefa outcrop illustrating the various associated geobodies responsible for the facies accumulation. (After Amer et al., 2018)

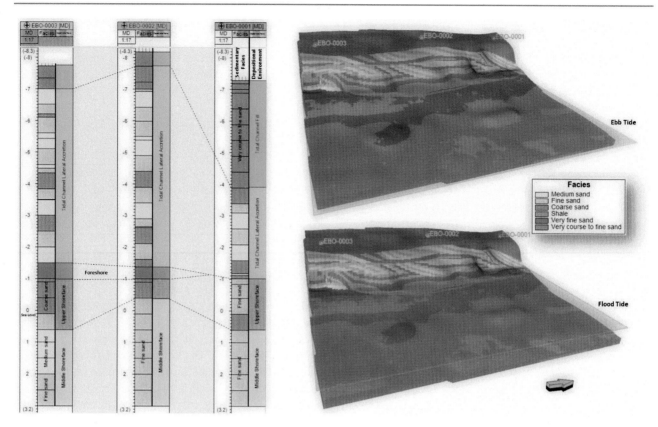

Fig. 1.17 Three measured sections along with the Enjefa Beach outcrop model. Note the inclined lateral accretion surfaces, the erosional nature of the capping tidal channel facies, and the present-day tidal range. After Amer et al. (2019b)

accretion and accommodation space. The model is capped by the tidal channel facies that express an erosional base nature, especially towards the northern end of the model. The authors also developed a synthetic seismic 3D model for the Enjefa Beach outcrop and used it for the base to understand similar sedimentary processes in the subsurface, making the Enjefa Beach a good analogue for subsurface shoreface and tidal facies.

1.8 Conclusion

Despite the numerous previous geological works and research conducted throughout the past decades, Kuwait's surface geology and stratigraphy are still poorly illustrated. Further modern and sophisticated studies are needed to enhance these domains. Several of the cited research in this chapter that demonstrates the fundamentals of the surface geology of Kuwait in various locations are relatively outdated. These researches were conducted during the past few decades of the past century (60s, 70s, 80s, and 90s). Therefore, it is essential to revisit and refine these previous studies using modern methods and technologies unavailable in the past. Despite the few recent attempts that were made to overcome this matter (e.g., Khalaf et al., 2017, 2019;

Amer et al., 2019b; Al-Hajeri et al., 2020), many early discussed areas are still in great need of further advanced and thorough studies covering all aspects of Kuwait's surface geology. Furthermore, the surface stratigraphic nomenclature of Kuwait rock units is another aspect that requires refining and re-establishing. The stratigraphic nomenclature of surface rock units of Kuwait—as discussed earlier—was established based on lithostratigraphic correlation to neighboring Saudi Arabia, Iraq, and Iran. This correlation must reassess, and more accurate modern methods must be applied to establish a better well-defined correlation and nomenclature. The contacts between the Kuwait Group's formations are also not well assigned and require more reevaluations. The naming of these formations (e.g., Ghar, Lower Fars, and Dibdibba) is also uncertain as the definitions of these formations' contacts are unclear and not well assigned. Several authors used different names for the same formations and different ages as well. Additionally, the age dating of most of Kuwait's surface rock units lack accurate estimation. Very few recent attempts were made to age date some of these formations to constrain a better understanding of Kuwait's surface and near-surface geology and stratigraphy (e.g., Amer et al., 2017, 2019b; Amer & Al-Hajeri 2019; Tanoli et al., 2019); however, many of the previously discussed outcrops are not yet accurately age dated in Kuwait.

Furthermore, the linkage between the exposed rock units and their presumed subsurface equivalent units is still not well established and requires further studies.

References

Al-Ameri et al., 2011 Al-Ameri, T. K., Pitman, J., & Naser, M. E., et al (2011). Programmed oil generation of the Zubair Formation, Southern Iraq oil fields: results from Petromod software modeling and geochemical analysis. *Arabian Journal of Geosciences* 4, 1239–1259.

Al-Anzi, M. (1995). Stratigraphy and Structure of the Bahrah Field, Kuwait. In: The Middle East petroleum geosciences (Geo94): Selected Middle East papers from the Middle East Geoscience Conference: Bahrain, Gulf PetroLink. pp 53–64.

Al-Asfour, T. (1982). Changing sea-level along the North Coast of Kuwait Bay. Kegan Pual, London, p. 310

Al-Awadi, E., Al-Ruwaih, F., Al-Rawdan, Z., & Ozkaya, I. (1997). The stratigraphy of the Middle Eocene-Pleistocene sediments in Kuwait. *Journal of Arid Environments, 37*, 1–22.

Al-Hajeri, M., Amer, A., Djawair, D., Green, D., & Al-Naqi, M. (2020). Origin of enigmatic sand injectites outcrops associated with non-tectonic force-folding structure in Bahrah area, northern Kuwait Bay. *Marine and Petroleum Geology, 155*, 1–20.

Al-Hurban, A., & Al-Sulaimi, J. (2009). Recent sediments and landforms of the southern area of Kuwait. *European Journal of Scientific Research, 38*(2), 272–295.

Al-Juboury, A., Al-Gharear, J., & Al-Rubaii, M. (2010). Petrography and diagenetic characteristics of the upper Oligocene–lower Miocene Ghar formation SE Iraq. *J. Petrol. Geol., 33*(1), 67–86.

Al-Sarawi, M. (1982). The origin of the Jal-Az Zor escarpment. *Journal of the University of Kuwait (science), 9*, 151–162.

Al-Sarawi, M., Al-Zamel, A., & Al-Rifaiy, I. (1993). Late Pleistocene and Holocene of the Khiran area (South Kuwait). *Journal of the University of Kuwait (science), 20*, 146–156.

AlShuaibi, A., & Khalaf, F. (2011). Development and lithogenesis of the palustrine and calcrete deposits of the Dibdibba Alluvial Fan, Kuwait. *Journal of Asian Earth Sciences, 42*, 423–439.

Al-Sulaimi, J., & Mukhopadhyay, A. (2000). An overview of the surface and near-surface geology, geomorphology and natural resources of Kuwait. *Earth Science Reviews, 50*, 227–267.

Amer, A., & Al-Hajeri, M. (2020). The Jal Az-Zor escarpment as a product of complex duplex folding and strike-slip tectonics; A new study in Kuwait, northeastern Arabian Peninsula. *Journal of Structural Geology, 135*, 1–14.

Amer, A., & Al-Hajeri, M. (2019). Strontium isotope radiometric dating reveals the late Eocene and Oligocene successions in northern Kuwait. *Arabian Journal of Geosciences, Springer, 12*, 288.

Amer, A., Al-Hajeri, M., Najem, A., & Al-Qattan, F. (2019a). Facies architecture of Lower Fars Formation at Jal Az-Zor escarpment, Kuwait. *Arabian Journal of Geosciences* 12, 502.

Amer, A., Al-Wadi, M., & Salem, H., (2019b). Geological modelling of the Enjefa Beach Marginal Marine Outcrop; A comparison between Holocene and Cretaceous Tidal Channel Complexes. Society of petroleum engineers. SPE-194895-MS. pp. 1–9.

Amer, A., Al-Wadi, M., Abu-Habbial, H., & Sajer, A. (2018). Tectonostratigraphic evolution of Enjefa's marginal marine Holocene deposits: highlighting the final stage of Kuwait's emergence. *Arabian Journal of Geosciences, 11*, 712. https://doi.org/10.1007/s12517-018-4070-9

Amer, A. (2017). New approach to modeling your reservoir: A technique based on understanding modern deposits, outcrops, and well-log data. In: SPE Middle East Oil & Gas Show and Conference. Society of Petroleum Engineers.

Amer, A., Al-Hajri, M., Najem, A., & Al-Qattan, F. (2017). New insights for Kuwait's lower fars formation exposure over Jal Az-Zor ESCARPMENT: the importance of linking surface to subsurface. AAPG Datapages Middle East Region Abstract #90299.

Benham, P., Freeman, M., Zhang, I., Choudhary, P., Spring, L., Warrlich, G., Ahmed, K., Al-Haqqan, H., Al-Boloushi, A., Jha, M., Sanwoolu, A., Shen, C., & Al-Rabah, A. (2018). The function of baffles within heavy oil reservoir and their impact on field development: A case study from Kuwait. 2018 SPE HOCE, Kuwait Dec10 -12th 2018. SPE-193775.

Bergstrom, R. E., & Aten, R. E. (1965). Natural recharge and localization of fresh groundwater in Kuwait. *Journal of Hydrology., 2*, 213–231.

Bou-Rabee, F., & Kleinkopf, D. (1994). Crustal structure of Kuwait: Constraints from gravity anomalies. US Department of the Interior, US Geological Survey.

Brun, J.-P., Sokoutis, D., & Van Den Driessche, J. (1994). Analogue modeling of detachment fault systems and core complexes. *Geology, 22*, 319–322.

Burdon, D. J., & Al-Sharhan, A. (1968). The problem of the Paleokarstic Dammam limestone aquifer in Kuwait. *Journal of Hydrology, 6*, 385–404.

Carman, G. J. (1996). Structural elements of onshore Kuwait. *GeoArabia, 1*, 239–266.

Cox, P., & Rhodes, R. (1935). The geology and oil prospects of Kuwait territory.-Kuwait Oil Company, unpublished report, Kuwait.

Dalongeville, R., & Sanlaville, P. (1987). Confrontation des datation isotopiques avec les donnees gemorphologiques et archeologies a propos des variation relative du niveau marin sur la rive arabe du Golfe persique. In: Aurenche, O., Evin, J., Houra, F. (eds), Chronologies in the Near East, vol. 379. Bar Int. Series, pp. 567–583 (ii).

Duane, M. J., Reinink-Smith, L., Eastoe, C., & Al-Mishwat, A. T. (2015). Mud volcanoes and evaporite seismites in a tidal flat of northern Kuwait—implications for fluid flow in sabkhas of the Persian (Arabian) Gulf. *Geo-Marine Letters, 35*, 237–246.

Fuchs, W., Gattinger, T. E., & Holzer, H. F. (1968). *Explanatory text to the synoptic geologic map of Kuwait: A surface geology of Kuwait and the neutral zone.* Geological Survey of Austria.

Gischler, E., & Lomando, A. (2005). Offshore sedimentary facies of a modern carbonate ramp, Kuwait, northwestern Arabian-Persian Gulf. *Facies, 50*, 443–462.

Gregory, J. W. (1929). The structure of Asia. Methuen & co ltd.

Gunatilaka, A. (1986). Kuwait and the Northern Arabian Gulf: A study in quaternary sedimentation. *Episodes, 9*(4), 223–231.

Jolly, R., & Lonergan, L. (2002). Mechanisms and controls on the formation of sand intrusions. *Journal of the Geological Society, London. 159*, 605–617. Kassler, P., (1973). The structural and geomorphic evolution of the Persian Gulf. In: The Persian Gulf. Springer, pp. 11–32.

Khalaf, F. I., Gharib, I. M., & Alkadi, A. S. (1982). Sources and genesis of the Pleistocene gravelly deposits in northern Kuwait. *Sedimentary Geology, 31*, 101–117.

Khalaf, F., Gharib, I., & Al-Hashash, M. (1984). Types and characteristics of the recent surface deposits of Kuwait. Arabian Gulf. *Journal of Arid Environments, 7*(1), 9–16.

Khalaf, F. I. (1988). Quaternary calcareous hard rocks and the associated sediments in the intertidal and offshore zones of Kuwait. *Marine Geology, 80*, 1–27.

Khalaf, F., & El-Sayed, M. I. (1989). Fossil cyclic calcrete in the Kuwait group clastic deposits (Mio-Pleistocene) of Kuwait, Arabian Gulf. *Geologische Rundschau, 78*, 525–536.

Khalaf, F., Mukhopadhyay, A., Naji, M., Sayed, M., Shublaq, W., Al-Otaibi, M., Hadi, K., Siwek, Z., & Saleh, N. (1989). Geological assessment of the Eocene and Post-Eocene aquifers of Umm Gudair, Kuwait. Kuwait Inst Sci Res Rep (EES-91), KISR3176.

Khalaf, F. I. (2011). Occurrence of diagenetic pseudobreccias within the paleokarst zone of the upper Dammam Formation in Kuwait, Arabian Gulf. *Arabian Journal of Geoscience, 4*, 703–718.

Khalaf, F. I., & Abdullah, F. A. (2013). Petrography and diagenesis of cavity-fill dolocretes, Kuwait. *Geoderma, 207–208*, 58–65. https://doi.org/10.1016/j.geoderma.2013.05.002

Khalaf, F. I., & Abdullah, F. A. (2014). Occurrence of diagenetic alunites within karst cavity infill of the Dammam Formation, Ahmadi, Kuwait: an indicator of hydrocarbon gas seeps. *Arabian Journal of Geosciences, 8*(3), 1549–1556.

Khalaf, F. I., Abdullah, F. A., & Gharib, I. M. (2017). Petrography, diagenesis, and isotope geochemistry of dolostones and dolocretes in the Eocene Dammam Formation, Kuwait. *Arabian Gulf. Carbonates and Evaporites, 33*(1), 87–105.

Khalaf, F., Abdel-Hamid, M., & Al-Naqi, M. (2019). Occurrence and genesis of the exposed Oligo-Miocene Ghar Formation in Kuwait, Arabian Gulf. *Journal of African Earth Sciences, 152*, 151–170.

Lababidi, M. M., & Hamdan, A. N. (1985). Preliminary lithostratigraphic correlation study in OAPEC member countries. Kuwait: Organization of Arab Petroleum Exporting Countries, Energy Resources Department. 171 pp.

Macfadyen, W. I. (1938). Water supplies in Iraq, Iraq Geol. Dept., Pub. No. 1, Baghdad.

Milton, D. I. (1967). Geology of the Arabian Peninsula. Kuwait: US geological survey, professional paper.

McClay, K. R. (1990). Extensional fault systems in sedimentary basins: A review of analogue model studies. *Marine and Petroleum Geology, 7*, 206–233.

Mukhopadhyay, A., Al-Sulaimi, J., Al-Awadi, E., & Al-Ruwaih, F. (1996). An overview of the Tertiary geology and hydrogeology of the northern part of the Arabian Gulf region with special reference to Kuwait. *Earth-Science Reviews, 40*, 259–295.

Nieuwland, D. A. (2003). Introduction: New insights into structural interpretation and modeling. *Geological Society, London, Special Publications, 212*, 1–5.

Owen, R.M.S. and Nasr, S.N., (1958). Stratigraphy of the Kuwait-Basra Area: Middle East. 'Habitat of Oil' American Association Petroleum Geologist Memoir 1, pp. 1252–1278.

Picha, F., & Saleh, A. (1977). Quaternary sediments in Kuwait. *Journal of the University of Kuwait (science), 4*, 170–184.

Picha, F. (1978). Depositional and diagenetic history of pleistocene and holocene oolitic sediments and Sabkhas in Kuwait, Persian Gulf. *Sedimentology, 25*, 427–450.

Saleh, A. (1975). Pleistocene and Holocene oolitic sediments in Al-Khiran area, Kuwait. M.Sc. thesis, Kuwait University.

Salman, A. S. (1979). Geology of the Jal Az-Zor, Al-Liyah area, Kuwait. Unpublished, M Sc Thesis, Kuwait University 128.

Schlumberger. (1972). Log interpretation, Vol. I—Principles. Houston: Schlumberger Ltd. 113.

Sharland, P. R., Archer, R., & Casey, D. M., et al. (2001). Arabian plate sequence stratigraphy. GeoArabia, Spec Publ 2, Gulf PetroLink. Oriental Press, Manama, Bahrain, p 371

Sharland, P. R., Casey, D. M., Davies, R. B., et al. (2004). Arabian plate sequence stratigraphy–revisions to SP2. *GeoArabia, 9*, 199–214.

Singh, P., Husain, R., Al-Kandary, A., & Al-Fares, A. (2011). Basement configuration and its impact on permian–triassic prospectivity in Kuwait. Kuwait City, Kuwait.

Tanoli, S. K., Al-Fares, A., & Al-Sahlan, G. (2012). The Enjefa beach exposure in Kuwait, Northern Gulf: Evidence of Late Holocene Regression. In: GEO 2012.

Tanoli, S.F., Husain, R., & Al-Khamiss, A. (2015). Geological handbook of Kuwait. Exploration Studies Team, Exploration Group, Kuwait Oil Company.

Tanoli, S. K. (2015). Sedimentological evidence for the Late Holocene sea-level change at the Enjefa Beach exposures of Kuwait, NW Arabian Gulf. *Arabian Journal of Geosciences, 8*, 6063–6074. https://doi.org/10.1007/s12517-014-1577-6

Tanoli, S. K., Youssef, A. H., Al-Bloushi, A., & Ahmad, K. (2019). Depositional pattern in the lower to middle miocene Jal Az-Zor formation from subsurface of North Kuwait. In H. R. AlAnzi, R. A. Rahmani, R. J. Steel, O. M. Soliman (eds), Siliciclastic reservoirs of the Arabian plate: AAPG Memoir 116, pp. 383–406

Veizer, J., Ala, D., Azmy, K., et al. (1999). 87Sr/86Sr, δ13C and δ18O evolution of Phanerozoic seawater. *Chemical Geology, 161*, 59–88. https://doi.org/10.1016/S0009-2541(99)00081-9

Ziegler, M. A. (2001). Late Permian to Holocene paleofacies evolution of the Arabian Plate and its hydrocarbon occurrences GeoArabia. 6(3), 445–504.

Subsurface Stratigraphy of Kuwait

Anwar Al-Helal, Yaqoub AlRefai, Abdullah AlKandari, and Mohammad Abdullah

Abstract

This chapter reviews the subsurface stratigraphy of Kuwait targeting geosciences educators. The lithostratigraphy and chronostratigraphy of the reviewed formations (association of rocks whose components are paragenetically related to each other, both vertically and laterally) followed the formal stratigraphic nomenclature in Kuwait. The exposed stratigraphic formations of the Miocene–Pleistocene epochs represented by the Dibdibba, Lower Fars, and Ghar clastic sediments (Kuwait Group) were reviewed in the previous chapter as part of near-surface geology. In this chapter, the description of these formations is based mainly on their subsurface presence. The description of the subsurface stratigraphic formations in Kuwait followed published academic papers and technical reports related to Kuwait's geology or analog (GCC countries, Iraq and Iran) either from the oil and gas industry or from different research institutions in Kuwait and abroad. It is also true that studies related to groundwater aquifer systems also contribute to our understanding of the subsurface stratigraphy of Kuwait for the shallower formations. The majority of the published data were covered the onshore section of Kuwait. The subsurface stratigraphic nomenclature

description is based on thickness, depositional environment, sequence stratigraphy, the nature of the sequence boundaries, biostratigraphy, and age. The sedimentary strata reflect the depositional environment in which the rocks were formed. Understanding the characteristics of the sedimentary rocks will help understand many geologic events in the past, such as sea-level fluctuation, global climatic changes, tectonic processes, geochemical cycles, and more, depending on the research question. The succession of changing lithological sequences is controlled by three main factors; sea-level change (eustatic sea level), sediment supply, and accommodation space controlled by regional and local tectonics influences. Several authors have developed theoretical methods, established conceptual models, and produced several paleofacies maps to interpret Kuwait's stratigraphic sequence based on the data collected over time intervals from the Late Permian to Quaternary to reconstruct the depositional history of the Arabian Plate in general and of Kuwait to understand the characteristics of oil and gas reservoirs.

2.1 Introduction

Kuwait is located within the depocentre of the Arabian platform carbonates basin and is covered by approximately integrated shallow marine Phanerozoic sedimentary succession with a thickness of more than 35,000 ft (10,688 km). This Phanerozoic section is interrupted by significant episodes of uplift and erosion or non-deposition manifested by well-recognized unconformities (Fig. 2.1). The sedimentary succession is influenced by large, gentle folds of different shapes and sizes associated with differential regional subsidence or uplifting along basement faults. These deep-seated structures cause lateral and vertical variations in sedimentary successions (Al-Sharhan & Nairn, 2003).

A. Al-Helal (✉)
Public Authority for Applied Education and Training, Collage of Basic Education, Science Department, P.O. Box 23167 13092 Safat, Kuwait
e-mail: ab.alhelal@paaet.edu.kw

Y. AlRefai
Earth and Environmental Sciences Department, Kuwait University, P.O. Box 5969 13060 Safat, Kuwait

A. AlKandari
Kuwait Oil Company, Exploration Group-Prospect Evaluation Team III, P.O Box 9758 61008 Ahmadi, Kuwait

M. Abdullah
Kuwait Foreign Petroleum Exploration Company, Administrative Shuwaikh Area 4, Street 102, Building No. 9, P.O. Box 5291 13053 Safat, Kuwait

© The Author(s) 2023
A. el-aziz K. Abd el-aal et al. (eds.), *The Geology of Kuwait*, Regional Geology Reviews,
https://doi.org/10.1007/978-3-031-16727-0_2

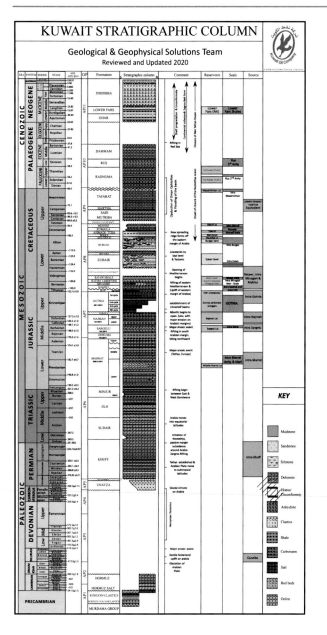

Fig. 2.1 Kuwait chronostratigraphic column. AP refers to the Arabian Plate tectonostratigraphic megasequence, which will not be discussed in this chapter in detail

The Phanerozoic succession is not exposed at the surface in Kuwait; hence, it is relatively poorly documented across its entire thickness. Most of the vital information about the subsurface formations in Kuwait are obtained from the great history of petroleum industry-related hydrocarbon drilling wells in the state of Kuwait (Fig. 2.2). Remarkable information was derived from deep wells drilled onshore, and few wells have reached the ancient basement (Al-Hajeri & Bowden, 2018). Hence, this chapter's authors found some uncertainties and rare information interpreting the Pre-Kuff

(Pre-Upper Permian) formations in Kuwait and excluded it from the discussion.

In an attempt to understand the vast amount of stratigraphic information, several authors published regional lithostratigraphic studies. The depositional environment interpretations of the identified sequences are based on sedimentary stacking patterns, core observations, lithology, physical and biogenic sedimentary structures, facies and their associations, facies boundary characteristics, etc. Ziegler (2001) reconstructs the depositional history of the Arabian Plate by generating paleofacies maps for given time intervals between the Late Permian and Holocene. Sharland et al. (2001) published the first chronostratigraphic interpretation of the rock units of the Arabian Plate. Beydoun (1988) and Al-Sharhan and Nairn (2003) have published regional lithostratigraphic reviews to understand the substantial amount of stratigraphic information. Pasyanos et al. (2007) used a limited amount of seismic data from the Kuwait National Seismic Network (KNSN) to estimate the lithospheric structure of Kuwait. They combined surface wave and receiver functions to develop a KUW1 model, presenting a sedimentary cover with (8 km) thick and crustal thickness of 45 km.

2.2 Late Paleozoic–Early Mesozoic

The Late Paleozoic–Early Mesozoic period in this chapter is referred to as the Permian to Triassic time, describing the potential source rocks formations including the oldest Khuff Formation (Late Permian–Early Triassic), Sudair Formation (Early Triassic), Jilh Formation (Middle-Late Triassic), and Minjur Formation (Late Triassic). This Permo-Triassic section in the subsurface of Kuwait is poorly understood, and very limited published reliable data are available. The formations that belong to this section gained less attention in drilling and exploration from oil companies due to the following reasons: The greater depths of these formations, the lack of high-resolution seismic and core data, the lack of hydrocarbon reservoirs within these formations, and most importantly, the higher cost of drilling (Abdullah et al., 2017; Husain et al., 2013; Pöppelreiter & Marshall, 2013; Singh et al., 2013).

2.2.1 Khuff Formation

The Khuff Formation cropped out in Central Saudi Arabia near the Riyadh–Jiddah Road and was named Khuff according to 'Ayn Khuff (Khuff Spring) (Steineke et al., 1958). The Khuff Formation is considered one of the largest

Fig. 2.2 Map illustrating the location of the oil and gas field from which the stratigraphic column of Kuwait was interpreted and constructed

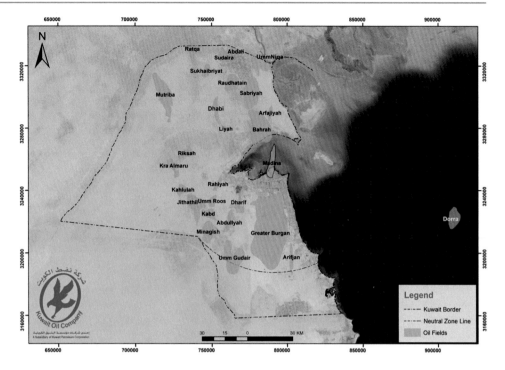

gas reservoirs in the Arabian Gulf region and worldwide. It was deposited in one of the most extensive shallow-water carbonate ramp environments in Earth's history, that extended for 3.7 square million kilometers along the eastern side of the Arabian Plate (the passive northeastern margin of Neo-Tethys) during the Permo–Triassic time (Zieglar, 2001; Pöppelreiter & Marshall, 2013). The Late Permian age is assigned for Khuff Formation due to the presence of foraminifera and algae, a fauna that includes *Globovalvulina vonderschmitti, Pachyphloia* cf. *ovata, Hemigordius* cf. *ovatus, Mizzia velebitana,* algae, echinoids, bryozoa, gastropods, and ostracods (Khan, 1989).

Information about Khuff Formation in Kuwait is deficient due to the greater depths of the formation (ranges between 15,000 and 24,000 ft (4572–7315 m)), which increases the drilling operations costs (Abdullah et al., 2017; Husain et al., 2013; Singh et al., 2013). Furthermore, the seismic data resolution for the Khuff Formation and deeper formations in Kuwait's subsurface are limited. Moreover, the presence of better and shallower hydrocarbon reservoirs above the Khuff Formation makes it less desirable for further exploration (Al-Sharhan et al., 2014; Husain et al., 2013).

Very few wells had penetrated entirely the Khuff Formation in Kuwait (six wells in Northwest Raudhatain, Burgan, and Umm Gudair structures) to the top of the Unayzah Formation that underling the Khuff Formation (Abdullah et al., 2017; Al-Sharhan & Nairn, 2003; Husain et al., 2013). Other wells were drilled only up to the upper parts of the Khuff Formation without full penetration (e.g., those in Sabriyah, Mutriba, and Kra Al Maru structures) (Husain et al., 2013).

In Kuwait, Khuff Formation is around 1800 ft (549 m) thick on average, and it underlies the Sudair Formation with uncomfortable to gradational(?) contact (Husain et al., 2013). The lower contact (base of AP6) of the Khuff Formation varies through different locations in Kuwait. In the Raudhatain Field, Khuff Formation lies unconformably on top of the Unayzah Formation, while it overlies a regional well-recognized unconformity consists of a clastic sequence of unknown age elsewhere in Kuwait (Al-Sharhan & Nairn, 2003; Husain et al., 2013; Strohmenger et al., 2003; Tanoli et al., 2008).

The Khuff Formation was divided into; Khuff-A, B, C, and D units (Husain et al., 2013; Singh et al., 2013). Another subdivision is widely used for Khuff Formation in Kuwait, and it subdivides the formation into; Upper Khuff (i.e., Khuff-A, Khuff-B, and Khuff-C) and Lower Khuff (i.e., Khuff-D). The Upper and the Lower Khuff are separated by a distinctive, extensive, and thick layer of anhydrite known as the Median Anhydrite (Husain et al., 2013; Khan, 1989).

Lithologically, Khuff Formation generally consists of dolomite and limestone (dolomudstone, dolowackestone, dolopackstone, and dolograinstone) with minor shale and anhydrite that were deposited in shallow subtidal to intertidal and lagoonal depositional settings (Husain et al., 2013; Khan, 1989; Singh et al., 2013). Both Khuff-D and Khuff-B are carbonates-rich in the lower sections with anhydrite-dominated in the upper sections. In Comparison, Khuff-C and Khuff-A are mainly carbonate-rich throughout their entire section (Husain et al., 2013). The early and late stages of diagenesis (e.g., micritization, dolomitization, cementation, stylolites, leaching, and dedolomitization) extensively affected the formation of lithology (Husain et al., 2013).

2.2.2 Sudair Formation

The Sudair Formation (informally called Sudair shale) is named for Khashm Sudayr in Saudi Arabia, where the lower part of the unit is well exposed (Powers et al., 1966). The Early Triassic Sudair Formation is considered a significant seal for the Khuff Reservoir along the Arabian Plate (Abdul Malek et al., 2005; Pöppelreiter & Obermaier, 2013). Lithology-wise, Sudair Formation consists mainly of shale and fine-grained siliciclastics with dolomite and anhydrite in most parts of the Arabian Plate, but it shows lateral facies variation when it is regionally correlated (Davies et al., 2019; Pöppelreiter & Obermaier, 2013).

Sudair Formation is composed of cycles of dolomudstones, laminated to bedded anhydrites and dolomitic shales deposited in an inner ramp and lagoon environments (Abdul Malek et al., 2005; Al-Sharhan et al., 2014). Despite the lack of oil discovery within the Sudair Formation, it has some potential for oil and gas discovery, especially in the West of Kuwait, where potential gas-condensate discoveries within the Sudair Formation were detected in the Mutriba area (Abdul Malek et al., 2005).

The Sudair Formation is unconformably/gradationally overlying the Khuff Formation and conformably underlies the Jilh Formation. The formation ranges in thickness from 98 ft (30 m) in Sabriyah Field, north of Kuwait, to 1201 ft (366 m) in Kra Al-Maru Field west of Kuwait (Al-Sharhan et al., 2014).

Sudair Formation was divided into three major units (from oldest to youngest) (Khan et al., 2010; Al-Sharhan et al., 2014):

Sudair Unit C: consists of argillaceous dolomite with interbedded shale and thin anhydrite.

Sudair Unit B: comprises dolomite, anhydrite, and minor shale.

Sudair Unit A: consists of dolomudastone, anhydrite, and shale. This unit can be found all over the subsurface of Kuwait, and it pinch-out towards the northeast (Khan et al., 2010).

2.2.3 Jilh Formation

The Jilh Formation was named for Jilh al 'Ishãr, a low escarpment in Saudi Arabia across which the section type was measured (Powers et al., 1966). The Jilh Formation is of Middle Triassic in age (Ladinian to Anisian) according to the presence of *Myophoria* sp. (Sharland et al., 2001; Al-Sharhan & Nairn, 2003). It is conformably overlying Sudair Formation and unconformably underlies the Minjur Formation. The thickness of the Jilh Formation in Kuwait ranges between 803 and 1804 ft (245–550 m) (Khan et al., 2010; Al-Sharhan et al., 2014). The formation is mainly composed of dolomite, anhydrite, minor limestone, and shale. The hypothetical depositional environment suggested hypersaline sabkha-dominated, shallow marine environments and intertidal/subtidal to continental influenced sabkha (Husain et al., 2009; Khan et al., 2010; Al-Sharhan et al., 2014).

The Jilh Formation can be subdivided into three units (from oldest to youngest) (Khan et al., 2010; Al-Sharhan et al., 2014; Arasu et al., 2018):

Jilh Unit C (Lower Jilh Carbonate Unit): Consists of altering various carbonates and dolostone (dolomudstone, dolowackestone, and dolopackstone) with anhydrite that indicates coastal marine and sabkha depositional settings.

Jilh Unit B (Jilh Dolomite): Comprises alternations of dolomudstone to dolowackestone with dolomitic shale and some thin-bedded salt at the top section of the unit. The lithology of this unit indicates a hypersaline shallow marine depositional setting.

Jilh Unit A (Upper Jilh Carbonate Unit): Consists of dolomudstone, anhydrite, with rare dolowackestone and dolopackstone. This unit represents more terrestrial sabkha-dominated depositional settings.

2.2.4 Minjur Formation

The Minjur Formation was named after Khashm al Manjûr in Saudi Arabia, where the upper part of the formation is exposed (Powers et al., 1966). In Kuwait, the Minjur Formation comprises mixed siliciclastics, carbonates, and evaporite deposits. It mainly consists of alternating dolomite, anhydrite, and sandstone with intercalation of siltstone and shale indicating siliciclastic input on top of the shallow Arabian shelf (Khan et al., 2010; Al-Sharhan et al., 2014; Davies et al., 2019).

As stated in Al-Sharhan and Nairn (2003), the Minjur Formation is estimated to have been deposited during the Late Triassic age (Carnian to Rhaetian) in agreement with the microhabitat of *Estheria minuta* Alberti and *Lingula tenuissima* var. *zenkeri* Alberta. The Minjur Formation is conformably overlying the Jilh Formation and unconformably underlies the Jurassic Marrat Formation. In Kuwait's subsurface, the Minjur Formation ranges between 524 and 1066 ft (160–325 m) in thickness and it was classified into five units by Al-Sharhan et al., (2014):

Minjur Unit 5: consists mainly of shale and siltstone with alternating sandstone, dolomite, and anhydrite.

Minjur Unit 4: predominantly composed of siliciclastics. It consists of sandstone with intercalated shale and siltstone.

Minjur Unit 3: consists mainly of dolomite with anhydrite and minor shale, siltstone, and sandstone.

Minjur Unit 2: comprises shale and siltstone with minor alternating dolomite, anhydrite, and sandstone. Signs of an arid environment paleosol were found in this unit's cores, indicating an arid floodplain and palaya depositional environments (Al-Sharhan et al., 2014).

Minjur Unit 1: predominantly consists of dolomite and anhydrite, intercalation of shale and siltstone, with minor limestone and sandstone.

2.3 Mesozoic

Kuwait was part of a carbonate shelf environment during the majority of the Mesozoic (Sharland et al., 2001; Ziegler, 2001). This shelf formed part of the Neo-Tethys Sea, which started during the Late Triassic due to the divergence between the central Iranian plate (s) and Arabian plate (Koop & Stoneley, 1982).

The Mesozoic period of the Jurassic and Cretaceous Kuwait stratigraphic column is well studied as part of oil exploration and production. The subsurface lithostratigraphic units have been described to demonstrate the close relationship between lithofacies, depositional environments, geological structures, and petroleum reservoirs. The reviewed Jurassic and Cretaceous stratigraphic columns in this chapter reflect the summary of these studies.

2.3.1 Jurassic

The Jurassic Formations of Kuwait from oldest to youngest are Marrat, Dhruma, Sargelu, Najmah, Jubaila, Gotnia, and Hith. The Marrat, Dhruma, Jubaila, and Hith formations are identified in surface exposures in Saudi Arabia (Powers et al., 1966), while the Sargelu, Najmah, and Gotnia formations were described in wells and outcrops in Iraq (Bellen et al., 1959).

2.3.1.1 Marrat Formation

The Marrat Formation was named and described in outcrop sections in Saudi Arabia (Powers et al., 1966; Steineke et al., 1958). Yousif and Nouman (1997) first published the term "Marrat Formation" in Kuwait.

The Marrat Formation is barren of nannofossils (Kadar et al., 2015; Al-Moraikhi et al., 2014). However, the deposition age was estimated between Late Sinemurian–Early Bajocian (Early–Middle Jurassic) based on isotope studies done by Al-Sahlan (2005) and Al-Moraikhi et al. (2014).

The Marrat Formation is predominantly composed of Carbonates. It mainly consists of micritic limestones with wackestones, packstones, and oolitic grainstones, interbedded with anhydrite, dolomite, and rare shale (Al-Sharhan et al., 2014; Yousif & Nouman, 1997). It conformably underlies Dhruma Shale Formation and unconformably overlies the Triassic Minjur Formation (Al-Sharhan et al., 2014; Neog et al., 2011). The Marrat Formation can reach up to 2001 ft (610 m) in thickness at Burgan field, where the formation is the thickest (Abdullah, 2001; Al-Sharhan & Nairn, 2003). It was deposited in a shallow marine carbonate platform to peritidal and Sabkha environments (Yousif & Nouman, 1997; Al-Eidan et al., 2009; Neog et al., 2011; Al-Sharhan et al., 2014).

The Marrat Formation was informally subdivided into five units: Unit A, B, C, D, and E by Yousif and Nouman (1997). Later it was subdivided into three members: The Lower, the Middle, and the Upper Marrat Members.

The Lower Marrat Member is composed of mixed lithologies of thinly interbedded carbonate, dolomite, argillaceous carbonates, and anhydrite (Al-Sharhan et al., 2014; Neog et al., 2011). It thickens towards the southeast of Kuwait (Neog et al., 2011). The Lower Marrat Member can be divided into seven sequences. Each of these sequences contains anhydrite that is either scattered through the formation or concentrated in the upper part, indicating peritidal, tidal flat, and sabkha environment settings (Al-Sharhan et al., 2014; Kadar et al., 2015; Neog et al., 2011).

The Middle Marrat Member consists of cleaner carbonates deposited in a tidal flat to open marine settings separated by a high-energy barrier (Kadar et al., 2015; Neog et al., 2011). The middle Marrat Member base consists of argillaceous limestone and debris flow deposits, while massive anhydrites fill the lagoons behind the barriers. The lagoonal deposits were later replaced by tidal flats and sabkha deposits (Al-Sharhan et al., 2014). The uppermost part of the Middle Marrat Member is dominated by carbonates and argillaceous limestone (Kadar et al., 2015). In the North of Kuwait, the Middle Marrat Member can be divided into three major cycles consisting of an alternation between carbonates and evaporites (Kadar et al., 2015; Neog et al., 2011) as follows:

1. Supratidal sabkha deposits of enterolithic anhydrite with thinly laminated carbonate beds interbedded with chicken-wire anhydrites.
2. Lagoonal evaporites of gypsum and anhydrite interbedded with carbonates related to cyanobacteria mats.
3. Shallow basin deposits consist of massive bedded anhydrite with dolomite partings, carbonate, and shale interbedded with layers of enormous halite.

The Upper Marrat Member consists of mixed lithology (generally more argillaceous lithofacies) that contains several thin alternating cycles of sabkha, tidal flat, lagoon, and carbonate platform deposits (Al-Sharhan et al., 2014; Kadar et al., 2015; Neog et al., 2011). The contact that separates the Middle and the Upper Members of the Marrat Formation is marked by a subaerial exposure surface (Davies et al., 2019; Kadar et al., 2015). The thickness of the Upper Marrat Member increases uniformly across the basin from 361 to 558 ft (110 m–170 m), indicating an absence of tectonic activities during the time of deposition (Al-Sharhan et al., 2014). The lower part of the Upper Marrat Member is composed mainly of evaporites (mainly anhydrite), with laminated reddish-brown silty dolomudstone, microbial laminites, argillaceous dolomudstone, and skeletal-peloidal, wavy bedded mudstone and wackestone representing sabkha, tidal flat, tidal channel, and shallow lagoon depositional settings. The uppermost part of the Upper Marrat Member consists of more open marine carbonates deepening upward into shale-dominated Dhruma Formation (Al-Eidan et al.,2009; Neog et al., 2011; Al-Sharhan et al., 2014).

2.3.1.2 Dhruma Formation

The Dhruma Formation was named and described concerning outcrops in Saudi Arabia near the town of Durmã (Steineke et al., 1958; Powers, 1966).

Kadar et al. (2015) analyzed the Dhruma Formation in Minagish-27 Well located in Minagish field; they estimated its lithological thickness of 138 ft (42 m), and they found that it consists of shale and calcareous shale with occasional interbeds of argillaceous lime mudstone and wackestone. The formation is thickening towards the South of Kuwait with a maximum of 175 ft (53 m) and (ca. 60 ft, 18 m) in North Kuwait (Cabrera et al., 2019; Kadar et al., 2015).

Based on calcareous nannofossil identifications, Kadar et al. (2015) estimated that Dhruma Formation was deposited during Middle Jurassic, within the age extending from Middle or Late Aalenian to early Late Bajocian.

The lower boundary of the Dhruma Formation with the Marrat Formation appears to be conformable, and it is usually detected with high gamma-ray (an e-log used to measure the gamma-ray radiation from the rocks). It was delineated where the Dhruma basinal shales replaced the interbedded shale and limestone beds of the Upper Marrat Member, suggesting a gradual upward deepening towards the Dhruma Formation (Kadar et al., 2015). The upper boundary with the overlying Sargelu–Dhruma Transition Member of the Sargelu Formation is conformable and marks a noticeable change to a possible shelf deposition with typical storm beds (Cabrera et al., 2019).

The Dhruma Formation is considered an excellent caprock for the Marrat reservoir (Yousif & Nouman, 1997).

2.3.1.3 Sargelu Formation

The Sargelu Formation was named after surface exposures of the succession in the Surdash Anticline, northeastern Iraq (Dunnington, 1958). Sargelu Formation is the oldest formation of the Riyadh Group that comprises of five members as follows: Sargelu, Najmah, Jubaila, Gotnia, and Hith.

Kadar et al. (2015) studied the Sargelu Formation in Minagish-27 Well between 13,166 ft (4,014 m) and 13,377 ft (4,077 m) and subdivided it into two members, a lower argillaceous member named "Sargelu-Dhruma Transition", and the overlying member named "Sargelu Limestone". The Sargelu Formation is the thickest in the country's southern region, adjacent to the Rimthan Arch near Umm Gudair Field, where it reaches a maximum thickness of more than 220 ft ($\sim$67.1 m). It thins consistently northward, parallel to the Burgan Arch, to less than 60 ft ($\sim$18.3 m) along the border with Iraq (Cabrera et al., 2019).

The Sargelu Formation, as interpreted by Kadar et al. (2015), Cabrera et al. (2019) is predominantly an aggradational sequence. The Sargelu–Dhruma Transition comprises extensively burrowed, interbedded argillaceous limestones and skeletal wackestones, with a decreasing number of calcareous nannofossils. The suggested conceptual model for the Sargelu–Dhruma Transition is an outer-ramp, slope, and basin margin. However, the Sargelu Limestone has been found to contain only thin shale interbeds that shallow upward into packstones consisting of cortoids, coated grains, and skeletal fragments with no Flora in the limestone. The conceptual model of the Sargelu Limestone is an open marine shelf.

Kadar et al. (2015) made the Nannofossil recovery in the Sargelu–Dhruma Transition. It was good and suggested Late Bajocian (Middle Jurassic) age. On the other hand, samples from the Sargelu Limestone were barren of nannofossils. Accordingly, the age was not determined, and they place lower Bathonian MFS (maximum flooding surface) near its base.

The Sargelu Formation conformably overlies the Dhruma Formation. Yousif and Nouman (1997) identified the contact between the Sargelu Formation with the underlying Dhruma Formation by the change from predominantly limestone with subordinate shale interbeds to mainly calcareous shale with subordinate limestone interbeds. The contact with the overlying Najmah Formation is interpreted to be conformable and transitional (Kadar et al., 2015; Yousif & Nouman, 1997).

The Sargelu Formation is one of the main carbonate reservoirs in the Jurassic section in Kuwait that includes Najmah, Sargelu, and Marrat Formations. (Singh et al., 2009).

2.3.1.4 Najma Formation

In Kuwait, the Najmah Formation is equivalent to that drilled on the Najmah–Quayarah Anticline southeast of Mosul located in northern Iraq (Bellen et al., 1959). It is the second formation of the Riyadh Group.

Kadar et al. (2015) divided the lithostratigraphy of the Najmah Formation penetrated in Minagish-27 Well into Najmah Shale and Najmah Limestone with further subdivision of Najmah Shale into three subunits:

1. Barren Najmah–Sargelu Transition.
2. Lower Najmah Shale developed during the middle Jurassic, between Late Bathonian and Middle Callovian.
3. Upper Najmah Shale deposited during Middle Callovian to Middle Oxfordian (i.e., Late middle Jurassic–Early-Late Jurassic).

The Najmah Limestone was interpreted to be deposited during Late Oxfordian (Late Jurassic). Kadar et al. (2015) approximated the thickness of the Najmah Formation ranges from 140 ft (42.7 m) in the north to 220 ft (67.0 m) in the south. The three subdivisions of Najmah Shale have correlated throughout Kuwait stratigraphic sections and were recognized by the sharp contact boundaries detected via electrical well-logs due to the variations in limestone lithofacies of strata that are interbedded with fissile, organic-rich, argillaceous lime mudstones reflecting different environmental settings (Fig. 2.3). (see Kadar et al., 2015 for more details).

The Najmah–Sargelu Transition, according to Kadar et al. (2015) does not contain calcareous nannofossils. Nevertheless, it consists of interbedded microlaminated organic-rich argillaceous lime mudstone (ORAM), skeletal microbial wackestones, and mud-dominated packstone beds with cortoids, coated grains, and various skeletal fragments. Calcareous nannofossils. This accumulation hypothesized transportation downslope from an outer-shelf or ramp environment and the microbial skeletal wackestones believed to signify an oxygen-minimum zone (OMZ) environment along the basin margin (Kadar et al., 2015).

The Lower Najmah Shale, as described by Kadar et al. (2015), is a lithological unit of low diversity of nannofossils. It contains a distinguishing association of filamentous microbial wackestones, tiny articulated bivalves, and juvenile ammonites, interbedded with ORAM. These beds were assumed to have been deposited in-situ in a base-of-slope environment along the basin margin.

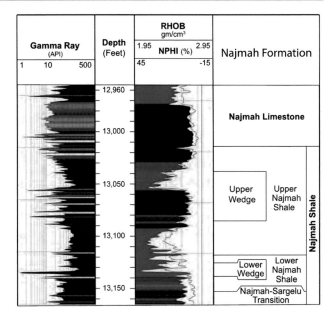

Fig. 2.3 Log display of the Najmah formation in Minagish-27 Well (Modified after Kadar et al., 2015)

Kadar et al. (2015) defined the Upper Najmah Shale as it was lacking burrowing and containing thick accumulations of the pelagic bivalve with abundant calcareous nannofossil assemblies suggesting a more basin-like environment. Furthermore, they found that it consists of predominantly ORAM with a few pure lime mudstone beds that could characterize very distal subaqueous flows.

The Najmah Limestone is characterized by Kadar et al. (2015) by the absence-to-abundance of nannoflora. They also found the basal beds overlie a condensed zone representing an outer shelf to the upper slope. The succession of lithofacies and ichnofacies could suggest shallowing upward and probable progression to the inner shelf or lower shoreface depositional environment.

The underlying boundary of the Najmah Formation with the Sargelu Formation is placed at the highest grain-supported beds. It is transitional, with thin organic-rich (kerogen) beds appearing in the uppermost Sargelu Formation and increasing upward. (Kadar et al., 2015).

The Najmah Limestone lies conformably on the Najmah Shale and unconformably overlain by a thin unit of shale, calcareous shale, and peloidal limestone beds referred to as Jubaila Formation. This unconformity surface has been correlated by Kadar et al. (2015) in eight cores as a minimum, with noticeable karsting and an indication of surface exposure. Erosion induces differences in the thickness of Najmah Limestone at this unconformity (Cabrera et al., 2019; Kadar et al., 2015).

2.3.1.5 Jubaila Formation

The Jubaila Formation type section is defined in Saudi Arabia (Powers et al., 1966). In Kuwait, Kadar et al., 2015 introduce for the first time the name of Jubaila Formation in Minagish-27 Well, referring to the lithological section with 46 ft ($\sim$ 14 m) thick and is thinnest towards the north to ca 8 ft (2.4 m) thick in Dhabi area (area in North of Kuwait). In the Sabriyah and Raudhatain Field area, it may be vanished or only a few inches to one or two feet in thickness (Carbrera et al., 2019).

The Jubaila Formation corresponds to the lower part of Najmah Unit 1 that Yousif and Nouman (1997) recognized, and below the fourth Anhydrite unit of the Gotnia Formation. The sediments of the Jubaila Formation are generally dark, reddish-brown to nearly black, and consist predominantly of dolomitic limestone, argillaceous limestone, and calcareous shale (Kadar et al., 2015).

Lithologically, the Jubaila Formation is a sandwich between two surfaces of unconformities that distinguished it from the overlain Gotnia Formation and the underlain Najmah limestone (Fig. 2.1) (Kadar et al., 2015).

Kadar et al. (2015) investigated the lithofacies of the Jubaila Formation across Kuwait. They predicted that the depositional setting in North Kuwait could be peritidal to shallow subtidal environments concerning sediment contents that consist of generally argillaceous, laminated to thin-bedded peloidal and mudstones, mud-dominated packstones, and wackestones. They also noticed that the formation had become more argillaceous and less microbial and had become burrowed with thickenings southward and westward. This type of sedimentation was interpreted to represent an inner ramp/shelf environment.

Kadar et al. (2015) identified extensive calcareous nannofossils in the argillaceous mudstone in some wells located in the south, northwest, and north of Kuwait. They also found Coccospheres within the Jubaila Formation, reflecting a low-energy, open-marine, depositional environment. Kadar et al. (2015) identified nannofossil assemblages in Jubaila Formation related to the Kimmeridgian age.

The high gamma-ray values and the low density between the top of the Najmah and the base of the Gotnia Formation were used to exclude the Jubaila unit from logs, indicating that it could be a source rock (Cabrera et al., 2019).

2.3.1.6 Gotnia Formation

Yousif and Nouman (1997) correlated the salt and anhydrite section penetrated in Burgan-113 Well in Kuwait with the Gotnia Anhydrite Formation in central Iraq in the Awasil-5 Well (Bellen et al., 1959). The Gotnia is equivalent to the Arab Formation in the Kingdom of Saudi Arabia (Cabrera et al., 2019).

The Gotnia Formation consists of alternating layers of cycles of salt and anhydrite units with a maximum of four cycles (Cabrera et al., 2019; Kadar et al., 2015). These anhydrite interbeds are mostly interlaminated with limestones, shales, and bituminous limestones.

The maximum thickness of the Gotnia Formation encountered in southern Kuwait in Minagish Field as it reaches 1,397 ft (426 m) thick (Kadar et al., 2015) and is thinning towards northeast Kuwait due to a gradual pinch out of the basal Gotnia units (Yousif & Nouman, 1997).

Kadar et al. (2015) couldn't date the evaporitic Gotnia and Hith Formations since they were barren of the calcareous nannofossils. However, they depend on Al-Sahlan's (2005) report based on the strontium-isotope analysis in West Kuwait, which assigned that the Hith Formation is of the approximate age of Kimmeridgian–Tithonian, and this would confine the Gotnia Formation to Kimmeridgian.

The Gotnia Formation was anticipated to be deposited in the restricted evaporitic Gotnia Basin. The basin was bordered to the south by the NE-trending Rimthan Arch, around the Kuwait–Saudi Arabian boundary (see Fig. 11 in Ziegler, 2001), and to the east by the Burgan Arch (Kadar et al., 2015; Wharton, 2017). This hypersaline environment occurred during the Late Jurassic (see Fig. 2 in Yousif & Nouman, 1997). The limestones and dolomites in the Gotnia anhydrite units are laminated to thin-bedded peloidal and microbial mudstones to packstones. Occasionally, the anhydrites conserve the geometry of stromatolites as well, suggesting deposition under more constrained conditions of a sabkha, tidal flats, or very shallow subtidal environments (Kadar et al., 2015).

Kadar et al. (2015) identified four anhydrite stratigraphic succession of the Gotnia Formation in many localities throughout Kuwait. They recognized that the basal fourth anhydrite unit characterized by microbial, evaporitic, and un-burrowed is unconformably overlain the Jubaila Formation. The upper boundary with the Hith Formation signifies the top of AP7 TMS (seventh Arabian Plate Tectonostratigraphic megasequence) (Fig. 2.1) that is defined by the widespread early Tithonian unconformity hiatus surface, which overlies the Late Jurassic evaporite deposits of the Gotnia and Hith formations (Sharland et al., 2001; Cabrera et al., 2019).

2.3.1.7 Hith Formation

The Hith Formation is the uppermost formation in the Jurassic section and the youngest member of the Riyadh group. The Hith Formation is exposed in a large anhydrite solution cavity called Dahal Heet (Dahl Hit; Ain Heet Cave), situates at the face of Mount Al Jubayl in Wadi As Sulay in a small village called Heet between Riyadh city and Alkharj, Saudia Arabia, Fig. 2.4 (Steineke et al., 1958; Powers et al., 1966; Splendid Arabia website, 2020). The cave sinkholes serve as an entrance to a large groundwater reservoir of Arab Formation (Gotnia Formation in Kuwait) that is

Fig. 2.4 Dahal Heet (Ain Heet Cave), the Blue line delineates the contact between the Hith Formation (below) and the Sulay Formation (up). Photo (**a**), the sinkhole developed due to the dissolution of Hith anhydrite and the collapse of the above Sulay (photo by Dr. Mohammad Al-Mahmood, Ex Geologist in Aramco). Photo (**b**) Copyright https://saudiarabiatours.net/trip-to-heet-cave/ showing the groundwater in Heet Cave

a

b

situated under the surface at a great depth, and it is the only place in Saudi Arabia where you can find the exposed Hith Formation (Powers et al., 1966).

Kadar et al. (2015) analyzed the core and interpreted the logs from Minagish-27 Well and described the Lithofacies of the Hith Formation; in such, it is mainly composed of interbedded massive to nodular anhydrite and peloidal to stromatolitic microbial limestones and dolomites; without salt beds and it becomes more carbonate-dominated upward. The Formation was truncated at 11,199 ft (3,413 m) depth with a thickness of 333 ft (101.5 m) in Minagish-27 Well, and it thins northward Kuwait. Hith Formation varies in thickness from 61 m (200 ft) in the North of Kuwait to 335 m (1,100 ft) in the Southwest of Kuwait. (Al-Sharhan & Nairn, 2003).

As in the case of the Gotnia Formation, the Hith Formation is also assumed to be deposited in sabkha, tidal flat, and shallow subtidal environments. (Kadar et al., 2015; Yousif & Nouman, 1997).

The contact between the Hith Formation with the underlying Gotnia Formation, as described by Kadar et al. (2015), is sharp and conformable. While the upper contact, with the Makhul (Sulaiy) Formation, is assumed unconformity, as they notice a profound shift in facies developed from the shallow, confined carbonate-evaporite cycles of the Hith Formation to deep, open-marine, basinal argillaceous mudstones interrupted by thin, probably distal turbidites that defined the Makhul Formation.

Al-Sahlan (2005) delineated the revised ages of formations using new biostratigraphic and strontium isotope data in a figure illustrating a hiatus in the Tithonian separating the

Hith Formation from the overlying Makhul Formation. From Sr isotope analysis in West Kuwait, The Hith Formation was assigned an age of about Tithonian/Kimmeridgian, which would confine the Hith–Gotnia section to Kimmeridgian.

The Gotnia and Hith formations are excellent caprocks for the underlying reservoirs (Kadar et al., 2015).

2.3.1.8 Makhul Formation

The Makhul (Sulaiy) Formation was located in Makhul-1 Well of northern Iraq (Bellen et al., 1959). According to Powers et al. (1966), the formation was first defined on the stable shelf in Saudi Arabia (Al-Sharhan & Nairn, 2003). Makhul Formation is the oldest sedimentary sequence of the Thamama Group in Kuwait. The Thamama Group accommodate Makhul, Minagish, Ratawi, Zubair and Shuaiba.

The Makhul Formation lithofacies were divided by Arasu et al. (2012) into upper, middle, and lower zones. The upper zone is an argillaceous lime mudstone/wackestone, interbedded with slightly dolomitic limestone. The middle zone comprises of gray mudstone to wackestone with distinctive packstone. The lower Makhul zone host kerogen-rich layers with alternating laminae of mudstone and wackestone. Arasu et al. (2012) suggested that it can act as a source rock.

The thickness of the formation varies in different locations: in Raudhatain Field (ca. 570 ft, 173,7 m), in northwest Raudhatain Field (ca. 780 ft, 237.7 m), and Minagish Field (650 ft, 198.1 m) (see Fig. 2.2 for locations) (Kadar et al., 2015).

Kadar et al. (2015) interpreted that the Makhul Formation demonstrates a significant deepening upward setting from the Hith Formation and gradually shoals upward to the Minagish Oolite.

The Jurassic/Cretaceous (Tithonian/Berriasian) boundary positioned in the argillaceous mudstone of the Makhul Formation in the Minagish Field was deposited above the Hith anhydrite during the Late Jurassic (late Tithonian)– Early Cretaceous (Early Berriasian) (Arasu et al., 2012; Kadar et al., 2015). The upper contact with the overlying Minagish Formation is described by Kadar et al. (2015) as diachronous.

2.3.2 Cretaceous

2.3.2.1 Minagish Formation

Kuwait Oil Company adopted the Minagish Formation name concerning well Minagish-8 in the Minagish Oil Field (Fig. 2.2) (Al-Sharhan & Nairn, 2003).

The Minagish Formation is the second sedimentary sequence of the Thamama Group in Kuwait overlying Makhul Formation. It is of the Early Cretaceous age (Berriasian– Valanginian), and it was divided into three members: Upper, Middle, and Lower. The Middle Oolitic Member is the main reservoir unit capped with the Upper Member, which comprises carbonate mudstone. The lithofacies of the Middle Member base and the Lower Member are composed of fine-grained, bioturbated, peloidal lime packstones (Abdullah & El Gezeery, 2016). The thickness of the Formation in Kuwait varies from 160 m (528 ft) in the south to nearly 360 m (1,188 ft) in the north. (Abdullah & El Gezeery, 2016).

The Oolitic limestone of the Middle Member composed of peloidal, bioclastic oolitic grainstone and packstone has been deposited on a broad, prograding carbonate ramp (Nath et al., 2014). The Middle Member is confined between a hard, dense, bioclastic micritic limestone. The Minagish Formation is fossiliferous, comprising skeletal fragments, ostracods, miliolids, echinoderms, calpionellids, and benthonic foraminifera, suggesting deposition in a shallow-water carbonate shelf environment characterized by the development of ooid shoals (Al-Sharhan & Nairn, 2003).

The Minagish Formation is one of the major oil reservoirs in South Kuwait (Abdullah & El Gezeery, 2016).

2.3.2.2 Ratawi Formation

The Ratawi Formation overlies the Minagish Formation and underlies the Zubair Formation. The formation was first described in the Ratawi Field in southern Iraq. It is identified in Kuwait, Bahrain, and Qatar, and it is equivalent to Buwaib in Saudi Arabia and Habshan and Lekhwair formations in the U.A.E. and Oman, respectively (Al-Sharhan & Nairn, 2003).

There is uncertainty about the age of the Ratawi Limestone in Kuwait, but it was inferred from Abu Dhabi and Yemen regarding Maximum Flood Surface (MFS) that associated with age diagnostic fauna like the *Salpingoporella pygmaea Biozone* in the upper Habshan Formation of the United Arab Emirates (Aziz & El-Sattar, 1997) and indicates an early Valanginian age (Sharland et al., 2001).

The Ratawi formation was divided into an upper Ratawi Shale Member and a lower Ratawi Limestone Member (Ratawi Oolite), reflecting the proportion of lime mud-wackestone, calcareous shales, and marls in each unit (Al-Fares et al., 1998). The formation varies in thickness from about 951 ft (290 m) in northern Kuwait to 426 ft (130 m) in the south. The upper part shaley unit seals off significant oil and gas accumulations in the lower Ratawi limestone and the underlying Minagish limestone reservoirs. (Al-Sharhan & Nairn, 2003).

The Ratawi Shale Member host three separate sandstone beds deposited as isolated sand bodies of limited extent in a shallow-marine setting. The Ratawi limestone is poorly interbedded with dense intervals of about 20% shale and is suggested to be deposited in intra-shelf and shallow carbonate environments. (Abdullah & Connan, 2002; Al-Sharhan & Nairn, 2003). Ratawi Formation contains benthonic foraminifera, algae, calpionellids, and other skeletal fragments, suggesting deposition on a shallow marine, shelf environment (Al-Sharhan & Nairn, 2003).

The upper boundary of the Ratawi with the Zubair Formation is referred to as the "Late Valanginian Unconformity" ("LVU"). It reflects a significant stratigraphic hiatus and erosion of the late Valanginian and possibly some early Hauterivian strata (Al-Fares et al., 1998). The lower Contact with Minagish formation is less conformable, as delineated from collected well data while drilling.

2.3.2.3 Zubair Formation

The Zubair Formation was deposited with a noticeable unconformity above the top of the Ratawi Shale (Al-Fares et al., 1998). It is almost identical to the Riyadh Formation in Saudi Arabia (Al-Sharhan & Nairn, 2003), and it is roughly equivalent to the lower and middle parts of the Zubair Formation in Iraq (Al-Sharhan & Nairn, 2003; Buday, 1980).

The Zubair Formation is a significant siliciclastic wedge in the Northern Arabian Gulf zone, deposited between the Ratawi Shale and the overlying Shuaiba carbonate formations. The lithostratigraphic successions of the Barremian– Lower Aptian (Early Cretaceous) Zubair Formation thickens from south to north, and it ranges in thickness from 1,158 ft (353 m) in the south to about 1,476 ft (450 m) in the north (Al-Sharhan & Nairn, 2003; Owen & Nasr, 1958).

The Formation predominantly consists of clastic sandstone interbedded with gray to black, thin laminated siltstone, and shale. In some places, a minor amount of limestone is present at the base of the formation (Owen & Nasr, 1958; Al-Sharhan & Nairn, 2003; Nath et al., 2014).

The regional geological correlation and litho-facies investigations designate that the Zubair Formation has been deposited on a fluvial delta complex that episodically displayed an estuarine character to a coastal plain environment with more marine (deeper) influence further eastward (Al-Sharhan & Nairn, 2003; Nath et al., 2015). Abdul Azim et al. (2019) discussed the impact of the depositional environment and the sequence stratigraphy and the structure on developing Zubair reservoirs in North Kuwait.

The shale layer within the Zubair formation forms a good seal for the oil and gas reservoirs in interbedded Zubair sand units. (Al-Sharhan & Nairn, 2003).

2.3.2.4 Shuaiba Formation

The Shuaiba Formation is the youngest unit of the Thamama Group. It was deposited during the Valanginian to Aptian stages (Early Cretaceous) and extended regionally through the Arabian Gulf region (Sharland et al., 2001).

The Shuaiba Formation in Kuwait thickens towards the north from 197 ft (60 m) in the south to 262.5 ft (80 m) in the north. It comprises coarse crystalline, porous, heavily fractured, and cavernous, dolomitized limestone with rare thin shale suggested to be deposited in a low-energy, shallow, lagoonal environment during the Aptian period (Al-Sharhan & Nairn, 2003).

From field observations and the data collected while drilling (i.e., cutting samples or core data) in Kuwait oil fields, the upper and the lower contact of the Shuaiba Formation is confirmable as it is a carbonate sandwich between two clastic formations.

The Shuaiba Formation carbonates are significant hydrocarbon reservoirs in some Arabian Gulf countries (Hohman et al., 2005).

2.3.2.5 Burgan Formation

The Burgan Formation nomenclature accredits the Burgan field, the greatest oil field in Kuwait. The Burgan field was placed on an elliptical-shaped anticline dome transected by multiple radial faults.

Burgan Formation is the oldest member of the Wasia group representing the Middle Cretaceous that overlies the Lower Cretaceous Thamama Group of the Arabian plate. Wasia Group in Kuwait is constructed of six members: Burgan, Mauddud, Wara, Ahmadi, Rumaila, and Mishrif. Burgan Formation was deposited during lower to middle Albian, and it is equal to the Nahr Umr formation of southeastern Iraq (Al-Sharhan & Nairn, 2003).

As described in the Burgan field, the Burgan Formation is composed of siliciclastic sediments characterized by well-bedded, well-sorted, rounded, medium- to coarse-grained littoral sands deposited near a delta front on a gradually sinking shelf (Al-Sharhan & Nairn, 2003). Figure 2.5 represents the present-day Mackenzie delta in Canada, a possible analog to the Burgan paleoenvironment. The Burgan Formation thickness ranges from approximately

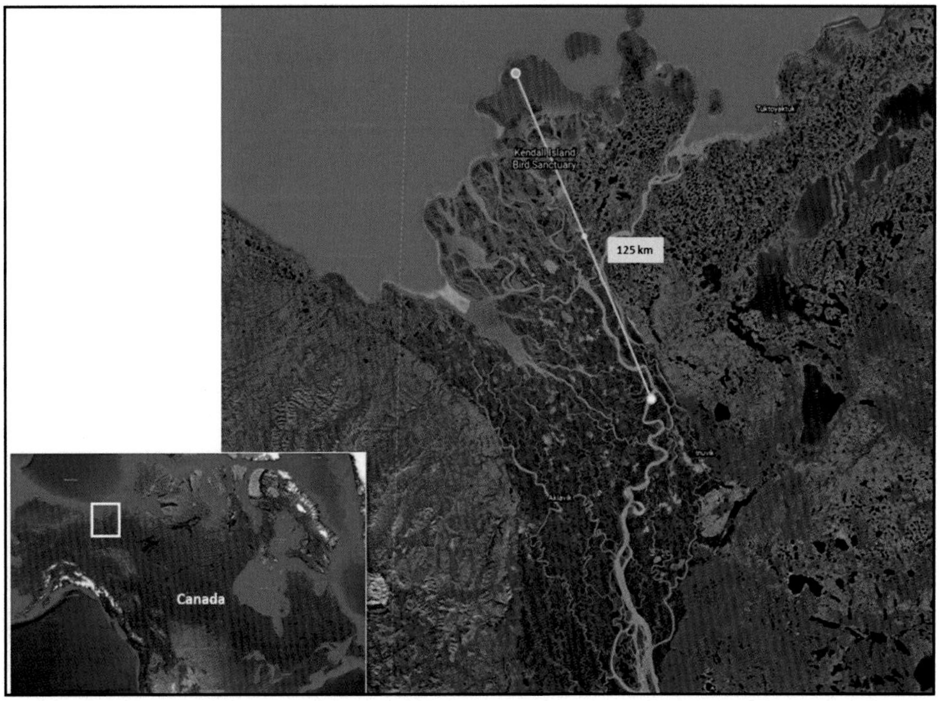

Fig. 2.5 Mackenzie delta in Canada, present-day analog to Burgan paleoenvironment (Google earth V 7.3. Mackenzie delta, Canada. 68° 51′ 11.61″ N, 134° 39′ 12.89″ W, Eye alt 265.16 miles. IBCAO 2021. https://earth.google.com [March 13, 2021])

1250 ft (380 m) at the Greater Burgan field area to nearly 900 ft (275 m) at Raudhatain and Sabriyah fields area (Bou-Rabee, 1996). It is grading upward into alternating fine-grained sandstone and siltstone. The shale is estuarine and contains abundant plant remains, Lignite, amber, and glauconite that exist all over the sequence with no foraminifera (Al-Sharhan, 1994; Brennan, 1990; Owen & Nasr, 1958).

Owen and Naser (1958) distinguished two units in the Burgan formation: the "Third" and "Fourth" Sand units. The "Third Sand Unit" (comparable to the Safaniya Member in Saudi Arabia) is approximately 476 ft (145 m) of glauconitic sand in the lower part, with an upper part of interbedded dark-gray shale. The middle section is almost pure quartz sand with very little secondary cementation and includes amber and lignite traces. The "Fourth Sand Unit" (equivalent to the Khafji Member in Saudi Arabia) has a total thickness of about 675 ft (206 m), of clean, well-sorted sand, with little secondary cementation and some traces of lignite, amber, and plant residues. Since no microfossil content in the sandstone, the Burgan Formation was dated by its apparent time equivalent with the Nahr Umr Formation in Iraq (Al-Sharhan & Nairn, 2003).

Strohmenger et al. (2006) discussed the sequence-stratigraphic framework and reservoir architecture of the Burgan and the overlying Mauddud Formation in Kuwait. The Burgan Formation is the producing reservoir in the Burgan, Ahmadi, Bahrah, Minagish, Raudhatain, and Sabriyah fields in Kuwait (Al-Sharhan & Nairn, 2003).

2.3.2.6 Mauddud Formation

The Mauddud Formation, first identified in Qatar (Sugden & Standring, 1975), is a common term used in the Middle East (Kuwait, Saudi Arabia, the UAE, and Bahrain) to the shallow marine carbonate-dominated succession resulting from the transgression that followed the deposition of the fluvial-deltaic-related clastic-dominated sediments of Burgan Formation (Cross et al., 2010).

To understand Mauddud reservoir heterogeneity, Cross et al. (2010) investigate detailed, integrated sedimentological and biostratigraphic, and dynamic reservoir data collected from Raudhatain and Sabiriyah Fields in northern Kuwait. The average thickness of the formation recorded was about 427 ft (130 m) and comprised a lower, interbedded mixed carbonate and clastic succession overlain by an upper, bio-turbated carbonate-dominated interval in which oil is accumulated. The foraminifera (e.g., orbitolina seifini) recorded within Mauddud Formation as recognized by Cross et al. (2010), indicating deposition during the Middle Cretaceous of the Late Albian Age.

The hypothetical depositional model described by Strohmenger et al. (2006) for the Mauddud Formation of north Kuwait as deposition on an open, northward-deepening carbonate ramp or low-angle shelf across which there were intermittent proximal siliciclastic incursions from the retreating Burgan that inhibited carbonate productivity (Cross et al., 2010) (Fig. 2.6).

Siliciclastic incursions punctuate the lower carbonate succession of the Mauddud reservoir assumed to be deposited in a high-energy inner to middle ramp environment. Compared to the Upper Mauddud, it is deposited in low energy inner to middle carbonate ramp (Cross et al., 2010).

Significant post-depositional erosion occurs at the contact between the Mauddud carbonate and the overlying Ceno-manian Wara Shale (Strohmenger et al., 2006). On the other hand, Sharland et al. (2001) documented a sharp contrast of the offshore mudrocks and the outer ramp wackestones at the base of the Mauddud Formation with the shoreface

Fig. 2.6 Hypothetical carbonate ramp depositional model of Muaddud Formation in Raudhatain and Sabiriyah Fields (modified after Cross et al., 2010)

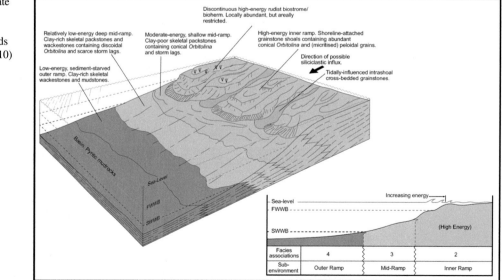

sandstones of the underlying Upper Burgan Formation that interpreted to indicate a substantial decrease in the supply of siliciclastic sediment from the Arabian Shield following the regional flooding event.

Mauddud Formation is highly affected by carbonate diagenesis, which enhances the hydrocarbon reservoir properties, especially on Kuwait's northern side. These include interparticle porosities created by allochems dissolution and fracturing (Cross et al., 2010).

The Mauddud carbonate reservoir is considered one of the primary oil producers in the domal structures of Raudhatain and Sabiriyah Fields, located in northern Kuwait (Cross et al., 2010).

2.3.2.7 Wara Formation

The Wara Formation consists of sandstone and interbedded shale. Carbonates are developed in the upper part of the formation in the north and northeast, whereas shale is dominant in the south and southwest of Kuwait (Al-Sharhan & Nairn, 2003).

The formation is a shallow marine clastic deposited in the progradation deltaic system of the offshore environment. The average thickness of Wara Formation from south to north is 151 ft (46 m)–299ft (91 m), respectively, and in the Greater Burgan Field, it ranges in thickness from 140–180ft (40–50 m) of which up to 60% of the total thickness comprises reservoir sand (Al-Sharhan & Nairn, 2003).

Palynological data helped Al-Enezi et al. (2011) reconstruct the relative sea level for the Burgan, Mauddod, and Wara Formations in the Greater Burgan field. They demonstrate that Burgan Formation was deposited during most of the Albian, whereas Mauddud is of the Late Albian age. Wara shales are early Cenomanian, while Wara sand and basal Ahmadi are of the Cenomanian age.

The Wara Formation is conformably overlain by the Ahmadi Formation, while it lies with slight disconformity on the Mauddud reservoir, forming a suitable hydrocarbon sealing rock (mudstone) (Bellen et al., 1959).

2.3.2.8 Ahmadi Formation

The transgressive Ahmadi Formation was deposited during the Early Cretaceous, Early Cenomanian age (Sharland et al., 2001). The sedimentary facies of the Ahmadi Formation were divided into discrete members, the upper and the lower. The upper member comprises clastic shale, while the lower member comprises wackestone and packstone carbonate rocks. The Lower Ahmadi carbonate is interbedded between Ahmadi shale at the top and Wara shale below (Zaidi et al., 2009). Ahmadi Formation is deposited in the middle ramp to the offshore environment. The thickness ranges from 203 ft (62 m) to 266 ft (81 m) (Al-Sharhan & Nairn, 2003).

A precise sequence boundary was identified between the non-calcareous lagoonal shale of the upper Ahmadi Formation and the highly fossiliferous, calcareous marine shale at the base of the Rumaila Formation (Youssef et al., 2014).

2.3.2.9 Rumaila Formation

Rumaila Formation is the second member of the Wasia Group deposited earlier than the Mishrif Formation. It was deposited in the Early Cenomanian age (Early Late Cretaceous). The Rumaila comprises mudstone, partially dolomitized bioclastic wackestone, calcareous shale, and marl with abundant microfossils and nanofossils (foraminifera and ostracods) (Youssef et al., 2014). The Rumaila Shale is described by Jaber (1972) as representing "the time of maximum Cenomanian transgression".

The proposed depositional environment ranges between the middle and the outer ramp with occasional inner ramp facies (Yossef et al., 2014). Correlation of the formation between the wells indicates that the thickness of Rumaila varies from 140 ft (43 m) in southeastern Kuwait to about 450 ft (137 m) in northern Kuwait (Al-Sharhan & Nairn, 2003).

The Rumaila Formation lies above Ahmadi Formation, with clear contact from non-calcareous lagoonal shale to highly fossiliferous, calcareous marine shale at the bottom of the Rumaila formation. The upper contact with Mishrif Formation was identified by the lithofacies' variation representing the difference in the deposition environments. The deeper facies of the Rumaila were marked by a much richer planktonic mudstone/wackestone. On the Other hand, the shallower facies of the overlying Mishrif Formation were observed by the recrystallization of the dominant packstone and occasional grainstone mainly enriched by praealveolinids, algal debree, permocalculus, with shell fragments, gastropods, and bivalves enriched upward with rudistid and coralline components (Youssef et al., 2014).

The Rumaila is meant to be a vital hydrocarbon seal rock in Kuwait, while the Mishrif is considered a satisfactory reservoir towards the south (Youssef et al., 2014).

2.3.2.10 Mishrif Formation

Mishrif is the youngest formation in the Wasia Group, spanning from the latest Aptian, Albian, and Cenomanian to the earliest Turonian ages of the Middle Cretaceous period (Youssef et al., 2014). The Wasia Group in Kuwait is constructed of six members arranging from the oldest to the youngest: Burgan, Mauddud, Wara, Ahmadi, Rumaila, and Mishrif. Mishrif was deposited during the Late Cretaceous, late Cenomanian, to possibly the early Turonian age (Al-Fares et al., 1998).

The Mishrif Formation is comprising mainly of bioclastic wacke/mudstone and contains various fossils, planktonic and benthonic foraminifera, ostracods, miliolids, and rudist bioherms with a chalky limestone fabric (Youssef et al., 2014; Ziegler, 2001). There is an unambiguous fossil changing upward of the Mishrif Formation. The lower Mishrif has abundant planktonic and benthonic foraminifera, while the upper part comprises many rudists and shell fragments (Yossef et al., 2014; Ziegler, 2001). This change is caused mainly by the middle ramp's regression to the inner ramp depositional environment (back reef, lagoon, and shoal) (Youssef et al., 2014).

The thickness of Mishrif is about 249 ft (76 m), and it was entirely eroded over the Burgan and Khafji–Nowruz Arches due to the pre-Aruma regional erosional surface (Al-Sharhan & Nairn, 2003; Ziegler, 2001).

The contact between the Mishrif Formation with the underlying Rumaila Formation is sharp and clear, showing regression from the deeper water facies of Rumaila to the shallower water facies of Mishrif (Yossef et al., 2014). The upper boundary marked hiatus followed marine deposits of the Mishrif Formation developed unconformity (Sharland et al., 2001). It is a tectonostratigraphic megasequence boundary (represents upper AP8) formed during Middle Turonian, causing variously erosive unconformity between the base of the Mutriba Formation and the top of Mishrif Formation Fig. (1) (Sharland et al., 2001; Ziegler, 2001; Youssef et al., 2014).

The diagenesis in Mishrif Formation has enhanced the hydrocarbon reservoir properties with vuggy porosity and fractures (Sharland et al., 2001; Youssef et al., 2014).

2.3.2.11 Mutriba

The Mutriba Formation is the oldest member of the Aruma Group in Kuwait. This group is constructed from four formations ranging from oldest to youngest as Mutriba, Sadi, Hartha, and Tayarat (Al-Sharhan & Nairn, 2003).

Mutriba consists of white to gray, dense detrital limestone, interbedded with shale horizons near its base. Lateral/geographic and vertical/stratigraphic wells correlations indicate that the formation is thinnest to the south, with thickness ranges from 928 ft (283 m) on the northern side of Kuwait to 79ft (24 m) on the southern side (Al-Sharhan & Nairn, 2003; Sharland et al., 2001).

Mutriba Formation lying above the Middle Turonian unconformity developed between the Aruma (above) and Wasia (below) lithostratigraphic groups resulting from the start of ophiolite obduction along the eastern margin of the Arabian Plate (Ziegler, 2001). This formation was deposited in the Early Cretaceous during the Santonian age. The upper Boundary with the Sadi formation is conformable.

2.3.2.12 Sadi Formation

The Sadi Formation is the second oldest Member of the Aruma Group, overlying the Mutriba Formation. Sadi Formation was deposited during the Late Cretaceous Period from the Early Santonian to Campanian age, and it lies conformably above Mutriba and unconformably overlain by Hartha Formation. The formation the thickest towards the north, from 39 ft (12 m) in the south to around 997 ft (304 m) in the north. The lithology consists mainly of fossiliferous lime mudstone interbedded with shale and dolomite (Al-Sharhan & Nairn, 2003; Han et al., 2015).

The hypothetical depositional environment of the Sadi Formation in Kuwait is correlated with that of southeast Iraq. Han et al. (2015) analyzed the characteristics and the genesis of the carbonate Sadi Formation reservoir in southeast Iraq. They found that Sadi Formation represents a third-order depositional cycle consisting of limestones and marlstones containing planktonic foraminifera and bioclastic debris as well as significantly bioturbated argillaceous rocks, believed to be deposited on a carbonate ramp setting.

Sadi and the overlying Hartha Formations are relatively shallow reservoirs that occur at depths ranging from 3500 ft (1067 m) to 4000ft (1219 m) in the southern part of the Burgan Field (Singavarapu et al., 2015).

2.3.2.13 Hartha Formation

The Hartha Formation is the third Member of the Aruma Group, on top of the Sadi Formation. There is no apparent contact between Sadi and Hartha, except that Hartha is composed of organic-rich limestone.

Hartha was deposited in the Late Cretaceous (Late Campanian to Early Maastrichtian) (Al-Sharhan & Nairn, 2003). Singavarapu et al. (2015) stated that the lithology of the Hartha Formation is limestone interbedded with shale. The thickness ranges from zero thickness over the crestal part of the great Burgan anticline to about 200ft (61 m), with a distribution that closely follows the structural trend. Singavarapu et al. (2015) assumed the depositional environment for the Hartha Formation as observed from the Minagish Field's cores to be distal inner to the middle ramp.

The formation is subdivided into three depositional units: upper, middle, and lower. The upper unit is typically composed of shaly limestone, and the middle unit is mostly wackestone to packstone in other areas. The lower unit is primarily wackestone grading upward into packstone (Singavarapu et al., 2015).

2.3.2.14 Tayarat Formation

Tayarat Formation represents the upper Cretaceous succession in Kuwait's stratigraphic column (Fig. 2.1) (Al-Sharhan & Nairn, 2003). The formations dips towards north and northeast cause variation in thickness from north to south of Kuwait, where it is around 656ft (200 m) in the south and about 1148 ft (350 m) in the north (Al-Sharhan & Nairn, 2003; Hayat et al., 2018). The Tayarat Formation in the Burgan field is roughly 800 ft (about 244 m) thick (Al-Hajeri & Bowden, 2018).

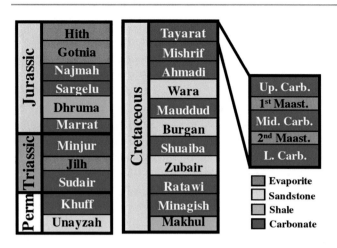

Fig. 2.7 Simplified stratigraphic column of Kuwait showing the five geological units of Tayarat Formation (After Al-Hajeri and Bowden, 2018)

Tayarat Formation is estimated to be a part of a shallow-water carbonate platform complex deposited across the Arabian Shield's interior during the late Maastrichtian stage of the Late Cretaceous (Al-Sharhan & Nairn, 2003; Dunnington, 1958; Owen & Nasr, 1958).

The Tayarat Formation comprises five geological units, three carbonate units interbedded with two thin carbonaceous shale zones, classified from top to bottom as follows: (1) the upper carbonate unit composed mainly of wackestone; (2) first Maastrichtian carbonaceous shale; (3) middle carbonate unit comprises of mudstone intercalated with wackestone; (4) second Maastrichtian carbonaceous shale; and (5) lower carbonate unit composed of packstone (reservoir pay zone), floatstones, and argillaceous floatstones intercalated with mudstones and shales (Fig. 2.7) (Al-Hajeri & Bowden, 2018).

The upper boundary of the Tayarat Formation represents the top of the tectonostratigraphic megasequence (AP9) that corresponds to the unconformity between the Mesozoic and Cenozoic. In Arabian Peninsula, this Pre-Cenozoic unconformity marks the end of the last phase of major Cretaceous ophiolite obduction along the northern plate margin and results in widespread regression (Fig. 2.1; Beydoun, 1991; Shalrland et al., 2001). Mud-logging data detected a thin shale separating the Tayarat Formation from the overlying Radhuma Formation (Taqi et al., 2018).

Tayarat and Radhuma Formations are shallow carbonate reservoirs in Kuwait (Al-Hajeri and Bowden, 2018). The Tayarat Formation truncated in the Burgan Field is highly dolomitized because of its complex depositional history and subsequent diagenetic processes (Buza, 2007).

2.4 Cenozoic

The Cenozoic Era is divided into Tertiary and Quaternary Periods; the Tertiary is classified into Paleogene and Neogene. The Paleogene is represented by Paleocene, Eocene, and Oligocene Epochs and consists of Radhuma, Rus, and Dammam Formations of the Hasa Group. The Neogene Period is designated by Miocene and Pliocene Epochs and consists of Ghar, Lower Fars, and Dibdibba Formations of the Kuwait Group.

Tertiary sediments are the primary source of usable groundwater in Kuwait. The Tertiary sedimentation began with a marine transgression in the Paleocene. Shallow marine to sabkha conditions prevailed in the area until the end of the Eocene; during this period, a carbonate-evaporite sequence (Radhuma, Rus, and Dammam Formations) was deposited (Mukhopadhyay et al., 1996).

The sea regressed at the end of Eocene, and a widespread unconformity, causing the loss of Oligocene deposits over most of the area. During this period, the karstification of the Dammam Limestone Formation produced localized accessible pathways for groundwater (Mukhopadhyay et al., 1996).

In the early Miocene, the deposition of the Kuwait Group's clastic sediments and its equivalents on the stable shelf started under mostly continental conditions. Occasional rainstorms recharge the Tertiary aquifers in Kuwait from the outcrops in Saudi Arabia and Iraq (Mukhopadhyay et al., 1996).

2.4.1 Paleogene

2.4.1.1 Paleocene
Radhuma

Radhuma Formation or Um Er Radhuma Formation (Radhuma is the commonly used name in Kuwait), named after the Umm Er-Radhuma water well in Saudi Arabia (Power et al., 1966).

Rudhuma Formation is the oldest known formation in the Hasa group. It comprises three carbonate-evaporite sequences: Radhuma, Rus, and Dammam Formations (AL-Sharhan & Nairn, 2003).

Based on carbon isotope dating, Youssef (2016) estimated that the formation is Paleocene–Eocene in age, which is confirmed by Dirks et al. (2018) and Mukhopadhyay et al. (1996). Radhuma formation is bounded by Rus Formation at the top, and it rests at erosional unconformity surface developed at the top of the Tayarat Formation. It is composed mainly of carbonate rocks, fine to coarse-grained dominant dolostone, and limestone with minor anhydrite streaks interbedded with dolostone. Shale is present at the top of the formation. In addition, lenses of lignitic, gypsiferous dolomite, and silicified anhydrite (Dirks et al., 2018; Mukhopadhyay et al., 1996; Youssef, 2016).

The formation is well developed and correlated all over Kuwait. Its thickness varies from 426 to 610 m in the North of Kuwait and shoreline areas (Mukhopadhyay et al., 1996).

Radhuma Formation was deposited in major transgression covering the Arabian Peninsula (Dirks et al., 2018). The predicted depositional environment of the Radhuma Formation is a shallow environment within an inner and middle ramp. It ranges from intertidal to supratidal and lagoon environments. There is shoals' development in some areas (Dirks et al., 2018; Mukhopadhyay et al., 1996). The fossil contents in the formation are mainly shallow-water fossils composed of mainly planktonic foraminifers, echinoderms, and Mollusca. In Sabriyah Field, some gastropods were present (Youssef, 2016).

From field observation and data collected while drilling (i.e., cutting samples or core data) in Kuwait oil fields, the contact between the Rus and the Radhuma Formations was recognized as sharp because Rus is mainly anhydrite while Radhuma is dolomite dominated. The contact between Radhuma and the underlying Tayarat Formation is sharp due to the thin shale layer present at the top of the Tayarat formation (Taqi et al., 2018).

The Radhuma Formation is considered as one of the major and significant aquafers in Oman (Mukhopadhyay, 1995) and Saudi Arabia (Dirks et al., 2018), while in Kuwait, it is not exploited (Mukhopadhyay et al., 1996). The water produced from the formation is used for agricultural and industrial use with partial domestic use but not for drinking as the salinity is high (UN-ESCWA and BGR, 2013).

2.4.1.2 Eocene
Rus Formation

Saudi ARAMCO geologist Bramkamp named Rus Formation after Um Er Rus. It is a small hill located on the southeastern flank of the Dammam dome in east Saudi Arabia (Power et al., 1966). Rus Formation was deposited during the Early Eocene, and it is dominantly evaporitic. The dominant rock type is white anhydrite and nonfossiliferous limestone with minor marls and shale. The formation thickness ranges from 450 ft (150 m) in the Northern area to 650 ft (200 m) in the offshore and southern regions (Al-Sharhan & Nairn, 2003).

The Rus Formation consists of two shallowing upward successions, each expected to represent a peritidal environment, i.e., starting with subtidal followed by intertidal and ultimately ending with supratidal setting (Tamar-Agha & Saleh, 2016).

The sharp contact between Rus and the above Dammam formation shows apparent changes from limestone to anhydrite, and this is also true for the contact between Rus and Radhuma as the lithology changes sharply from anhydrite to dolomite (Tanoli & Al-Bloushi, 2017).

Dammam Formation

Dammam Formation is the youngest carbonate unit developed in Kuwait, and it was named after the Dammam dome based in the Dammam Peninsula on the eastern side of the Kingdom of Saudi Arabia (Al-Sharhan & Nairn, 2003). In Kuwait, the upper Dammam Formation is exposed to the surface due to industrial activities in a quarry at the Al-Ahmadi area located in southeast Kuwait on the Al-Ahmadi ridge that extends parallel to the east coast of Kuwait and encounters within the Greater Burgan oil field (Fig. 2.8). On the other hand, the aquifer is exploited in the country's central and southern parts (Mukhopadhyay, 1995).

Several authors investigated the Dammam Formation in Kuwait, and they interpreted the geology, petrography, diageneses, lithostratigraphy, biofacies, geochemistry, hydrology, and the depositional history of the Dammam Formation (e.g., Burdon & Al-Sharhan, 1968; Mukhopadhyay, 1995; Al-Awadi et al., 1997, 1998a, 1998b; Mukhopadhyay & Al-Otaibi, 2002; Khalaf & Abdullah, 2015; Khalaf et al., 2018; Tanoli & Al-Bloushi, 2017).

The Dammam Formation represents the upper half of the sequence stratigraphic Palaeocene–Eocene megasequence AP10 of Sharland et al. (2001). The lower half of this megasequence includes the Rus and the Radhuma formations. Owen and Naser (1958) estimated that the Dammam Formation was deposited during the Middle Eocene Epoch. Ziegler (2001) also documented that the Dammam Formation contains nummulites and dated from its foraminiferal content as upper Ypresian to Priabonian (Middle Eocene Epoch).

The Dammam Limestone is found at depths ranging from the surface et al.-Ahmadi quarry to about 1,200 ft (366 m) in northeastern Kuwait. It dips at about 9 ft (2.7 m) per mile from southwest to northeast unless interrupted by post-Eocene structures (Milton, 1967).

The thickness of the Dammam Formation in Kuwait varies from about 492ft (150 m) in the southeast to about

Fig. 2.8 Location map of the Al-Ahamdi Quarry (black circle) (after Khalaf et al., 2018)

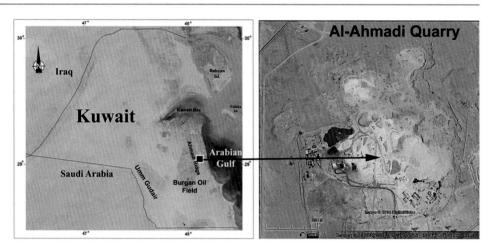

902 ft (275 m) in the northeast (Khalaf et al., 2018), and it consists mainly of dolomitized limestone chalky fossiliferous limestone and shale. It is represented by cycles of limestone and dolostone. Dammam Formation was subdivided into three members based on lithology and biofacies (Burdon & Al-Sharhan, 1968; Khalaf et al., 1989; Al-Awadi et al., 1997, 1998a, b).

The upper member range in thickness between 60 and 90 m (Al-Awadi et al., 1998a, b) and consists of white, very friable, porous dolomite with thin chert lenses and nodules. This unit's top is distinct by a karstified and cherty zone below the disconformity (Khalaf, 2011). Khalaf et al. (1989) further subdivided this unit into four lithotypes, from top to bottom are (a) vuggy chertified dolomicrite, (b) dense chalky dolomicrite, (c) massive earthly dolomicrite, and (d) vuggy dolomite. Al-Awadi et al., (1998a, b) identified an abundant number of molds and casts of bivalves and gastropods, tests of foraminifera, echinoid spheres, and ostracods in this friable chalky dolomite.

The middle member ranges in thickness from 30 to 40 m and is essentially composed of laminated biomicrite, dolomicrite, and lignitic seams and lenses (Al-Awadi et al., 1998b). Khalaf (2011) subdivided this unit into six sub-members, named as follows: (1) algal limestone, (2) fossiliferous limestone, (3) dolomite, (4) dolomitic limestone, (5) lignite, and (6) dolomitized limestone. This unit is rich in elongated or round yellow crystals of phosphates (Al-Awadi et al., 1998b).

The thickness of the lower Dammam Formation ranges from 50 to 70 m, and it is mainly composed of nummulitic limestone with phosphate-rich layers of dolomite (Al-Awadi et al., 1998a, b). It was also documented by Khalaf (2011) that it consists of shale interlayers at the base and grading into fossiliferous limestone at the top. Table 2.2 in Al-Awadi et al., (1998a, b) described in detail the different biofacies that characterized the three major lithostratigraphic units of

the Dammam Formation in the Umm Guadair area, located in southwest Kuwait as identified by Khalaf et al. (1989).

The Dammam sediments were assumed to have been deposited in a shallow marine inner shelf environment that has gradually become shallower with time. Regional regression and surface exposure to weathering and erosion during the Late Eocene Epoch marked the sedimentation's end (Al-Sharhan & Nairn, 2003; Al-Awadi et al., 1997, 1998 a,b; Mukhopadhyay, 1995; Khalaf et al., 2018; Tanoli & Al-Bloushi, 2017).

The dolomitic Dammam Formation is laying conformably on the anhydritic Rus Formation (Fig. 2.9), and it is capped by a paleokarstic zone delineating the regional disconformity with the overlying Ghar Formation, the oldest formation of the Neogene clastic deposits in Kuwait (Al-Awadi et al., 1998a, 1998b; Khalaf, 2011). Burdon and Al-Sharhan (1968) were the first to recognize this zone in the Al-Ahmadi quarry.

The pre-Neogene unconformity developed on the top of the dolomitic Dammam Formation was interpreted as a result of the tectonic uplift associated with the compression occurring on the northeast margin of the Arabian Plate (Goff et al., 1995). The top of the formation has pronounced karstification associated with this unconformity. The depositional history of the Dammam Formation in Kuwait is well discussed by Al-Awadi et al., (1998a, 1998b) and Tanoli and Al-Bloushi (2017).

Data yield from drilled core and cutting in Kuwait oil fields helped to specify the sharp contact between the dolomitized limestone of Dammam Formation (base of AP11) and the overlying sandstone with minor shale characterizing the Ghar formation. It is also true for the underlying Rus formation characterized by anhydrite with minor dolomite, limestone, and shale.

The Dammam limestone Formation is one of the principal aquifers of the Arabian Gulf region. In Kuwait, it is

Fig. 2.9 Core Photographs under white light illustrating the sharp contact between the Dammam Formation and the underlying Rus Formation. The contact between the two formations is the red arrow at around 713 ft. From 696 ft. to 707 ft. is the Dammam gray to lack mudstone Facies 2. After that, it converts to Facies 1, which is Dammam brown limestone. Rus Anhydrite starts from 713 ft. (After Tanoli & Al-Bloushi, 2017)

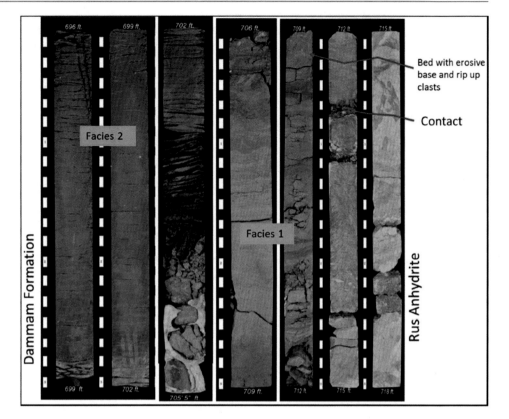

considered one of the significant brackish water aquifers (Al-Awadi et al., 1998a, 1998b). This formation and the overlying the Mio-Pleistocene clastic deposits form the important aquifer system containing compatible quality of the brackish water (Khalaf, 2011).

2.4.1.3 Oligocene

The opening of the Red Sea and the Gulf of Aden in the west of the Arabian plate and the closure of the Neo-Tethys in the east of the Arabian plate begins at the late Eocene and passing through Oligocene to the present time, causes pronounced fall in the sea level to expose almost the entire Arabian Plate. A significant unconformity and sedimentary hiatus mark the boundary between megasequences AP10 and AP11 (Fig. 2.1) (Ellis et al., 1996; Sharland et al., 2001; Ziegler, 2001; Albaroot et al., 2016).

2.4.2 Neogene

2.4.2.1 Miocene

Ghar formation

Ghar formation was named after a locality in the Al-Basra area of Iraq. It constitutes the basal part of the Kuwait Group siliciclastic sequence that unconformably overly the Eocene

Dammam Formation in southern Iraq and Kuwait (Owen and Naser, 1958). Power et al. (1966) correlate Ghar Formation with that of the Hadrukh Formation (Early Miocene) in Saudi Arabia, where its base also rests unconformably upon Eocene Dammam limestone. Owen and Naser (1958) accredit the Ghar Formation of Oligocene to Early Miocene in age as they only assigned its deposition time concerning stratigraphic considerations.

The Kuwait Group was subdivided into three formations in northern Kuwait, from bottom to top: Ghar, Lower Far, and Dibdibba Formations. This stratigraphic subdivision is based on the presence of the fossiliferous clay bed that correlated with the Lower Fars Formation in Iran. These beds were deposited during the Early to Middle Miocene and separated the overlying Dibdibba Formation from the underlying Ghar Formation; both comprise essentially marginal to nonmarine coarse clastics that lack age-diagnostic fossils (Owen & Naser, 1958; Milton, 1967; Picha & Saleh, 1977). In southern Kuwait, however, no such subdivision is possible, as the fossiliferous Lower Fars Formation is absent, and the Kuwait Group is defined by an undifferentiated sequence of clastic deposits (Mukhopadhyay et al., 1996).

In Bahra1 Well (Northeast Kuwait), the thickness of the Ghar Formation is 800ft (244 m) (Milton, 1967), and it is around 984 ft (300 m) in Um-Gudair (Southwest Kuwait) (Al-Rawdan et al., 1998). Al-Awadi et al. (1997) and Al-Rawdan et al. (1998) studied and identified the subsurface stratigraphy of the Kuwait Group of Miocene to Pleistocene age in different locations and linked it with the surface exposure in JalAz-zor escarpment (Northeast Kuwait). Formation boundaries were debatable because their studies lack field and fossils indicators. Milton (1967) was the first to link the lithology of an 800 ft (244 m) section characterized by subsurface Ghar formation encountered in Kuwait Oil Company Bahra 1 well with a 100 ft (30.5 m) thick sequence surface exposure in Jal Az-Zor escarpment in the Bahrah area. As discussed in the previous chapter, Khalaf et al. (2019) studied the occurrence and genesis of the exposed Oligo-Miocene Ghar Formation in Northern Kuwait.

The main composition of the Ghar Formation is clastic sequences comprising of non-fossiliferous cross-bedded, coarse-grained to pebbly sandstone intercalated with few green mud beds. These are interbedded with sandy limestone (calcareous sandstone), gypsum laminae, and minor shale (Al-Rawdan et al., 1998; Milton, 1967).

Ghar Formation was deposited in a fluvial environment characterized by cross-bedded fining-upward cycles of clastic sediments of alluvial fan and shallow marine deposition (Aqrawi et al., 2010). In the lower part, where shale is present, the proposed environment is a restricted low-energy environment (lagoonal) (Al-Rawdan et al., 1998).

The lower contact of the Ghar Formation with the Dammam Formation is unconformable, and the contact between Ghar and the overlying Lower Fars Formation is transitional and distinguished by sediment color and the presence of fossils (Owen and Naser, 1958; Al-Sharhan & Nairn, 2003).

Lower Fars

Owen and Naser (1958) referred to the Lower Fars Formation nomenclature defined from exposures in the Fars Province, southwest Iran. The formation was considered deposited during the Early to Middle Miocene age (Owen and Naser, 1958). Its age is amended in the light of more reliable biostratigraphic data from late Serravalian (12 Ma) to late Langhian (15.5 Ma) (Burdoun and Al-Sharhan, 1968; Sharland et al., 2004).

Lower Fars Formation ranges in thickness from 200ft (61 m) in the west of Kuwait to more than 600 ft (183 m) in the eastern offshore area (Al-Sharhan & Nairn, 2003). Between the depth of 345ft (105 m) to 704 ft (215 m) in the Raudhatain well No. 1 (Kuwait Oil Company), the Lower Fars Formation of 359 ft (109 m) thick was clearly identifiable. It mainly consists of anhydrites, gypsum, clays, marls, and shallow-water limestones. Remains of Ostrea

latimarginata, Clausinella, and Quinquiloculina sp. were identified (Owen and Naser, 1958). Milton, 1967 and Amer et al., 2019; described facies architecture of the Lower Fars Formation exposed at the Jal Az Zor escarpment mentioned in the previous chapter.

Ferdous et al. (2013) demonstrate that Lower Fars Formation truncated at a shallow depth of less than 800 ft (244m) and extended laterally to cover a vast area in northern Kuwait. It consists of two unconsolidated sandstone units separated by a middle shale unit. Intervening siltstone and shale units commonly divide the two sandstone units into four stratified reservoir units of Upper A and B and Lower A and B zones. Each zone varies in thickness from about 10–30 ft (3–9 m).

The depositional environment is predicted based on the vertical and lateral distribution of approximately 15 different Lithofacies types. This suggests a fluvio-deltaic complex with fluvial channels delivering fresh water and sediments from the south and west to the river's mouth in the north, where they fed and formed a series of delta lobes. The northern part of this fluvio-deltaic complex has been flooded and transgressed by shallow-marine deposits on an episodic basis. These transgressive events, which were relatively short-lived, transformed the delta complex's distal parts into estuarine to shoreface depositional environments (Ferdous et al., 2013).

The contact between the Lower Fars Formation and the overlying Dibdibba delineated by 20-30 ft (6–9 m) thick shale unit, sealing the unconsolidated sandstone reservoir system of the Lower Fars. The contact with the underlying Ghar Formation is gradational (Ferdous et al., 2013).

The Lower Fars Formation is a shallow-depth, multi-stacked sandstone oil reservoir covering a vast area in northern Kuwait (Ferdous et al., 2013). It is one of the essential oil-bearing formations in Ratqa, Raudhatain, Sabriyah, Bahrah, and Mutriba oil fields (Abdul Razak et al., 2018).

2.4.2.2 Pliocene

Dibdibba Formation

Dibdibba Formation is the youngest formation of the Kuwait Group. The name of Dibdibba was first used by Macfadyen (1938) to describe superficial gravel beds deposited in southern Iraq. According to Milton (1967), the Dibdibba Formation in Kuwait comprises all beds above the fossiliferous horizon of the Lower Fars Formation, except for the Recent sediments. The exposed Dibdibba Formation in Kuwait is discussed in the earlier chapter.

The formation is of Miocene to Pliocene in age (Milton, 1967; Mukhopadhyay et al., 1996). Dibdibba Formation is underlain by Lower Fars Formation and overlain by the surface sediments in some areas where it is not exposed to the surface.

Dibdibba Formation deposition took place in an alluvial fan setting (AlShuaibi & Khalaf, 2011; Milton, 1967). The subsurface Dibdibba Formation in Kuwait can be divided into two members; the lower member is of Mio-Pliocene age and consists of coarse-grained, poorly sorted gritty, and pebbly sandstone lithified by chalk and carbonate cement. The upper member is of the Plio-Pleistocene age and composed of gravelly sand and sandy gravel. Gypsiferous is a common cement, while calcareous cement is limited (Owen & Naser, 1958; Mukhopadhyay et al., 1996). The gravels deposited in sheets and trains originated from igneous and metamorphic rocks derived from Najed, Al-Hijaz, the Syrian Desert, and Northern Arabia. There are no index fossils documented in Dibdibba Formation (Owen & Naser, 1958; Milton, 1967 and reference therein; Picha & Saleh, 1977).

The contact with the underlying Lower Fars formation is gradational. The sand and gravel beds of the Dibdibba Formation are capped by coarse pebbles of rounded igneous, metamorphic, or sedimentary dreikanter (three-edged) shaped deposited on a typical desert floor (Owen & Naser, 1958). As mentioned earlier, in southern Kuwait, the fossiliferous Lower Fars Formation is absent, and the Kuwait Group is defined by an undifferentiated sequence of clastic deposits (Mukhopadhyay et al., 1996).

In northern Kuwait, the contact between Dibdibba and Lower Fars is sharp due to the presence of a 20–30 ft (6–9 m) thick shale at the base of the Dibdibba Formation, forming a regional barrier and a sealing unit for the shallow Lower Fars reservoir system (Ferdous et al., 2013).

A well drilled to approximately 100 ft (30.5 m) in northern Kuwait during 1960 and found an accumulation of potable water in the Dibdibba Formation. This well was drilled in a topographical low west of the Raudhatain oil field where the Raudhatain anticline has changed the regional dip (Milton, 1967).

2.4.3 Quaternary

2.4.3.1 Recent Sediments (Surface)

The Quaternary deposits in Kuwait are discussed in the previous chapter as part of the surface sediments of Kuwait, it characterized by an abundance of gravel. Dibdibba is the most extensive occurrence of these Quaternary gravels. The Dibdibba Formation covers most of the northern area of Kuwait (Al-Sulaimi et al., 1997). The exposed sediments in the surface of Kuwait are dated from the Early Miocene to

recent (Milton, 1967). These surface deposits have two major environments: desert-related depositions and costal related depositions (Al-Hurban, 2016).

2.5 Summary and Conclusions

This chapter presents informative information in a simple geological expression to entice geoscience educators to develop their knowledge in understanding subsurface geology and paleoenvironments that once dominated Kuwait's geologic past.

The subsurface stratigraphy of Kuwait introduced in this chapter, adopted published reports, conference proceeding articles, and research papers associated with the geological descriptions of the Phanerozoic succession of Kuwait and neighboring countries (Saudi Arabia, Iran, and Iraq). Those publications are based on oil industry activities, groundwater aquifer research, and linking surface exposure outcrops of some stratigraphic Formations to subsurface analog.

Most lithostratigraphic Formations in Kuwait consisted of carbonate rocks deposited in a shallow water environments through the Phanerozoic succession. Exception goes to Lower Cretaceous, where the deltaic system dominated the area and deposited clastic sediments. One of the significant challenges in this chapter is the lack of publication for non-hydrocarbon, non-aquifer, and deeper Formations.

Understanding all possible integrated geologic characterization and features enabled many researchers to interpret Kuwait's surface and subsurface lithological strata and understand past geologic events, such as sea-level fluctuation, climate changes, and tectonic events. Accordingly, paleoenvironments and depositional history were reconstructed, and a successful Kuwait stratigraphic column was established.

Acknowledgements Praise is to Allah, the Almighty, Who by His blessings all righteous things are completed.

The authors would like to acknowledge the Ministry of Oil (MOO) of the State of Kuwait and Kuwait Oil Company (KOC) for their explicit support. It is crucial to gratitude to many staff in KOC for their persistent help and support.

A sincere thanks must go to Exploration Group Manager, Mr. Mohammad Al-Ajmi, Prospect Evaluation III Team leader Mr. Mashari Al-Awadi and Mr. Riyasat Hussain for their support.

The authors also appreciated the assistance of the Geological & Geophysical Solutions Team leader, Mr. Jarrah Al-Jenaie, and the Sedimentology and Stratigraphy Unit, directed by Ms. Ghaida Al-Sahlan, through their provision of the updated Kuwait Stratigraphic column 2020.

Appreciation is given to the Exploration Operations Team Leader Mr. Bader Al-Ajmi and Mr. Jalal Dashti to provide and edit the map of Kuwait oil fields and confirm the formation tops and the contacts between the formations.

Indeed, the authors would like to thank Dr. Mohammad Al-Mahmood, "Director & Owner of Dhahran Geological Consulting", for his persistent help and advice.

References

Abdul Azim, S., Kostic, B., Al-Anzi, S., Abou-Qammaz, L., Al-Blayees, M., Al-Ajmi, M.F., Al-Saad, B., & Hoppe, M. (2019). Impact of depositional environment, sequence stratigraphy, and structure on developing Zubair reservoirs in North Kuwait. In H. R. Al.Anzi, R. A. Rahmani, R. J. Steel, & O. M. Soliman (Eds.), *Siliciclastic reservoirs of the Arabian plate:* AAPG Memoir 116, pp. 185–218.

Abdul Razak, M. H., Al-Jenaie, J., & Moubarak, H. (2018). Understanding the Reservoir Architecture of Lower Fars Formation in North Kuwait through seismic reservoir characterization. *GEO 2018 13th Middle east geosciences conference and exhibition, Manama, Bahrain.*5–8 March 2018.

Abdul Malek, S., Bahattacharya, S., Husain, R., Sajer, A., Lau, K. (2005). Gas-condensate discovery from the Sudair Formation of Kuwait-exploration implications for Triassic plays (Abstract). *AAPG international conference and exhibition*, Paris, France.

Abdullah, F. (2001). A preliminary Evolution of Jurassic Source Rock Potential in Kuwait. *Journal of Petroleum Geology, 24*(3), 361–378.

Abdullah, F., & Connan, J. (2002). Geochemical study of some cretaceous rocks from Kuwait: Comparison with oils from cretaceous and Jurassic reservoirs. *Organic Geochemistry, 33,* 125–148.

Abdullah, F., Shaaban, F., Khalaf, F., Bahman, F., & Akbar, B. (2017). *Journal of Asian Earth Sciences, 148,* 105–120.

Abdullah, F. H., & El Gezeery, T. (2016). Organic geochemical evaluation of hydrocarbons in Lower Cretaceous Middle Minagish reservoir. *Kuwait. Marine and Petroleum Geology., 71,* 41–54.

Al-Awadi, E., Al-Ruwaih, F., Al-Rawdan, Z., & Ozkaya, I. (1997). The stratigraphy of the Middle Eocene- Pleistocene sediments in Kuwait. *Journal of Arid Environments, 37,* 1–22.

Al-Awadi, E., Al-Ruwaih, F., and Ozkaya, I. (1998a). Stratigraphy of the Dammam Formation in Umm-Gudair area, Kuwait. *Géologie Méditerranéenne, 25*(2), 105–116.

Al-Awadi, E., Mukhopadhyay, A., & Al-Senafy, M. (1998b). Geology and hydrogeology of the Dammam Formation in Kuwait. *Hydrogeology Journal, 6*(2), 302–314.

Albaroot, M., Ahmad, A. H. M., Al-Areeq, N., & Sultan, M. (2016). Tectonostratigraphy of Yemen and geological evolution: A new prospective. *International Journal of New Technology and Research, 2*(2), 19–33.

Al-Eidan, A., Neog, N., Narhari, S., Al-Darmi, A., Al-Mayyas, R., De Keyser, T., & Perrin, C. (2009). Carbonate facies and depositional environments of the Marrat formation (Lower Jurassic), North of Kuwait. *AAPG annual convention and exhibition, Denver, Colorado*. Search and Discovery Article #50223.7–10 June 2009.

Al-Enezi, B., Burman, K., Datta, K., Le Guerroué, E.,& Filak, J.-M. (2011). Reconstruction of relative sea level for Burgan-Wara complex in greater Burgan field using an innovative approach. *International petroleum technology conference.* Bangkok, Thailand, 7–9 February 2012.

Al-Fares, A. A., Bouman, M., & Jeans, P. (1998). A new look at the middle to lower cretaceous stratigraphy. *Offshore Kuwait Geoarabia, 3*(4), 543–560.

Al-Hajeri, M., & Bowden, S. (2018). Origin of oil geochemical compositional heterogeneity in the Radhuma and Tayarat formations heavy oil carbonate reservoirs of Burgan Field, south Kuwait. *Arabian Journal of Geosciences, 11,* 649.

Al-Hurban, A. (2016). Sedimentomorphic evolution of recent surface deposits of Kuwait using Remote Sensing and GIS applications. *International Journal of Environment & Water, 5*(2). 55–75.

Al-Moraikhi, R., Verma, N., Mishra, P., Houben, A. J. P., van Hoof, T., & Verreussel, R. (2014). An updated chronostratigraphic framework for the Jurassic of the Arabian Platform: Towards a regional stratigraphic standard. *Search and Discovery Article* #30333.

Al-Rawdan, Z., Al-Ruwaih, F., & Ozkaya, I. (1998). Stratigraphy of Kuwait Group in Umm-Gudair and Surrounding areas, Kuwait. *Geologie Mediterraneenne, 25*(1), 3–18.8.

Al-Sahlan, G. (2005). Letter to the Editor. *GeoArabia, 10*(3), 193–194.

Al-Sharhan, A., & Nairn, A. (2003). *Sedimentary basins and petroleum geology of the middle east* (2nd ed., p. 878). Elsevier.

Al-Sharhan, A., Strohmenger, C., Abdullah, F., & Al Sahlan, G. (2014). Mesozoic stratigraphy evolution and hydrocarbon Habitats of Kuwait. In L. Marlow, C. Kendall, & L. Yose (Eds.), *Petroleum systems of Tethyan Region: AAPG Memoir*, Vol. 106, p. 451–611.

Al-Sharhan, A. S. (1994). Albian clastics in the western Arabian Gulf region: A sedimentological and petroleum geological interpretation. *Joural of Petroleum Geology, 17,* 279–300.

AlShuaibi, A., & Khalaf, F. (2011). Development and lithogenesis of the palustrine and calcrete deposits of the Dibdibba Alluvial Fan. *Kuwait. Journal of Asian Earth Sciences, 42*(3), 423–439.

Al-Sulaimi, J., Khalaf, F., & Mukhopadhyay, A. (1997). Geomorphological analysis of paleo drainage systems and their environmental implications in the desert of Kuwait. *Environmental Geology, 29* (1/2), 94–111.

Amer, A., Al-Hajeri, M., Najem, A., & Al-Qattan, F. (2019). Facies architecture of Lower Fars Formation at Jal Az-Zor escarpment. *Kuwait. Arabian Journal of Geosciences, 12,* 502.

Aqrawi, A., Goff, J., Horbury, A., & Sadooi, F. (2010). The petroleum geology of Iraq. Beaconsfield (U.K.): Scientific Press, 435 p.

Arasu, R., Singh, S., Al-Adwani, T., Khan, B., Macadan, J., & Abu-Ghaneej, A., (2012). Hydrocarbon prospectivity of the Late Jurassic–Early Cretaceous Makhul Formation in north and northwestern Kuwait. *Fourth Arabian Plate Geology Workshop.* Late Jurassic/Early Cretaceous Evaporite-Carbonate-Siliciclastic Systems of the Arabian Plate. 9–12 December 2012, Abu Dhabi, UAE.

Arasu, R., Rao, B., Das, S., & Abu-Taleb, R. (2018). Seismo-stratigraphic characterisation of Triassic intra-platform play: Jilh Lower Carbonates, West of Kuwait. In M. C. Pöppelreiter (Ed.), *Lower Triassic to Middle Jurassic Sequence of the Arabian Plate.* European Association of Geoscientists & Engineers (EAGE). Netherlands, pp. 179–190.

Aziz, S. K., & El-Sattar, A. M. M. (1997). Sequence stratigraphic modeling of the lower Thamama group, east onshore Abu Dhabi. *United Arab Emirates. Geoarabia, 2*(2), 179–202.

Bellen, R.C., Van, Dunnington, H. V., Wetzel, R., & Morton, D. M. (1959). Iraq. Lexique Stratigraphique International, Paris, Vol. III, Asie, Fascicule, 10 a. Electronic version. http://paleopolis.rediris.es/LEXICON/IRAQ/.

Beydoun, Z. R. (1991). Arabian plate hydrocarbon geology and potential—A plate tectonic approach. *American Association of Petroleum Geologists, v*(33). 77 p.

Beydoun, Z. R. (1988). *The middle east: Regional geology and petroleum resources* (p. 292p). Scientific Press.

Brennan, P. (1990). Greater Burgan Field. *In TR: Structural Traps i: Tectonic Fold Traps, Treatise, AAPG Special Publication, 1,* 103–128.

Bou-Rabee, F. (1996). Geologic and tectonic history of Kuwait as inferred from seismic data. *Journal of Petroleum Science and Engineering, 16,* 151–167.

Buday, T. (1980). *The regional geology of iraq, stratigraphy and paleontology*. State Organization for Minerals Library, Baghdad, Iraq, 1445 p.

Burdon, D. J., & Al-Sharhan, A. (1968). The problem of the Palaeokarstic Dammam limestone aquifer in Kuwait. *Journal of Hydrology, 6,* 385–404.

Buza, J. (2007). Tayarat task force final report. Kuwait Oil Company internal report.

Cabrera, S. C., Keyser, Th., Al-Sahlan, Gh., Al-Wazzan, H, Kadar, A. P., & Karam, Kh. A. (2019). Middle and Upper Jurassic Strata of the Gotnia Basin, Onshore Kuwait: Sedimentology, Sequence Stratigraphy, Integrated Biostratigraphy and Palaeoenvironments, Part 1. *Stratigraphy,16*(3), 165–193.

Cross, N., Goodall, I., Hollis, C., Burchette, T., Al-Ajmi, H. Z., Johnson, I. G., Mukherjee, R., Simmons, M., & Davies, R. (2010). Reservoir description of a mid-Cretaceous siliciclastic-carbonate ramp reservoir: Mauddud Formation in the Raudhatain and Sabiriyah fields. *North Kuwait. Geoarabia, 15*(2), 17–50.

Davies, R., Simmons, M., Jewell, T., & Collins, J. (2019). Regional controls on siliciclastic input into mesozoic depositional systems of the arabian plate and their petroleum significance. In H. Al-Anzi, R. Rahmani, R. Steel, & O. Soliman, (Eds.), *Siliciclastic reservoirs in the Arabian Plate*: AAPG Memoir 116, pp. 103–140.

Dirks, H., Al-Ajmi, H., Kienast, P., & Rausch, R. (2018). Hydrogeology of the Umm Er Radhuma Aquifer (Arabian Peninsula). *Groundwater, 23*(1), 5–15.

Dunnington, H. V. (1958). Generation, migration, accumulation, and dissipation of oil in northern Iraq. In L. G. Weeks, (Ed.) *Habitat of Oil*. AAPG Special publication.

Ellis, A. C., Kerr, H. M., Cornwell, C. P., & Williams, D. O. (1996). A Tectono-stratigraphic Framework for Yemen and its Implications for Hydrocarbon Potential. *Petroleum Geoscience, 2*, 29–42.

Ferdous, H., Chaudhary, P., Ahmad, F., Abbas, F., Ahmed, K., Lierena, J., & Al-Sammak, I. (2013). Challenges to explore shallow sandstone reservoir for optimized unconventional development strategy in Kuwait. *AAPG 2013 annual convention and exhibition*. Pennsylvania. Search and Discovery (Article #90163).19–22 May 2013.

Goff, J. C., Jones, R. W., & Horbury, A. D. (1995). Cenozoic basin evolution of the northern part of the Arabian Plate and its control on hydrocarbon habitat. In M. I. Al-Husseini (Ed.), *Middle east petroleum geosciences, GEO'94*. Gulf PetroLink, Bahrain, 1, pp. 402–412.

Han, H., Zhao, L., Xu, X., & Lu, W. (2015). Characteristics and genesis of low to ultra-low permeability carbonate rocks: A case study in late cretaceous Sadi formation in H Oilfield, Southeast of Iraq. In *SPE reservoir characterization and simulation conference and exhibition*. Society of Petroleum Engineers. 14–16 September, Abu Dhabi, UAE.

Hayat, L., Al-Qattan, M., Al Jallad, O., Dernaika, M., Koronfol, S., Kayali, A., & Gonzalez, D. (2018). Geological and petrophysical evaluations of tayarat heavy oil carbonates in burgan field-Kuwait (SPE-193661-MS). *SPE international heavy oil conference and exhibition*, Kuwait, 10–12 December 2018.

Hohman, J. C., Al-Emadi, I. A. A., & Zahran, M. M. E. (2005). Sequence Stratigraphic analysis of the Shuaiba Formation: Implications for exploration potential in Qatar. *International petroleum technology conference*. Doha, Qatar, 21–23 November 2005.

Husain R., Sajer A., Al-Ammar, N., Khan, D. A., Rabie A., & Iqbal, M. K. (2009). Sequence stratigraphy of Triassic Jilh formation in Kuwait. *AAPG international conference and exhibition*, Rio de Janeiro, Brazil. AAPG Search and Discovery, Article # 90100.15–18 November 2009.

Husain, R., Khan, A., Sajer, N., Al-Ammar, N., & Al-Fares, A. (2013). Khuff formation in Kuwait: An overview. In M. Pöppelreiter (Ed.), *Permo-Triassic Sequence of the Arabian Plate* (pp. 303–326). EAGE Publication.

Jaber, A. S. (1972). Remarks on the distribution of the middle cretaceous succession in the Kuwait—Saudi Arabia Divided Neutral Zone Offshore Area. *Eighth Arab Petroleum Congress*, Paper 94 B, pp. 1–15.

Kadar, A. P., Keyser, Th., Neog, N., Karam, Kh. A. with contributions from Yves-Michel Le Nindre & Davies, R. B. (2015). Calcareous nannofossil zonation and sequence stratigraphy of the Jurassic System, onshore Kuwait. *GeoArabia, 20*(4), 125–180.

Khalaf, F. (2011). Occurrence of diagenetic pseudobreccias within the paleokarst zone of the upper Dammam Formation in Kuwait. *Arabian Gulf. Arabian Journal of Geosciences., 4*(5), 703–718.

Khalaf, F., Abdel-Hamid, M., & Al-Naqi, M. (2019). Occurrence and genesis of the exposed Oligo-Miocene Ghar Formation in Kuwait, Arabia Gulf. *Journal of African Earth Sciences, 152*, 151–170.

Khalaf, F., & Abdullah, F. (2015). Occurrence of diagenetic alunites within karst cavity infill of the Dammam Formation, Ahmadi, Kuwait: An indicator of hydrocarbon gas seeps. *Arabian Journal of Geosciences., 8*(3), 1549–1556.

Khalaf, F., Abdullah, F., & Gharib, I. (2018). Petrography, diagenesis, and isotope geochemistry of dolostones and dolocretes in the Eocene Dammam Formation, Kuwait. *Arabian Gulf. Carbonates and Evaporites, 33*(1), 87–105.

Khalaf, F., Mukhopadhyay, A., Naji, M., Sayed, M., Shublaq, W., Al-Otaibi, M., Hadi, K., Siwek, Z., & Saleh, N. (1989). Geological assessment of the Eocene and post-Eocene aquifers of Umm Gudair, Kuwait. Kuwait Institute for Scientific Research (KISR). unpublished report # KISR3176, 322 pp.

Khan, A. (1989). Stratigraphy and hydrocarbon potential of the Permo-Triassic sequence of rocks in the State of Kuwait. in Country reports and case studies presented at the seminar on deep formations in the Arab countries: *Hydrocarbon Potential and exploration techniques: OAPEC*, Kuwait. pp.E3-E29.

Khan., D., AL-Ajmi, M., Amar, N., & AL-Mukhaizeem, M. (2010). *Atlas on facies and sedimentological characters of rocks of Kuwait*. Exploration Studies, Exploration Group. Kuwait Oil Company, Kuwait.

Koop, W. J., & Stoneley, R. (1982). Subsidence history of the Middle East Zagros Basin, Permian to Recent. In *The evolution of sedimentary Basin*. Proceedings of a Royal Society discussion meeting, pp. 149–168.

Macfadyen, W. (1938). Water Supplies in Iraq, Iraq Geology Department., Pub. No. 1, Baghdad. 206 p.

Milton, D. I. (1967). Geology of the Arabian Peninsula; Kuwait. Washington: USGS. Geological Survey professional paper. 560-F, 14 p.

Mukhopadhyay, A. (1995). Distribution of Transmissivity in the Dammam Limestone Formation. *Kuwait. Groundwater, 33*(5), 801–805.

Mukhopadhyay, A., Al-Sulaimi, J., Al-Awadi, E., & Al-Ruwaih, F. (1996). An overview of the Tertiary geology and hydrogeology of the northern part of the Arabian Gulf region with special reference to Kuwait. *Earth-Science Reviews, 40*(3–4), 259–295.

Mukhopadhyay, A., & Al-Otaibi, M. (2002). Numerical simulation of freshwater storage in the Dammam Formation, Kuwait. *Arabian Journal for Science and Engineering, 27*(2B), 127–150.

Nath, P. K., Singh, S. K., Ye L., Al-Ajmi, A. S., Bhukta, S. K., & Al-Otaibi, A. H. (2014). Reservoir Characterization and Strati-structural Play of Minagish Formation, SE Kuwait. *International petroleum technology conference*. Doha, Qatar, 20–22 January 2014.

Nath, P. K., Singh, S., Al-Ajmi, A. S., Bhukta, S. K., & Al-Shehri, E. S. (2015). Integrated reservoir characterization and depositional model of Zubair formation in exploration phase, in Bahrah Area, Kuwait. Adapted from extended abstract prepared in conjunction with oral presentation at AAPG Annual Convention & Exhibition 2015, Denver, Colorado, May 31-June 3, 2015.

Neog, N., Rao, N., Al-mayyas, R., De Keyser, T., Perrin, C., & Kendall, C. (2011). Evaporite facies: A key to the mid Mesozoic sedimentary stratigraphy of North of Kuwait. AAPG International Convention and Exhibition, Calgary, Alberta, Canada, Search and Discovery Article #40682.15 February 2010.

Owen, R. M. S., & Nasr, S. N. (1958). Stratigraphy of the Kuwait-Basrah areas. In L. G. Weeks, (Ed.), *Habitat of oil, a symposium: American Association of Petroleum Geology*. AAPG Memoir (1), pp. 1252–1278.

Pasyanos, M. E., Tkalčić, H., Gök, R., Al-enezi, A., & Rodgers, A. J. (2007). Seismic structure of Kuwait. *Geophysical Journal International, 170*(1), 299–312.

Picha, F., & Saleh, A. (1977). Quaternary sediments in Kuwait. *Journal of Kuwait University (Science), 4*, 169–185.

Pöppelreiter, M., & Marshall, E. (2013). The Khuff formation: Play element and development history of an Epicontinental carbonate platform. In M. Pöppelreiter (Ed.), *Permo-Triassic Sequence of the Arabian Plate* (pp. 9–22). EAGE Publication.

Pöppelreiter, M., & Obermaier, M. (2013). The Khuff formation: Play element and development history of an Epicontinental carbonate platform. In M. Pöppelreiter (Ed.), *Permo-Triassic Sequence of the Arabian Plate* (pp. 387–400). EAGE Publication.

Powers, R. W., Ramirez, L. F., Redmond, C. D., & Elberg Jr., E.L. (1966). *Geology of the Arabian Peninsula: Sedimentary geology of Saudi Arabia*. United States Geological Survey Professional Paper 560-D, 147 p.

Sharland, P. R., Archer, R., Casey, D. M., Davies, R. B., Hall, S. H., Heward, A. P., Horbury, A. D., & Simmons, M. D. (2001). Arabian plate sequence stratigraphy. *GeoArabia Special Publication* (2), Gulf PetroLink, Bahrain, 371 p.

Sharland, P. R., Casey, D. M., Davies, R. B., Simmons, M. D., Sutcliffe, & Owen E. S. (2004). Arabian plate sequence stratigraphy—Revisions to SP2. *GeoArabia, 9*(1). Gulf PetroLink, Bahrain.

Singavarapu, A., Singh, S. K., Borgohain, B., Al-Ajmi, A., Roy, M., & Al-Busairi, A. (2015). Integrated seismic attribute analysis to characterize the Upper Cretaceous Hartha Reservoir, an emerging exploration play in Kuwait. *SEG New Orleans annual meeting*. SEG Technical Program Expanded Abstracts. Society of Exploration Geophysicists, pp. 3357–3361.

Singh, S. K., Akbar, M., Khan, B., Abu-Habbiel, H., Montaron, B., Sonneland, L., & Godfrey, R. (2009). Characterizing fracture corridors for a large carbonate field of Kuwait by integrating borehole data with the 3-D surface seismic. AAPG Convention, Denver, Colorado, June 7–10, 2009.

Singh, P., Husain, R., Al-Zuabi, Y., Al-Khaled, O., Rahaman, M., Mohammed, H., Ebrahim, M., Hafez, M., Al-Rashid, T., Al-Ghareeb, S., Al-Kandary, A., & Al-Fares, A. (2013). Basement configuration and its impact on Permo-Triassic prospectivity in Kuwait. In M. Pöppelreiter (Ed.), *Permo-Triassic Sequence of the Arabian Plate* (pp. 43–54). EAGE Publication.

Splendid Arabia. (2005–2020). Heet Cave. Travel portal of the Kingdom of Saudi Arabia (website). http://www.splendidarabia.com/destinations/riyadh-province/heet-cave/.

Steineke, M. R. A., Bramkamp, R. A., & Sander, N. J. (1958). Stratigraphic relations of Arabian Jurassic oil. In L. G. Weeks (Ed.),, *Habitat of Oil*. AAPG Symposium, pp. 1294–1329.

Strohmenger, C., Al-Anzi, M., Pevear, D., Ylagan, R., Kosanke, T., Ferguson, G., Cassiani, D., & Douban, A. (2003). Reservoir quality and K–Ar age dating of the pre-Khuff section of Kuwait. *GeoArabia, 8*(4), 601–620.

Strohmenger, C. J., Patterson, P. E., Al-Sahlan, G., Mitchell, J. C., H. R., Feldman, T. M., Demko, R. W., Wellner, P. J., Lehmann, G. G., McCrimmon, R. W., Broomhall, & Al-Ajmi, N. (2006). Sequence stratigraphy and reservoir architecture of the Burgan and Mauddud formations (Lower Cretaceous), Kuwait. In P. M. Harris, & L. J. Weber (Eds.), *Giant Hydrocarbon Reservoirs of the World: From Rocks to Reservoir Characterization and Modeling*: AAPG Memoir 88/SEPM Special Publication, pp. 213–245.

Sugden, W., & Standring, J. J. (1975). Qatar peninsula. *Lexique Stratigraphique International, v*(III), 10b3.

Tamar-Agha, M. Y., & Saleh, S. (2016). Facies analysis and depositional environment of the Rus and Jil Formations (l- Eocene) in Najaf and Samawa Areas, Southern Iraq. *Journal of Environment and Earth Science, 6*(4), 30–39.

Tanoli, S., & Al-Bloushi, A. (2017). Depositional History of the Eocene Dammam Formation in Kuwait. *SPE Kuwait Oil and Gas Show and Conference*, Kuwait .15–18 October 2017.

Tanoli, S., Husain, R., & Sajer, A. (2008). Facies in the Unayzah Formation and the Basal Khuff Clastics in subsurface, northern Kuwait. *GeoArabia, 13*(4), 15–40.

Taqi, F., Ahmed, K., Saika, P., Tyagi, A., Freeman, M., Ren, Z., & Al-Rabah, A. (2018). Integrated Petrophysical Evaluation of Tayarat Formation for Water Disposal Purpose. *SPE International Heavy Oil Conference and Exhibition*. Kuwait, 10–12 December 2018.

UN-ESCWA and BGR (United Nations Economic and Social Commission for Western Asia; Bundesanstalt für Geowissenschaften und Rohstoffe). (2013). Umm er Radhuma-Dammam Aquifer System (North). Inventory of Shared Water Resources in Western Asia. Beirut. https://waterinventory.org/sites/waterinventory.org/files/chapters/Chapter-16-Umm-er-Radhuma-Dammam-Aquifer-System-North-web.pdf (online version).

Youssef, A. (2016). Sequence Stratigraphy of Radhuma Section; Onshore Kuwait. *conference proceeding*, Sixth Arabian Plate Geology Workshop, European Association of Geoscientists & Engineers, Dec 2016, Volume 2016, pp 1–4.

Youssef, A. H., Kadar, A. P., Karam, K. A. (2014). Sequence stratigraphy framework of late early to middle cenomanian Rumaila and Late cenomanian to earliest turonian Mishrif formations, Onshore Kuwait. *International Petroleum Technology Conference*. IPTC-17296-MS. Doha, Qatar. 20–22 January 2014.

Yousif, S., & Nouman, G. (1997). Jurassic geology of Kuwait. *GeoArabia., 2*(1), 91–110.

Wharton, S.R. (2017). The Rimthan Arch, basin architecture, and stratigraphic trap potential in Saudi Arabia. *Interpretation, 5*(4), T563–T578.

Zaidi, S., Khan, A. N., Nair, S. R., Al-Ghadhban, A., Mishra, R. K., Chetri, H. B. & AL-Enzi, E. (2009). Sequence stratigraphy and reservoir architecture of Tuba in Sabriyah field, North Kuwait. *International petroleum technology conference*. Doha, Qatar. 7–9 December 2009.

Ziegler, M. (2001). Late Permian to Holocene paleofacies evolution of the Arabian Plate and its hydrocarbon occurrences. *GeoArabia, 6*(3), 445–504.

Sand Dunes in Kuwait, Morphometric and Chemical Characteristics

3

A. M. Al-Dousari, M. Al-Sahli, J. Al-Awadhi, A. K. Al-Enezi, N. Al-Dousari, and M. Ahmed

Abstract

There are around 2,304 sand dunes scattered within seven dune fields in Kuwait. The longest barchanoid chain dune in Kuwait is 2400 m and 900 m. All the dunes in Kuwait were mapped, sampled, and analyzed. This chapter passes through dunes in Kuwait covering the following:

- The dunes mapping.
- The dune migrations using satellite images.
- The chemical properties include
  - Mineralogical properties.
  - X-ray diffraction (major and minor trace elements).
  - Micro-inclusions within Aeolian particles.
- The physical properties include
  - The particle size analysis.
  - Statistical properties.
  - Particle micro-features.
  - BET surface area.
  - Perimeter and diameter.

This chapter covers all aspects regarding sand dunes in Kuwait; therefore, this chapter will be a key reference for future studies tackling aeolian landforms in Kuwait and regional areas.

A. M. Al-Dousari (✉) · A. K. Al-Enezi · N. Al-Dousari · M. Ahmed
Environmental and Life Sciences Research Center, Kuwait Institute for Scientific Research, Kuwait City, Kuwait
e-mail: adousari@kisr.edu.kw

M. Al-Sahli
Department of Geography, College of Social Sciences, Kuwait University, Kuwait City, Kuwait

J. Al-Awadhi
Department of Earth and Environmental Sciences, Faculty of Science, Kuwait University, Kuwait City, Kuwait

3.1 Introduction

Desert sand seas, or ergs, are contained beyond 95% of the world's aeolian sand. Of this 85% is contained in sand seas with an area exceeding 32,000 km^2 (Wilson, 1973). Most areas of sand sea accumulation lie in the old-world deserts such as the Sahara, Arabia, Central Asia, Australia, and South Africa, the sea cover is often less than 45% of the area's surface ranging between 20 and 45% that is classified as arid (Lancaster, 1989). Mobile sand is a common phenomenon in desert areas, and it can pose a threat to various urban and industrial activities. In Kuwait, it can affect various activities such as agriculture, urban development, and communication. Mobile sand and dust transported by wind, and the corresponding problems and impact on the environment, are very important.

Kuwait constitutes part of the northwestern coastal plain of the Arabian Gulf. It covers an area of about 17,818 km^2 spreading between 46° 33′ and 48° 36′ E, and latitudes 28° 30′ N and 30° 05′N. On the north and west, the state of Kuwait is bounded by Iraq, while on the southwest and south it is bordered by Saudi Arabia. The state has some offshore islands, the largest of which are Bubiyan, Failaka, and Warba. Only Failaka Island is inhabited.

The geological, environmental, and geographical characteristics of Kuwait and its surrounding areas contribute to the dynamism and occurrence of the migrating sand belt.

Sand and dust storms are common in Kuwait. Windblown sand forms one of the chief difficulties in the future development of Kuwait. The mobile sands and dust threaten the main roads, encroach cultivated lands and several inland areas. Scientific studies were conducted on the various types of sand deposits and the locations of the sediments in the Al-Huwaimiliyah and Al-Atraf areas (Khalaf et al., 1980; Abu-Eid et al., 1983). Some of the key findings of these studies are that the development activities in these areas are prone to severe impacts due to the migration of sand. They also suggest that the use of mobile sand belts should be established to help in the land-use planning in these regions.

Khalaf et al. (1980) identified four types of aeolian accumulation in Kuwait. These include sand sheets, aeolian wadi fill deposits, sand dunes, and sand drifts and surface sediment maps of Kuwait were produced.

There is a lack of information concerning the characteristics of the mobile surface sediments in Kuwait usually and in the study area particularly. This study focuses on a comprehensive examination of the morphology, dynamics, and sedimentology of numerous categories of free dunes (barchan, domal, barchanoid, and barchan complex), in addition to anchored dunes such as nabkhas and falling dunes. A sand dune belt dominates in northwest Kuwait. Also, a detailed dunes map has been developed using remote sensing to detect sand dunes within the Al-Huwaimiliyah–An-Nimritayn zone.

An assessment of baseline data and information about the main environmental conditions in the desert of Kuwait was an essential prerequisite for the implementation of this study. The following is a brief review of the most relevant literature that provided such data and information.

Fuchs et al. (1968) conducted a comprehensive geological study of the surface and near-surface formations in Kuwait. They prepared a surface geologic map with a scale of 1: 250,000. In 1981, the Kuwait Oil Company surveyed the surface geology of Kuwait and created a detailed geological map at a scale of 1: 50,000 (Warsi, 1990).

A comprehensive study was conducted by Khalaf et al. (1984) on the various geological formations in Kuwait. He produced a sedimentomorphic map showing the distribution and type of recent surface sediments and redeveloped by Misak (2000). A detailed geomorphological zonation of Kuwait's terrain was carried out by Al-Bakri (1988a, b). All these studies include the study area of this project.

Aeolian landforms have been widely discussed (Hunter, 1977; Howard et al, 1978; Chaudhri & Khan, 1981; Paisley et al., 1991; Nickling & Wolfe, 1994; Ramakrishna et al., 1994; Anthonsen et al, 1996; Neuman et al, 1997; Bullard et al, 1997; Blumberg, 1998).

Dune morphology and control lay factors were first mentioned by Beadnell (1910). This was followed by numerous subsequent discussions Folk (1971), Bowler (1973), Breed et al., (1979a, b, 1979b), Nielson et al., (1982), Tsoar (1983a, b, 1986), Greeley and Iversen (1985), Thomas (1986, 1988a, b, c, 1989), El-Baz (1986), Cooke et al., (1993). The genesis of the barchan dunes has been described by Bagnold (1941).

Sand dunes and other aeolian landforms in Kuwait were discussed by Abu-Eid et al. (1983) and many later investigators Foda et al., (1984), Vincent and Lancaster (1985),

Khalaf (1989), Gharib et al., (1985), Khalaf et al, (1995), Misak et al., (1996), Abdullah, (1988), Nayfeh (1990). In Kuwait, the dunes can be found as isolated, sporadically distributed, or clustered in the form of dune belts (Khalaf et al, 1984). Al-Dousari (1998) delineates six belts of barchan sand dunes extending NW–SE direction in the northwestern part of Kuwait.

At a wider regional scale, aeolian sediments and landforms within Iraq and Arabia also have been discussed widely. Several workers (Dougrameji, 1984; Skocek & Saadallah, 1972) discussed the dunes of southern Iraq. They found a close similarity in mineralogy and textural characteristics between the southern Iraq dune sediments and the Mesopotamian flood plain.

The southern Mesopotamian flood plain is frequently hit by dust storms in the Middle East. It is regarded as one of the main sources of dust in the world (Khalaf & Al-Ajmi, 1993). The textural characteristics of aeolian sediments and landforms in eastern Saudi Arabia have attracted the interest of many investigators (Al-Saud, 1986; Fryberger et al., 1984; Shehata et al., 1992). The application of different fixation methods was widely used on dunes and mobile sand in the eastern desert of Arabia (Abolkhair, 1986; Watson, 1990). The study of the sand dunes has been discussed by Embabi and Ashour (1993), and Al-Sheeb (1998) in Qatar, and Goudie et al. (2001), El-Sayed (1999) in the United Arab Emirates.

Most dune types can be accommodated in an extended morphological categorization of dunes, following Pye and Tsoar (1990). Misak et al. (1996) classify the dunes into three main categories, namely, free or mobile, anchored and stabilized dunes.

Al-Dabi et al. (1997) analyzed the distribution and quantity of free dunes in Kuwait. They found that the increase in the number of free dunes occurred from 1989 to 1992. The Gulf War's disruption of Kuwait's environmental desert surface contributed to the development of these dunes. They noticed that with the same wind velocity during the past three decades, the rate of dune formation suddenly increased from 31 dunes per year for the pre-war images (1985–1989) to 321 dunes per year for after-war images (1992–1994).

Barchan sand dunes are the prevailing form in the study zone. Individual dunes and groups of barchans cover more than 80% of some dune belt areas (Al-Dousari, 1998). The barchan dunes appear primarily close to their source in the peripheral zones of the sand sea, especially downwind, away from the advancing tips of linear dunes in Al-Najaf sand sea in Iraq. These barchan dunes extend from Al-Najaf downwind to the Al-Atraf area in Kuwait.

There are four major conditions for the formation of barchan dunes. They are.

i. sufficient continuous source of sand;

ii. dominant, unidirectional wind;

iii. relatively flat, hard surfaces, such as desert pavement; and

iv. sand trapping obstacles such as vegetation or abrupt changes in microtopography.

The conditions listed above were subjected to various conditions that were conducive to the formation of sand dunes. However, the initiation of sand dunes is still considered a poorly understood process in the aeolian process. According to Khalaf and Al-Ajmi (1993), the sand dunes in Kuwait can be formed by the accumulation of flat patches of sand around 80–100 m^2 of sand on an area measuring less than 30 cm high. These sand patches grow into mounds and continue growing to form barchan dunes. Bagnold (1941) and Mainguet (1984) asserted that pre-barchanic dunes are transformed over time into crescentic dunes. Al-Dousari (1998) mentioned that Landsat TM images using bands 2, 5, and 7 differentiate between thick and thin mobile sand sheets. He noticed that the thick mobile sands are usually formed around the dunes, while thin sand sheets are mainly initiated downwind directly at the end of the mobile sand dunes in the Al-Atraf area as a continuation of the wind corridor. This strongly suggests that the sand dunes are one of the major sources of sand sheets in Kuwait.

Philip (1968), Skocek and Saadallah (1972) found a close similarity between the mineralogical and chemical composition of sand dunes in Iraq and the Mesopotamian flood plain deposits. Omar et al. (1989) mentioned that the barchan sand dunes in the northwest of Kuwait have nearly the same mineralogical composition as the upwind dunes in Iraq.

The average annual amount of sand drift in Kuwait is calculated by Al-Awadhi and Cermak (1995) as 99.596 kg/(m.yr). The highest average net flux amount of sand transport is 18.620 kg/(m.mo) in June, while the lowest is 592 kg/(m.mo) in December, but the sand drift attains a maximum during summertime (Al-Awadhi & Cermak, 1995).

3.2 Materials and Methods

The aeolian deposits have been traced using remote sensing (satellite images or aerial photographs) and fieldwork.

A total of 1684 samples were collected from the top 5 cm of the surface in the seven major fields of dunes in Kuwait. A global positioning system was used to determine the random sample locations. The number of samples collected varied from one dune field to another depending on the field size (Table 3.1).

The sampling sites were randomly collected within the wind corridor from the top surface sediments of sand dunes of different forms in Kuwait. Samples cover all types of dunes (barchans, dome, nabkha, and falling). At each chosen site samples were collected from an undisturbed surface at a location that appeared to be representative of the aeolian surface sediments at each site location.

A global positioning system was used to determine the specimen locations (GPS). The crest, the slip face, the horn, the windward, and the middle area between the crest and the windward were all sampled separately from the dune surfaces.

3.3 Dunes Migration

- Satellite Imagery

The level-3A RapidEye and PlanetScope images from the Planet Labs Inc (https://www.planet.com) acquired on different dates (Table 3.2) were used to detect the yearly rate of dune migration in Kuwait. The Level-3A images were geometrically and radiometrically calibrated by the Planet Labs. The RapidEye satellite imagery consists of five spectral bands ranging from 440 to 850 nm at a spatial resolution of 5 m (https://www.satimagingcorp.com), whereas the PlanetScope satellite imagery consists of four spectral bands ranging from 455 to 860 nm at a spatial resolution of 3 m (https://www.planet.com).

Table 3.1 Sampling sites and number of samples collected

Dune field	Number of samples	Remarks
Huwaimiliyah–Al-Atraf	1225	NW Kuwait, 60 km long and 5 to 8 km wide
Um Niqa (NE Kuwait)	5	NE of Kuwait
Falling dunes in Jal Zur	500	Four main sites in Jal Zur
Um Eish sand dunes	30	Part of dunes in Sabah Al-Ahmed National Reserve
Ras Sabiyah dunes	78	
Kabd sand dunes	38	
Dhubaiyah coastal dunes	15	Oolitic dunes (high calcium)

Table 3.2 The dates of satellite images used to study dune migration

Area	Satellite	Date
Atraf	RapidEye	2011–09-15, 2016–04-09
Buhaith	RapidEye	2011–09-15
	PlanetScope	2018–10-30
Huwaimliyah	RapidEye	2011–09-15, 2016–04-09
	PlanetScope	2020–10-26
Kabd	PlanetScope	2016–11-16, 2020–10-29
Um Urta	RapidEye	2011–09-15, 2016–04-09

Fig. 3.1 Dune zones in Kuwait. Dunes are widely distributed in Huwaimliyah, Um Urta, and Atraf compared to the other areas

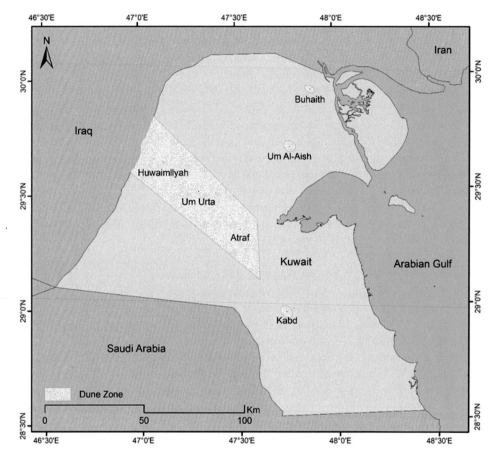

- Dune Delineation

Dunes were identified and delineated using ArcMap 10.6 based on a visual interpretation that incorporates linear and non-linear image contrast techniques (e.g., standard deviations, histogram equalization, and minimum–maximum stretch) (Jensen, 2015). A random sample of dunes was selected from five areas in Kuwait, i.e., Atraf, Buhaith, Huwaimliyah, Kabd, and Um Urta (Fig. 3.1). The selected dunes were varied in size and separated from other adjacent dunes. To estimate the dunes' migration, the selected dunes were observed two times over 4 years or longer. This period was selected to ensure that the magnitude of dune displacement was sufficient to be detected. The area and centroid of delineated dunes of the two times were calculated and saved in a shapefile.

- Dune Migration

The dune migration analysis was conducted using the R programming language. The dune shapefile was imported to RStudio, and its attribute table was stored as a data-frame object. The total dune displacement was calculated by determining the distance from the dune's centroid in time-1 to the dune's centroid in time-2 using Eq. (3.1):

$$d = \sqrt{(x_{t1} - x_{t2})^2 + (y_{t1} - y_{t2})^2} \qquad (3.1)$$

where d is total dune displacement, x_{t1} and y_{t1} represent the dune position in the time-1, and x_{t2} and y_{t2} represent the dune position in the time-2.

The dune migration yearly rate was then estimated by dividing d by the time difference in years from time-1 to time-2. Furthermore, the direction of dune migration in radians was calculated using the two-dimensional arctangent function available in R. These radian values were converted to degrees for the statistical analysis.

The dune displacement and its direction were statistically summarized. Descriptive statistics of these values within each area were also conducted to investigate the differences in dune migration in these areas. Dune migration differences in northwestern areas (i.e., Huwaimliyah, Um Urta, and Atraf) were investigated using one-way ANOVA analysis. The other areas were excluded due to some limitations in number of dune samples in these areas.

3.4 The Volume of the Dunes and Single Dune Sampling

The volume of the dunes was measured using two formulae:

- Barchan and domal dune volume = 4/3 * 3.14 * (Length * width * height).
- Falling dune volume = 1/6 * (Length * width * height).

A good estimation for the volume of all the dune fields was achieved using these equations.

Three different single dunes at three different locations were sampled to see the statistical variations within the same dunes.

Careful sampling, analysis, and data interpretation are essential to ensure those good management decisions are possible. Along with the quality of sampling, a sufficient sampling number of samples must be taken. In this study, the sample sites were chosen, randomly distributed areas all different kinds of mobile surface sediments. The collected samples were taken from the top 3 cm of both mobile and anchored surface sediments in all cases. The collected samples have been subjected to several types of laboratory analysis. The laboratory work is represented by trace and major elements (ICP, XRF), carbonate content, mineralogical surface area, and quartz morphometry analysis.

A total of 26 different micro-features were identified by the Scanning Electron Microscope (SEM). The chosen method used to describe the analysis features of the surface is considered as the finest technique due to the contributions of Krinsley and Doornkamp (1973), Bull (1981), Culver et al. (1983). The tabulation of the surface features on the quartz particles using percentages, numbers, or other statistical methods was also useful in distinguishing different samples. The roundness scale was also measured using a visual scale developed from Powers (1953) (Table 3.3 and Fig. 3.2).

Three different sets of aeolian samples along the main wind corridor were collected from upwind to downwind and one extra set was collected from the Dibdibba Formation for comparison (Fig. 3.1). With special care, samples were impregnated with epoxy, cut, and polished. The initial three set samples were gathered from upwind zones around Huwamiliyah area. One set of three samples D1, D4, and D9 represents the Dibdiba Formation. D1 is made up of fine sand particles (0.250 mm–0.180 mm), whereas the rest of the samples, whether from the Dibdiba Formation or aeolian sediments, are made up of medium sand fractions (0.5 mm–0.350 mm), which are the most prevailing size fractions. Samples preparation included the split and mix of both size fractions and placed in a resin block producing polished surface. Several grades of diamond abrasive are used in the grinding and polishing process of aeolian sediment samples.

Table 3.3 Methods and number of samples

Property	Particle morphometry	Roundness
Upwind samples	10 (21)[a]	10 (21)[a]
Downwind samples	9 (21)[a]	9 (21)[a]
Al-Dibdibba Formation	–	–
Measured items	26 Particles Micro-features	Particle Roundness
Analyzed fraction	Coarse sand	Coarse sand
Instrument used	JEOL super prob JXA-860MX SEM	Modified from Powers (1953)

[a]Values between brackets are the average number of studied particles

Fig. 3.2 Modified particle shape visual comparator scheme of Powers (1953), showing numerical indices and associated descriptive terminology

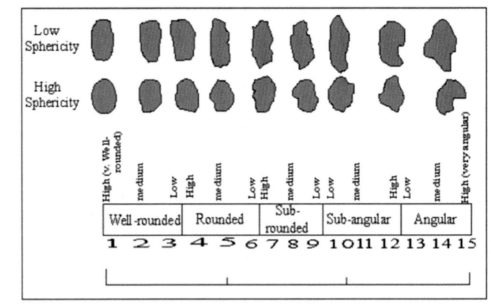

The samples were examined using a Hitachi 3000 variable pressure SEM equipped with BSE deflectors. The microscope can be used in a variety of ways, including

a. provide illustration in BSE or SEM mode of the sample;
b. using the EDX attachment, to analyze spots, areas, or line cross sections;
c. determine the map element distributions.

An intensive study was performed on various features within detected inclusions in quartz, feldspar, and other particles. The identification of some mineral species was challenging since there are also polymorphs that have similar chemical composition but have different structures. There are seven known polymorphs of titanium dioxide TiO_2 with the same chemical composition, but with different structures (e.g., rutile, anatase, and brookite). This makes it difficult to identify them using the crystals they are made from. For instance, rutile may have long hair-like needles, while other forms may have euhedral lath-like structures.

3.5 Results

Al-Dousari et al. (2008) identified and investigated six dune fields in Kuwait, namely, (Fig. 3.1).

- Al-Huwaimiliyah–Al-Atraf zone area (Fig. 3.3).
- Um Niqa (NE Kuwait) (Fig. 3.4).
- Falling dunes in Jal Al-Zur and Ras Sabiyah.
- Um Eish sand dunes (Fig. 3.4).
- Kabd sand dunes (Fig. 3.5).
- Dhubaiya coastal dunes (Fig. 3.5).

There are six major types of dunes (Fig. 3.6) in the dune fields, namely,

- Barchan or crescentic dunes.
- Domal dunes.
- Falling dunes.
- Climbing dunes.
- Barachanoid dunes.
- Nabkha dunes.

Most of the dunes in Kuwait are in the Huwaimiliyah (2026 dunes) and Jal Al-Zur (190 dunes) fields. About 36% of the dunes are barchans while only 10% are domal dunes in the Huwaimiliyah. The total volume of dunes in Kuwait is about 93,379,916 m^3 up to date. This volume is renewable, as the aeolian mobile sand is renewable especially in the summertime.

The longest barchanoid chain dune in Kuwait is with 2400 m and 900 m in Huwaimiliyah and Um Urta areas, respectively, shown in Fig. 3.7. Abu-Eid et al. (1983) showed dunes mapping used as a reference for our study that was collected in 1979 with a scale of 1:10,000 shown in Fig. 3.8. Dunes mapping during this study in the upwind area of the Huwaimiliyah-Atraf zone area is visible in Figs. 3.9, 3.10, and 3.11.

3.6 Dunes Migration

There are around 2304 sand dunes in Kuwait, all were sampled in this study. A sample of 35 dunes from five areas (i.e., Atraf, Buhaith, Huwaimiliyah, Kabd, Um Urta) were investigated. Um Al-Aish dunes were not investigated

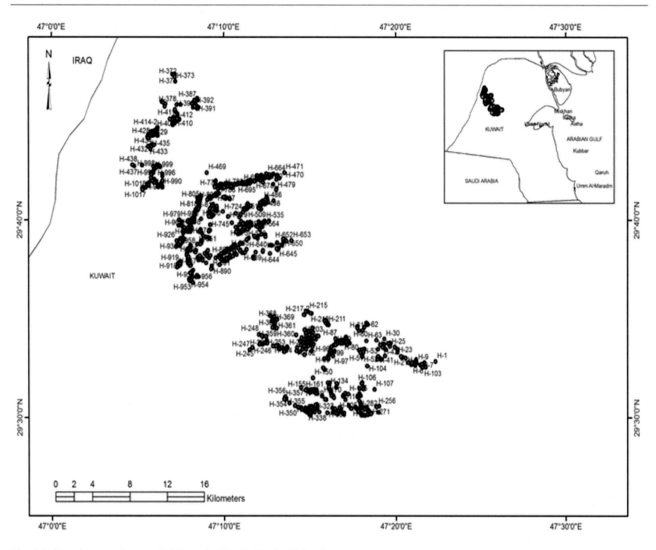

Fig. 3.3 Location map for sampled dunes in Huwaimiliyah–Al-Atraf zone area

because we could not find a clear site for them in the satellite images. The sample dunes varied in the area from about 2,300 m^2–40,000 m^2 (Table 3.4). The dune migration yearly rate ranged from 2 m in Buhaith area to 24 m in Um Urta area with an overall mean of about 14 m (Table 3.5) and (Fig. 3.12). The dunes were generally migrated toward the southeast enforced by the northwestern wind (the permanent wind). The dune migration direction, however, varied from 121° to 142° with a mean of 129° (Fig. 13a). We found one exception in Buhaith area: a dune displaced to the east (87°)

by 2 m yr^{-1}. The dune was the largest in the sample with an area of about 40,000 m^2. Generally, the dune migration rate had a negative correlation with the dune area ($r = -0.81$).

The dune migration rate in Buhaith area was different compared to the other areas (Fig. 3.13b). The dune migration in the northwestern areas exhibited some differences (Fig. 3.13). Um Urta and Atraf had the widest range of dune migration, whereas Huwaimiliyah dune migration rate had relatively a narrow range. This difference in dune migration rate among the three areas was significant (F = 4.15 and

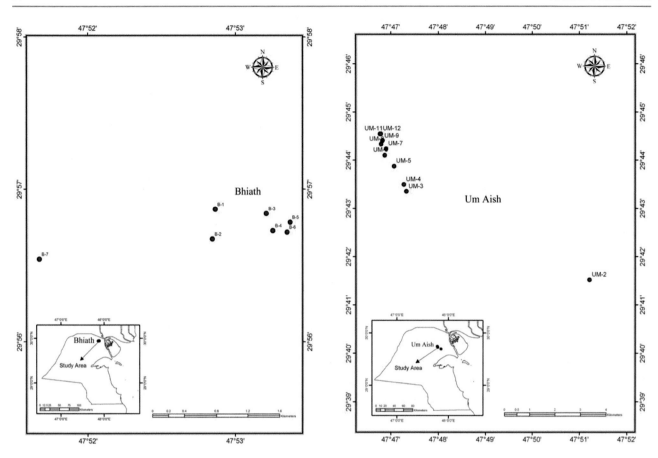

Fig. 3.4 Location map for sampled dunes in Um Niqa/Bhaith (left) and Um Eish

P-value = 0.008). The dune migration yearly rate significantly varied in northwestern areas in which Huwaimiliyah area had the narrowest range of dune migration yearly rate (Fig. 3.14).

3.6.1 Dunes Granulometric Characteristics

Most barchans in the region that extends from the Najaf Sand Sea toward Kuwait appear as isolated, occasionally dispersed in clusters or groups forming dune belts. The statistical parameters that were used to determine the barchan dune body's characteristics varied from one area to another. For instance, the mean particle size decreased from the upwind to the downwind edge, and the finest sand arising from the middle of the slip face. A significant variation in the border and mid-dune was present only in very fine sand with a p-value of 0.047. When relating particle

samples in the upwind of the barchan and climbing dune, all particle size classes presented significant variations except for the very coarse class (p-value = 0.44). Contrary to that, the particle sizes of the downwind barchans and falling dunes were remarkably different. For instance, sorting develops and improves in the direction of the border of the dune, while the horns and the mid of the dunes were poorly sorted compared to other sites. The difference in the sizes of the various particles in the two areas was also significant (Table 3.6). The average particle size demonstrates even and uniform dispersal with increased percentages of the medium, fine, and coarse sand size fractions, respectively. The upwind samples comprehend the largest percentages of coarse sand (68%) compared to mid and other dunes.

Kurtosis differs relatively with a vast range from 0.83 (platykurtic) to 1.6 (very leptokurtic) with an average of 1.12 (leptokurtic). The presence of several small populations on the crest could explain the decrease in particle sizes found in

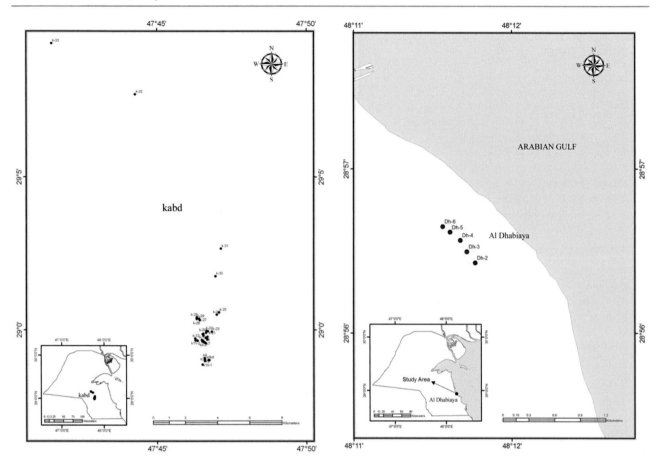

Fig. 3.5 Location map for sampled dunes in Kabd (left) and Dhubaiyah

Fig. 3.6 Pictures illustrating the types of studied dunes, barchan dunes (**a**), falling dunes (**b**), climbing dune (**c**), and nabkha (**d**)

(a) Barchans in Huwaimiliya area

(b) Falling dune in Jal Al-Zur

(c) Climbing dune in Kabd

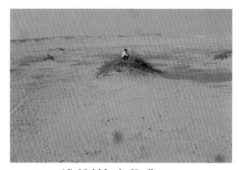

(d) Nabkha in Kadhma

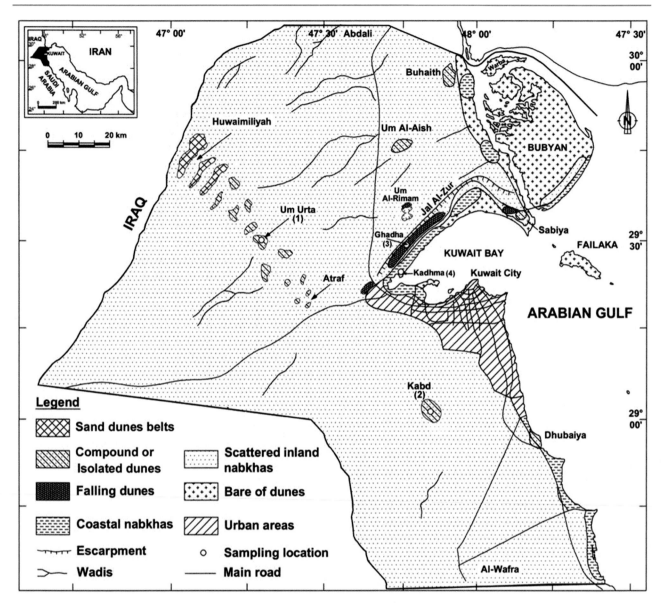

Fig. 3.7 Sampling locations of sand dunes in Kuwait and the Dibdiba Formation

border samples. The limited mixing of numerous particle size populations on the crest might be the indication for the decrease in the samples from the borders than those from the crest.

Statistical parameters of the studied dunes are variable regionally. The mean size diameter of dune sediments in the Mesopotamian Flood Plain is finer than other dunes on a regional scale. The dune sediments in Kuwait are coarser than the surrounding dunes in Iraq and Saudi, Qatar, and Emirates (Table 3.7). The mean grain size shows extensive

differences shifting between 1.3φ and 3.1φ (Tables 3.8 and 3.9). The barchans in Kuwait demonstrate a greater mean particle size, enhanced sorting, and more leptokurtic but similar skewness related to those upwind in Iraq and downwind in Saudi Arabia (Table 3.10). The location of Iraq's dunes near the Mesopotamian flood plain is known to have contributed to the higher levels of fine sediments found in these areas (Fig. 3.15). Also, climbing dunes and sabkhas contain much coarser sediments than comparable studies. Due to the limited literature on falling dunes, the available

Fig. 3.8 Dunes mapping in reference to aerial photos 1979 (Abu-Eid et al., 1983)

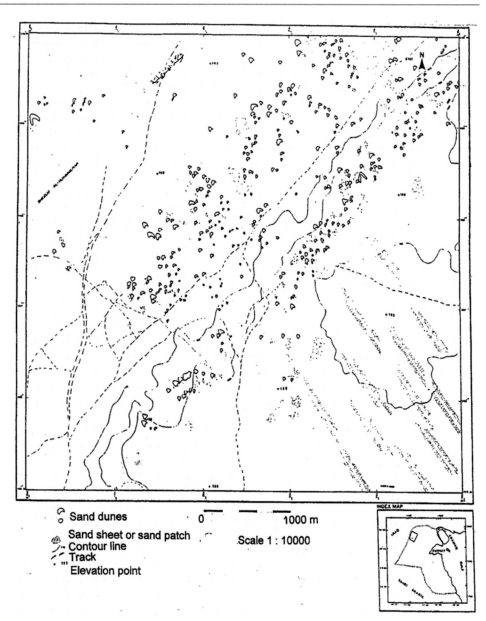

content on this matter was correlated with the size of the large linear dunes. According to the findings, the falling dunes had a greater mean grain size, better sorting, but similar skewness.

The textural features of different types of dunes in the same field global aspect were studied by Al-Dousari and Pye (2005). Considering the time factor, the statistical parameters of sediments across the dune fields in the study area were limited.

The results of the study showed that the skewness had negative effects on the overall dunes characteristics with fewer coarse particles among dunes with a tendency for coursing toward the dune margins. It made them more prone

to settling toward the margins (Table 3.11). The values range between −0.17 and 0.40 with an average of 0.14 (positively skewed). Skewness values of margins samples showed no significant variation compared to the samples from the crest of the same dune (Fig. 3.16).

3.7 Mineralogy

The mineral semi-quantitative percentages show lower quartz in upwind dunes in Iraq and the Mesopotamian Flood Plain. The highest purity quartz is present in two main areas: northwestern Arabia and Kuwait (Table 3.12).

Fig. 3.9 Dunes mapping during this study in the upwind area of the Huwaimiliyah-Atraf zone area

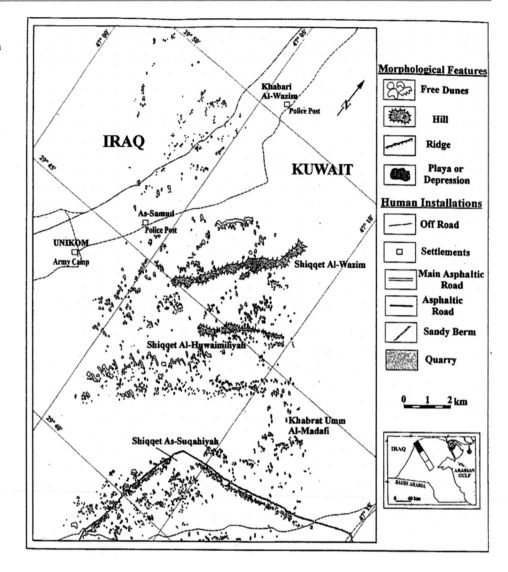

3.8 XRF Results

The XRF percentages show a lower concentration of silica in Dhubaiyah (56%) and Huwaimiliyah (82%), but Dhubaiyah dunes contain the highest amount of calcium oxides (39.2%) in comparison to all dune fields in Kuwait (Table 3.13). On the other hand, the RHUL results show percentage similar values to those of KISR before calibration (Table 3.12). The KISR results have been achieved with calibration, and are, therefore, more authentic.

Major oxide analyses of the bulk sand samples taken from the study area of aeolian sediments revealed the chemical interrelationships between the various components of the sand. The chemical differences between aeolian sands in the study area and Al-Dibdibba Formation sand are supported by mineralogical and ICP analyses result which show that the relative abundance of quartz and feldspar differs slightly between these two sand populations. The Al-Dibdibba Formation and Mesopotamian flood plain sand have been subjected to more extensive physical and chemical weathering. The major oxide values coincide the trace

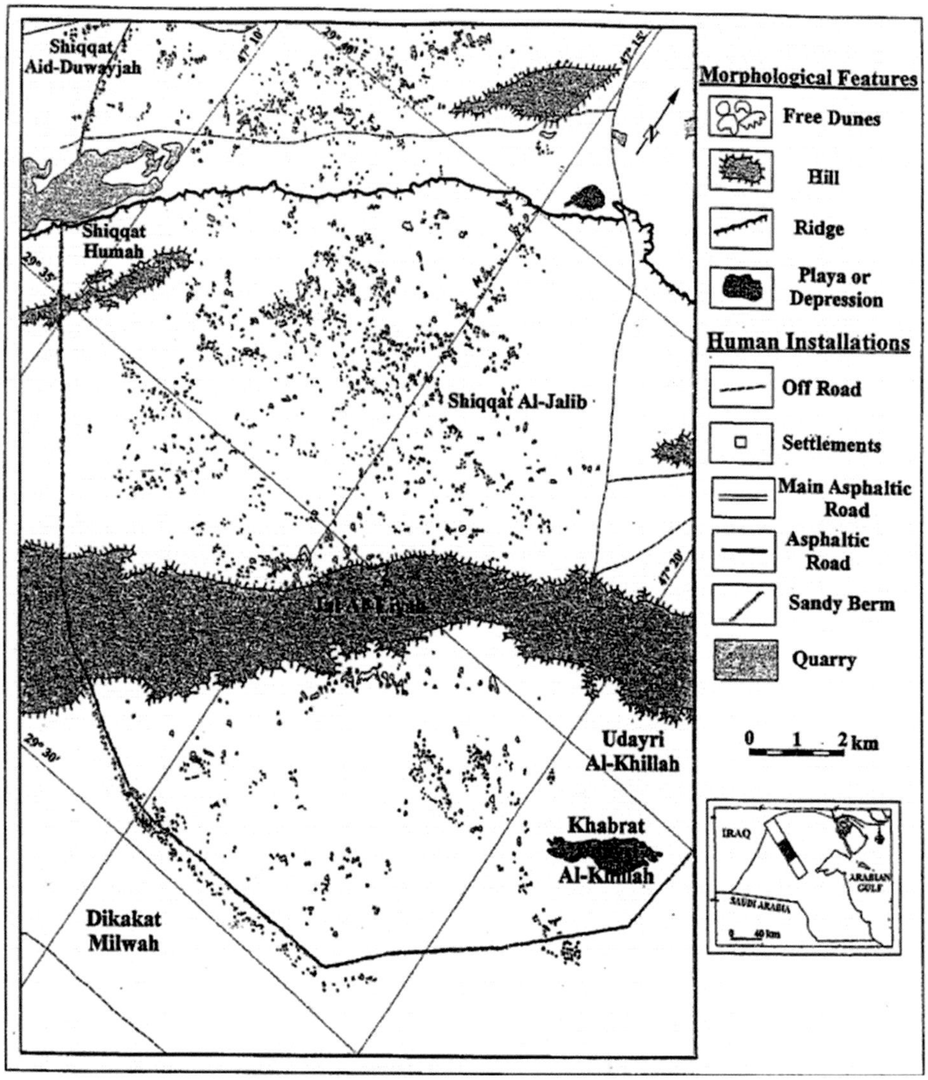

Fig. 3.10 Dunes mapping in the mid-area of the Huwaimiliyah-Atraf zone area

element results, which clearly differentiated the three sand populations. In addition, the close similarities in plot diagrams between the aeolian sediments of the research area and the Al-Dibdibba Formation sediments are well observed.

Differences between coarse and medium sand in terms of the concentration of major oxides are very limited. The aeolian sediments are characterized by high concentrations of SiO_2 (range from 86 to 97%) in both dominant size fractions. The high silica percentages reflect the high quartz proportion.

Although the major oxide results show close percentages between sediments from various regions, bivariate plots revealed clear distinctions. These results show that the aeolian sands in the study area are chemically related to the Al-Dibdibba Formation sand. Table 3.14 shows a very close interrelationship between Al_2O_3 and both K_2O and Fe_2O_3 indicating that the Al oxides are mostly associated with K-bearing minerals, principally K-feldspar. The aeolian sediments show slightly lower percentages of most of the major elements in comparison to the Al-Dibdibba sediments.

Fig. 3.11 Dunes mapping in downwind area of the Huwaimiliyah-Atraf zone area

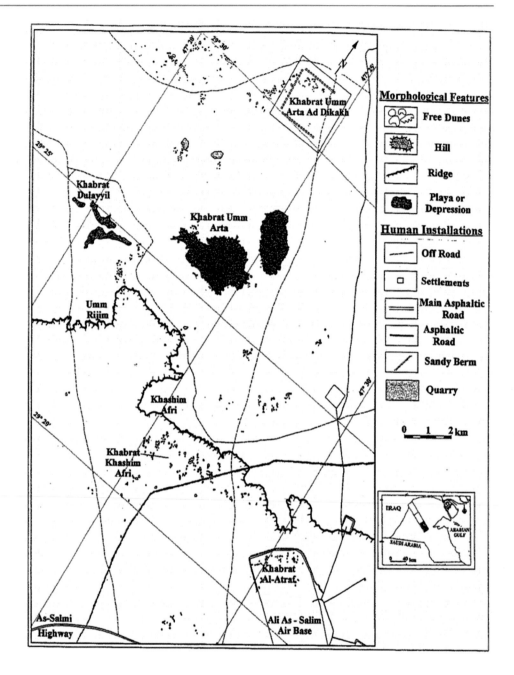

The division evident in the SiO_2 is also reflected in the variation of Al_2O_3 abundance. The high abundance of Al_2O_3 in the Al-Dibdibba sediments probably reflects an increased relative abundance of feldspar in these samples, indicating that they are less mature than the aeolian transported sands. Prolonged wind abrasion can reduce sand-sized K-feldspar and Fe-bearing minerals to silt sizes, causing silt is deflated away, and quartz-rich residue to form. The point counting results of coarse sand using binocular microscopy show that the average feldspar content is 2.37% in the total minerals of aeolian sand.

3.9 Micro-inclusions in Dunes Particles

Studying the physical and chemical properties of inclusions within aeolian particles is essential to use these properties as a fingerprint to compare them with global and regional sediments (Al-Dousari et al., 2020). Back-scattered electron microscopy was applied to examine the inclusions in 12 different samples from dunes (recent) and the Dibdiba Formation (Miocene to Pleistocene) sediments (Table 3.15). Samples were collected from a line across the main wind

Table 3.4 Number and percentages of dunes in Kuwait for each type with a total volume (m³) of sand

Area	Kabd	Bhiath	Dhabiaya	Um Aish	Huwaimiliyah	Jal Zur	Sabiya
Total dunes	33	7	6	12	2026	190	30
Barchans %	27.27	57.14	0.00	8.33	72.46	0	0
Dome dunes %	36.36	42.86	83.33	83.33	20.24	0	0
Barchanoid dunes %	6.06	0.00	0.00	0.00	4.54	0	0
Falling dunes %	27.27	0.00	16.67	8.33	0.39	100	100
Complex dunes %	3.03	0.00	0.00	0.00	2.37	0	0
Total volume (m³)	1,034,442	130,415	52,752	12,560	92,140,223	5384	4140

Table 3.5 Summary statistics of the dune sample in each area. The highest mean migration yearly rate was observed in Kabd

Site	n	Min	Max	Median	Mean	STD
Atraf	10	6	20	13	12.8	4.6
Bhaith	2	2	6	4	4	2.8
Huwaimiliyah	10	13	18	16.5	16.4	1.6
Kabd	3	14	20	16	16.7	3.1
Um Urta	10	7	24	14.5	15.5	5.9
Overall	**35**	**2**	**24**	**16**	**14.4**	**5.1**

corridor. The samples were analyzed using the electron microscopes to determine the presence of various elements within the deposits. The 12 aeolian samples were gathered from a line transect across the main wind corridor, representing the downwind, mid, and upwind areas (three in each set). All examined samples are observed in the major elements within minerals. Iron, aluminum, calcium, and zirconium, respectively, were the highly dominant elements found in the aeolian samples. Some variation was observed in the inclusion counts of various materials in the samples. For instance, in the downwind samples, the lower counts were observed in quartz and feldspars, while the aluminum percentages gradually increased upwind. The samples in the mid-area exhibited the presence of more sodium than in the downwind and upwind zones. The average percentages of barium found within the particles of quartz and feldspar indicate that the conditions in the samples have gradually increased downwind. The consistency across the various

elements in the samples from the Dibdiba Formation suggests that there is a strong relationship between the various elements.

The variations in the properties of the various elements found in the samples from the Dibdiba Formation are very limited. For instance, titanium is mainly produced through the natural solubility of smectite in cracks, fractures, and borders of quartz particles to form rutile (TiO_2) with needle-like or lath-like crystals. Although there were no further changes observed in the barium inclusions in quartz samples when compared with both the Dibdiba Formation (4.7%) and the aeolian (4.8%) samples. Other elements within the samples revealed a similar trend of resemblance. Different forms of titanium minerals were found in which (a) signify single particle with variations in brightness (Fig. 3.17). Aeolian particle enclosing heavy minerals inclusions (a) Smectite, inclusion made of iron, silica, titanium, and minor aluminum (Fig. 3.18).

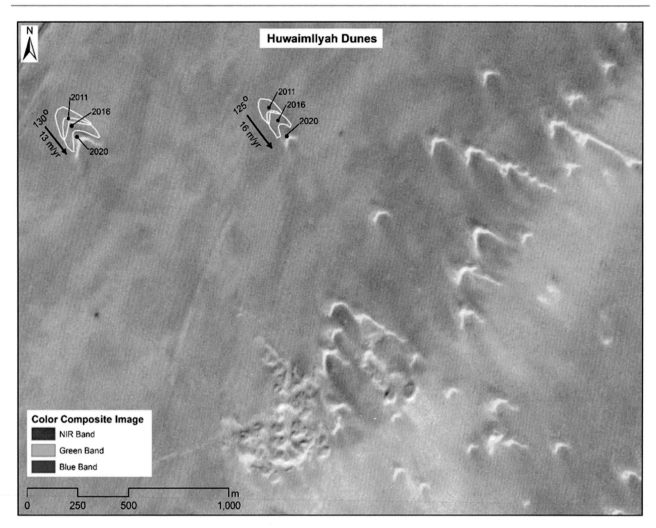

Fig. 3.12 The dune migration rate of two dunes in Huwaimiliyah. The dunes were different in size, migration rate, and direction

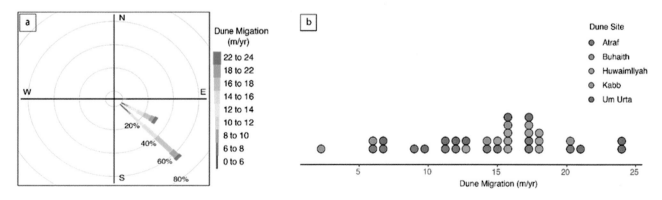

Fig. 3.13 **a** Rose diagram shows the migration direction and yearly rate. Most of the dunes (60%) were migrated to the southeast direction between 125° and 142°. **b** The distribution of dunes migration yearly rate. Colored dots show the variation of dune migration yearly rates within the dune areas

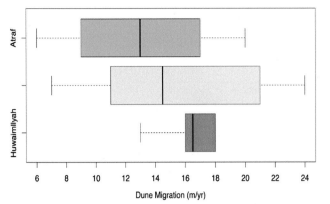

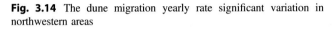

Fig. 3.14 The dune migration yearly rate significant variation in northwestern areas

3.10 Mechanical Micro-textures (Micro-features)

Mechanical micro-features are mainly mainly composed of various types of pits, such as crescentic pits, rounded pits, coalescing pits, V-shaped pits, dish-shaped depressions, curved sutures, straight sutures, grooves, upturned plates, meandering ridges, stepped cleavage planes, striations, and conchoidal fractures.

Most quartz grains have a dotted surface due to particle-to-particle coalescence caused by saltation. The most prominent features on the surfaces of the quartz particles were discovered to be V-shaped pits and dish-shaped depressions (Fig. 3.19a). The size of the rounded and

Table 3.6 Statistical parameters for the five sampling localities on barchan dunes in Kuwait

Locality	Statistical parameters				Grain size %		
	Mean (phi)	Sorting	Skewness	Kurtosis	C.S	M.S	F.S
Slip face	1.64	0.61	0.05	1.01	12.15	48.11	35.90
Horn	1.28	0.72	0.23	1.10	38.77	38.05	18.36
Crest	1.29	0.60	0.24	1.28	26.72	53.49	15.55
Mid	1.54	0.70	0.12	1.00	18.70	42.66	33.53
Windward	0.92	0.70	0.21	1.12	46.18	33.51	13.35
Average	1.34	0.67	0.17	1.10	28.51	43.16	23.34

Table 3.7 Barchan dune movement rates in the study area and other areas

Location	References	No. of Dunes (n)	Mean height (m)	Range in height (m)	Travel Distance (m/y)	Range in travel distance (m/y)
Saudi Arabia	Fryberger et al. (1981)	16	5.7	2.9–12.0	14.8	6.0–28.0
Saudi Arabia	Fryberger et al. (1984)	20	7.6	2.9–23	14.8	3.0–23
Saudi Arabia	Watson (1985)	67	8.54	3.2–25.1	14.64	0.2–44.4
Saudi Arabia	Shehata et al (1992)	3	9.67	7.0–11	21.5	9.7–27.7
Qatar	Al-Sheeb (1998)	100rds	–	1–45	20.2	7.5–40
Kuwait (NW)	Khalaf and Al-Ajmi (1993)	10	3.31	1.8–7.0	17.5[a]	7.0–27.1[a]
Kuwait	Present study (field)	2	3.09	2.1–4.1	24	16–31
Kuwait	Present study (photo's)	86	3.1	0.53–7.7[b]	24.3	3.7–59.7

[a] Only for 8 months
[b] n = 21(field measurements)

Table 3.8 Statistical parameters for five fields of sand dunes in Kuwait

Area	S.P	Methods of moments				
		Mean	Mean (mm)	Sorting	Skewness	Kurtosis
Huwaimiliyah	AVERAGE	1.5	0.4	0.6	1.2	4.7
	MAX	2.3	1.5	1.1	20.1	23.3
	MIN	0.6	0.2	0.2	-0.7	0.3
	STDEV	0.3	0.1	0.2	1.2	3.3
Kabd	AVERAGE	1.3	0.5	0.6	1.8	6.2
	MAX	2.0	0.5	0.9	3.4	13.0
	MIN	1.0	0.3	0.4	0.0	1.9
	STDEV	0.3	0.1	0.1	0.9	3.2
Um Aish	AVERAGE	1.2	0.5	0.7	2.0	6.7
	MAX	1.6	0.5	1.0	3.6	15.4
	MIN	1.0	0.4	0.5	0.8	1.9
	STDEV	0.2	0.0	0.1	0.8	3.9
Dhubaiyah	AVERAGE	1.0	0.5	0.4	2.8	13.1
	MAX	1.2	0.5	0.6	4.3	23.3
	MIN	0.9	0.5	0.3	1.5	5.0
	STDEV	0.1	0.0	0.1	1.0	6.8
Bhaith	AVERAGE	1.1	0.5	0.6	2.7	9.7
	MAX	1.2	0.5	0.7	3.3	13.8
	MIN	1.0	0.5	0.5	2.0	5.9
	STDEV	0.1	0.0	0.1	0.6	3.5

V-shaped pits ranges from 5 to 50 m. When quartz particles are exposed on the surface, they are subjected to weathering and abrasion during transportation. As a result, the surface of the chemically smoothed particles is dominated by crescentic, rounded, and V-shaped pits. Weathering causes adhering particles on the surface of quartz particles as well as dissolution and precipitation. Additionally, grooves and etchings were revealed on the surface of some particles (Fig. 3.19b). Some of the quartz particles had extensive fracturing, abrasion, and weathering (Fig. 3.19c). Some of the particles exhibited a smooth surface texture. However, some of them showed a variety of unusual features, such as Sutures and grooves that appeared mostly without any regularity in orientation. The angular quartz particles are also known to have multiple micro-features, such as sutures, cracks, and fractures either linear or conchoidal. Due to the dominance of other micro-features such as dish-shaped depressions and mechanical pits, they are suppressed in visual observation (Fig. 3.19d). Upturned plates were discovered to be abundant on the majority of the particles. Adhering particles were rare and uncommon on most of the majority of quartz particles, but the rounded quartz particles with smooth surfaces had none or very few particles compared with angular particles rich in micro-features. It is concluded that some quartz particles had a longer paleoenvironmental history than other quartz particles. It is worth noting that the pits, which were either V-shaped or crescentic and rounded, covered the majority of the particles and other surface micro-features, indicating that they were younger than other surface micro-features.

In terms of the relative effect of weathering and transport of the individual minerals, the quartz/feldspar ratio is a good indicator of the general maturity of sediments (Zimbelman & Williams, 2002).

Table 3.9 Particle size percentages for five fields of sand dunes in Kuwait

Area	Parameter	V.C.S	C.S	M.S	F.S	V.F.S	Mud
Huwaimiliyah	AVERAGE	0.3	43.2	36.8	10.3	2.3	0.0
	MAX	60.9	92.2	94.8	59.1	22.0	0.0
	MIN	0.0	0.0	0.0	0.0	0.0	0.0
	STDEV	3.9	27.8	26.9	6.4	2.1	0.0
Kabd	AVERAGE	2.1	58.8	22.5	5.2	2.2	0.0
	MAX	58.4	347.0	89.8	12.0	7.0	0.0
	MIN	0.0	0.0	0.0	0.0	0.0	0.0
	STDEV	9.6	58.5	24.5	3.4	2.1	0.0
Um Aish	AVERAGE	0.8	71.3	6.9	8.1	3.8	0.0
	MAX	1.6	88.2	17.2	19.1	8.7	0.0
	MIN	0.0	0.0	0.0	0.0	0.0	0.0
	STDEV	0.5	25.3	5.3	5.0	2.3	0.0
Dhubaiyah	AVERAGE	0.4	30.6	7.4	1.2	0.4	0.0
	MAX	1.8	84.6	25.7	3.9	1.6	0.0
	MIN	0.0	0.0	0.0	0.0	0.0	0.0
	STDEV	0.8	42.3	11.3	1.7	0.7	0.0
Bhaith	AVERAGE	0.1	48.7	6.4	3.0	1.8	0.0
	MAX	0.3	84.4	13.6	5.4	4.8	0.0
	MIN	0.0	0.0	0.0	0.0	0.0	0.0
	STDEV	0.2	44.5	6.2	2.7	2.0	0.0

Table 3.10 Studied dunes statistical parameters values compared to global and regional dunes values

Location	Mean (Phi)	Sorting	Skewness	Kurtosis	Dune type	Reference
NW-Kuwait	1.30	0.68	0.14	1.12	Barchan	Present study
Kabd-Kuwait	1.29	0.75	0.21	1.23	Climbing dune	Present study
Khadhma-Kuwait	0.64	0.89	1.44	0.70	Nabkha	Present study
Northern Kuwait Bay	3.20	1.50	0.12	1.21	Nabkha	Khalaf et al. (1989)
Um Nega-Kuwait	1.67	0.65	0.04	1.26	Barchan	Omar et al. (1989)
Najaf-Iraq	2.04	0.52	0.32	0.94	Barchan	Skocek and SaadAllah (1972)
Samawa-Iraq	2.05	0.73	0.04	1.11	Barchan	Skocek and SaadAllah (1972)
Beeji dunes-Iraq	1.85	0.82	0.27	1.39	Barchan	Dougrameji (1984)
Humar Lake-Iraq	2.60	0.93	-0.19	1.04	Barchan	Skocek and SaadAllah (1972)
Basrah-Iraq	1.95	0.93	0.25	0.92	Barchan	Khalaf (1989)
Musayab-Iraq	3.08	0.99	0.02	0.90	Barchan	Dougrameji (1984)
Nasriya-1-Iraq	2.42	0.35	0.04	1.32	Barchan	Skocek and SaadAllah (1972)
Nasriya-2-Iraq	1.91	0.73	0.31	1.53	Barchan	Dougrameji (1984)
Nafud Rumhat-Saudi	1.88	0.57	0.32	1.53	Barchan	Binda (1983)
Nejav-Palestine	2.54	0.47	…	…	Climbing dune	White and Tsoar (1998)
Mopti, Mali	3.58	2.38	…	…	Nabkha	Nickling and Wolfe (1994)
Molopo, S. Africa	1.91	1.18	-0.34		Nabkha	Dougill and Thomas (2002)
Ghadha-Kuwait	1.68	0.89	0.28	0.94	Falling dune	Present study
Jal Al-Zour-Kuwait	1.86	0.8	0.14	…	Falling dune	Al-Enezi (2001)
Australia	2.34	0.55	0.24	…	Linear dune	McKee (1979)
Australia	2.17	0.36	0.36	…	Linear dune	McKee (1979)
Arizona	2.04	0.61	0.15	…	Linear dune	McKee (1979)
Algeria	2.19	0.5	0.31	…	Linear dune	McKee (1979)
Tunisia	3.38	0.22	0.22	…	Linear dune	McKee (1979)
Libya	2.16	0.63	0.14	…	Linear dune	McKee (1979)
Namib Sand dunes	2.3	0.49	0.16	…	Linear dune	Lancaster (1981)
Kalahari	2.03	0.72	0.17	…	Linear dune	Livingstone et al. (1999)

3.11 Chemical Micro-textures

The only abundant chemical micro-feature found in the studied samples was irregular solution and silica precipitation. Other micro-features discovered to occur to a limited extent include silica plastering, deep surface solution, oriented V-shaped pits, disintegration by solution, and large-scale decomposition. Chemical surface features were discovered to be sporadic in distribution and varied in extent from particle to particle. Deep grooves were discovered, sometimes deeply etched, and were frequently overlapping by mechanical features such as conchoidal fractures and deep triangular pits (Fig. 3.19e). The surface of some quartz particles has a frosted appearance with large depressions. Deep grooving and etching occur more frequently in sutures, pits, or around corners of quartz particles (Fig. 3.19f).

3.12 Upwind and Downwind Variations

In general, upwind particles in Huwaimiliyah have more mechanical and chemical properties than southeastern particles (downwind). This could imply that the coarse quartz particles are primarily derived from local sources (Fig. 3.19a). The dominant elements in the upwind samples

Fig. 3.15 Particle size distribution histograms from upwind in Iraq to downwind in Kuwait (V.C.S, C.S. M.S., F.S., V.F.S.: very coarse, coarse, medium, fine, and very fine sand, respectively, S: silt)

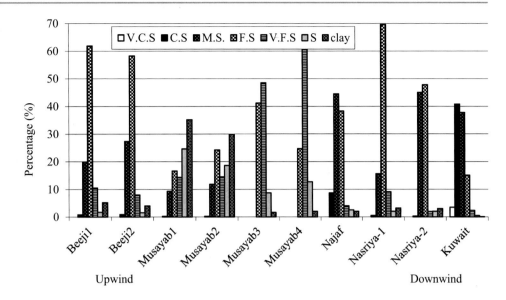

Table 3.11 Relative statistical parameters for barchan type in several sand seas

Location	Mean	Sorting	Skewness	Kurtosis
Humar Lake (Iraq)[a]	2.60	0.93	−0.19	1.04
Zubir (Iraq)[b]	1.95	0.93	0.25	0.92
Musayab (Iraq)[c]	3.08	0.99	0.02	0.90
Beeji dunes (Iraq)[c]	1.85	0.82	0.27	1.39
Najaf (Iraq)[d]	2.04	0.52	0.32	0.94
Nasriya-1 (Iraq)[d]	2.42	0.35	0.04	1.32
Samawa (Iraq)	2.05	0.73	0.04	1.11
Red Sea (Saudi)[e]	2.6	0.46	0.04	0.99
Jafurah Sand Sea (Saudi)[e]	2.02	0.95	0.35	1.12
Nafud Rumhat (Saudi)[e]	1.88	0.57	0.32	1.53
NW Kuwait	1.28	0.72	0.23	1.10
NE Kuwait	1.67	0.65	0.04	1.26
Qatar	1.93	0.48	0.09	0.72
Average	2.09	0.70	0.15	1.13

[a]Skocek and SaadAllah (1972)
[b]Al-Dousari et al. (2008)
[c]Dougrameji (1984)
[d]Skocek and SaadAllah (1972)
[e]Benaafi and Abdullatif (2015)

are upturned plates, dish-shaped depressions, V-shaped pits, and rounded and crescentic pits, while the dominant features in the downwind samples are upturned plates, irregular solution precipitation, rounded and crescentic pits, and V-shaped pits. Dish-shaped depressions were much less common in downwind samples than in upwind samples (Table 3.16).

3.13 Particle Area, Perimeter, and BET Surface Area

BET surface area varied in range from 1 m^2g^{-1} to 47 m^2g^{-1}. The closer the samples to the Mesopotamian flood Plain, the higher the BET surface area (Fig. 3.20). This can be

Fig. 3.16 Statistical parameters with sampling sites for barchan dune surface sediments in Um Urta

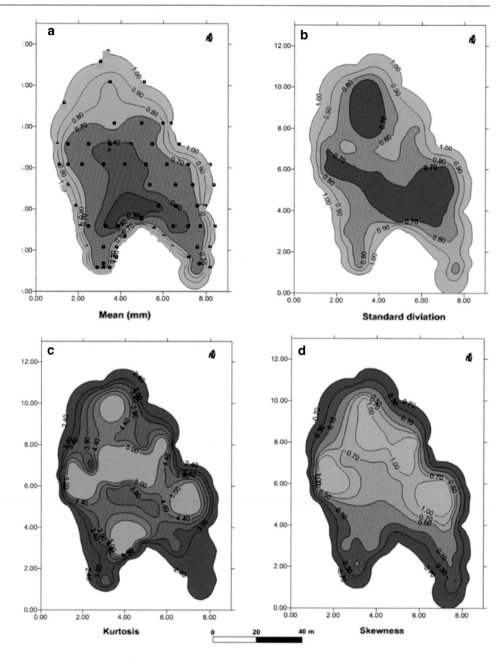

Table 3.12 The average mineral percentages for barchan dunes in the Middle East

Location	Reference	Quartz	Feldspar	Calcite	Dolomite	Gypsum	Others
North Iraq	IV	75.0	12.0	10.0	2.0	0.5	0.5
Southwest Iraq	IV	49.6	11.1	19.2	2.3	4.6	13.3
West Iraq	IV	64.4	22.5	5.0	1.0	0.5	6.6
East Iraq	IV	68.9	10.0	15.1	3.5	1.3	1.2
Northwest Kuwait	II	92.9	2.8	2.9	0.6	0.4	0.4
Northwest Saudi	I	100.0	0.0	0.0	0.0	0.0	0.0
East Saudi	II	87.7	5.8	6.0	0.5	0.0	0.0
Southeast Saudi	III	91.4	3.8	4.8	0.0	0.0	0.0
Southwest Saudi	III	78.5	2.6	18.1	1.8	0.0	0.0
Qatar	III	75.0	13.0	9.0	2.0	0.5	0.5
Emirates (Ain)	III	68.1	0.58	29.9	1.3	0.0	0.1
Average		77.4	7.7	10.9	1.4	0.7	2.0

Table 3.13 Samples analyzed by XRF Laboratory (number of samples analyzed for each location is given in brackets)

Location	XRF Percentages										
	Na_2O	K_2O	Al_2O_3	SiO_2	Fe_2O_3	SrO	TiO_2	MgO	CaO	MnO	BaO
Kabd (37)	1.24	1.44	5.51	86.8	0.56	0.01	0.24	0.87	3.33	0.02	0.05
Huwaimiliyah (56)	1.55	1.16	6.80	82.7	4.16	0.01	0.25	1.11	5.45	0.18	0.04
Huwaimiliyah (94)[a]	0.18	0.94	1.98	94.1	0.39	0.00	0.07	0.20	0.67	0.01	0.02
Dhuabiyah (6)	0.64	0.52	1.99	56.0	0.47	0.28	0.15	0.81	39.15		
Um Aish (12)	1.05	1.44	4.03	89.5	0.49	0.01	0.19	0.49	2.46	0.02	0.04
Bhaith (9)	0.69	0.71	2.83	92	0.41	0.01	0.13	0.51	2.68		
Jal Zur (190)[a]	0.51	1.40	2.70	91.9	0.47	0.01	0.06	0.16	1.13	0.01	0.03
Sabiya (30)[a]	0.29	1.44	2.52	92.1	0.46	0.01	0.06	0.20	1.03	0.01	0.03

[a]Samples analyzed by RHUL labs

Table 3.14 Average XRF analysis results of the Al-Huwaimiliyah-Atraf Dunes for two size fractions

Elements %	Average	
	Coarse sand	Medium sand
SiO_2	93.28	94.07
$Al2O_3$	2.52	1.97
K_2O	1.34	0.94
CaO	0.69	0.67
$Fe2O_3$	0.39	0.39
MgO	0.23	0.20
Na_2O	0.23	0.17
TiO_2	0.07	0.07
P_2O_5	0.01	0.01
MnO	0.01	0.01
Total	98.77	98.50
Trace elements (ppm)		
Ba	280.40	206.22
Sr	55.00	49.59
Zr	34.05	35.89
Rb	32.43	23.27
Ni	18.57	20.33
Cl	39.23	25.66
Zn	3.71	4.02
Ga	3.71	3.04
Pb	3.76	3.08
V	9.11	7.76
Cu	3.04	3.67
Cr	10.58	9.91
Sc	0.16	0.34
Nb	1.10	1.04
Th	0.14	0.16
Y	2.53	2.49
La	3.32	3.92
Ce	2.93	4.23
Nd (ppm)	2.79	3.14

Table 3.15 Frequency of detection of the mean elements in inclusions of aeolian and Dibdiba Formation particles

	Aeolian sample				Dibdiba Formation
Averages (in quartz particles)					
Elements	Average	Downwind	Mid-area	Upwind	Average
S	0.38	0.00	0.67	0.33	1.67
Mn	0.50	1.00	0.00	0.67	2.33
Mo	1.25	0.33	1.67	1.33	3.33
Ba	2.00	1.67	2.33	2.00	2.67
Ti	5.50	2.67	7.00	5.33	5.33
Zr	8.00	6.33	8.00	8.33	11.67
Ca	8.50	6.00	8.33	9.67	6.67
Al	11.00	4.33	12.67	13.00	10.67
Fe	11.13	5.00	12.67	12.33	9.67
In feldspar and other grains[a]					
Mn	0.63	0.67	0.67	0.67	4.00
Mo	1.00	2.33	0.67	0.67	3.00
Ba	1.13	2.33	0.67	1.00	2.33
Zr	1.25	0.33	2.33	0.67	3.33
Ti	1.63	0.33	2.33	1.67	8.33
Ca	4.00	0.67	7.33	2.67	12.67
Fe	5.00	2.33	8.00	3.33	8.67

[a]Other grains are clay, calcrete, heavy minerals, or intergrowth of quartz and feldspar (lithics)

attributed to the high content of mud size fractions, carbonates, and heavy minerals compared to samples in Arabia and Kuwait that contain more quartz.

The particle area and perimeter measurements of aeolian sediments (upwind and downwind) in Kuwait and the underneath or surrounding geological formation (Dibdiba Formation) indicate close similarities between the upwind and downwind. Using these values, the similarities are also observed when comparing the aeolian sediments and the Dibdiba Formation samples. The mean area and perimeter values of aeolian samples are slightly higher than the Dibdiba Formation samples. Also, upwind samples were slightly higher in the mean perimeter values than downwind samples, but the values overlap by one standard deviation (Fig. 3.21).

3.14 Conclusions

There are around 2304 sand dunes in Kuwait; all were sampled in this study. The sum up conclusion of this study is as follows:

Fig. 3.17 Titanium minerals in various forms, where **a** represents a single particle with varying brightness. The first is anatase, which is composed of titanium, iron, silica, and minor manganese (1), and the darker tone contains more iron than titanium without manganese (2), **b** veins of silica, calcium, titanium, and aluminum (ilmenite) within quartz particle, **c** lath-like crystals of anatase (1) composed of titanium and silica with minor aluminum within quartz particle (2), **d** rutile (1) formed in the border of quartz particle due to alteration from smectite (2), **e** albite particle (1) contains ilmenite (2) and epidote (3) inclusions, and **f** euhedral crystals within quartz particle in sample D9 at Dibdiba formation

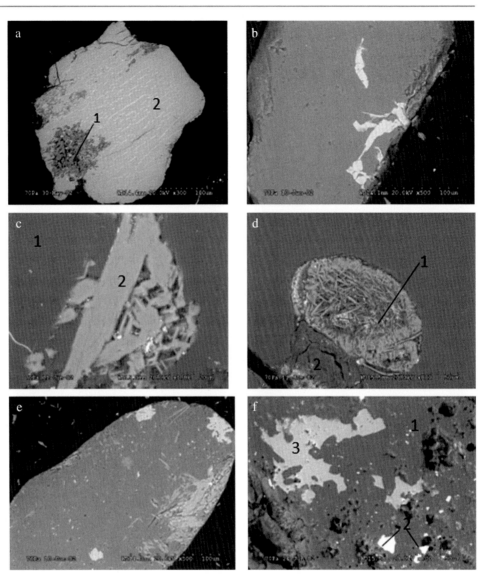

Fig. 3.18 Aeolian particle with enclosing heavy minerals inclusions **a** Smectite (1), bright inclusion composed of iron and silica (2) engulfing a sphene (3), **b** a matrix composed of iron, silica, titanium, and minor aluminum (1) containing variety of inclusions, silica, calcium, zircon, and titanium (2), pyroxene (3) and hornblende with altered titanium (4), **c** zircon (1) inserted around a cavity coated by silica, magnesium, and iron (2), **d** biotite mica as large inclusion within a quartz particle **e** hypersthene inclusion (1) containing small inclusion composed of silica, molybdenum, iron, calcium, and barium, **f** a feldspar particle full of inclusions of garnet minerals (almandite)

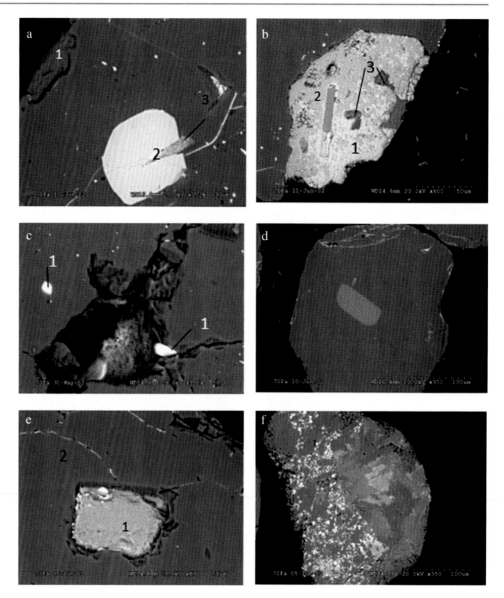

Fig. 3.19 Quartz sand particles shown in scanning electron micrographs (SEM).
a Well-rounded smoothed quartz particle with mechanical V-shaped, rounded, and crescentic pits and dish-shaped depressions. **b** Deep chemical pitting in a quartz particle. **c** An angular particle with a large number of mature conchoidal fractures. **d** Medium outline relief particle with a variety of micro-features such as meandering ridges (1), upturned plates (2), capping layers (3), conchoidal fractures (4), and adhering particles (5). **e** Extensive grooves, cracks, and chemically etched V-forms. **f** Generalized etching arrangement, solution disintegration, V-shaped sutures (1), and weathered feldspar-inclusion (2)

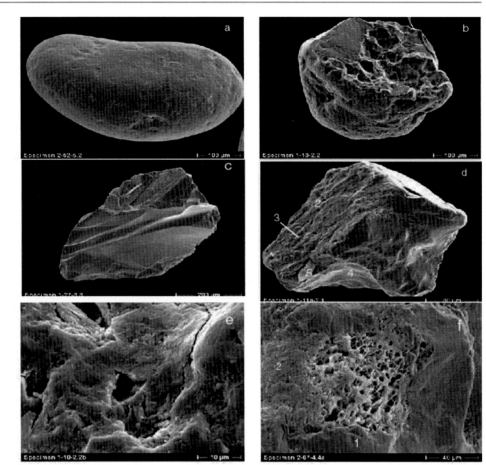

Table 3.16 The mean occurrence of 26 surface textural features on quartz sand particles from upwind and downwind aeolian samples in Al-Huwaimiliyah

Micro-features	Mean occurrence	
	Upwind	Downwind
Conchoidal fractures	2.17	1.99
Rounded and crescentic pits	5.98	5.80
Small striations	1.09	0.80
Upturned plates	7.31	6.25
Straight suture	2.91	2.91
Curved sutures	3.09	2.15
Dish-shaped depression	6.20	3.36
Coalescing pits	0.68	0.27
V-shaped pits	5.91	5.12
Stepped cleavage planes	1.69	1.39
Rounded particles[a]	7.44	6.81
Adhering particles	1.25	1.15

(continued)

Table 3.16 (continued)

Micro-features	Mean occurrence	
	Upwind	Downwind
Chemically etched V-forms	1.61	1.45
Straight/curved grooves	5.96	5.46
Smooth precipitation surface	1.90	1.54
Oriented V-shaped pits	3.16	2.80
Silica plastering	4.85	3.15
Deep surface solution	3.96	3.96
Dulled solution surface	0.37	0.28
Meandering ridges	2.97	2.01
Capping layer	0.06	0.03
Quartz crystal growth	0.15	0.00
Disintegration by solution	2.96	4.16
Irregular solution-precipitation	6.07	5.98
Silica coating structure	1.21	0.73
Large-scale decomposition	2.51	2.29

[a]Using the visual scale in Fig. 3.19

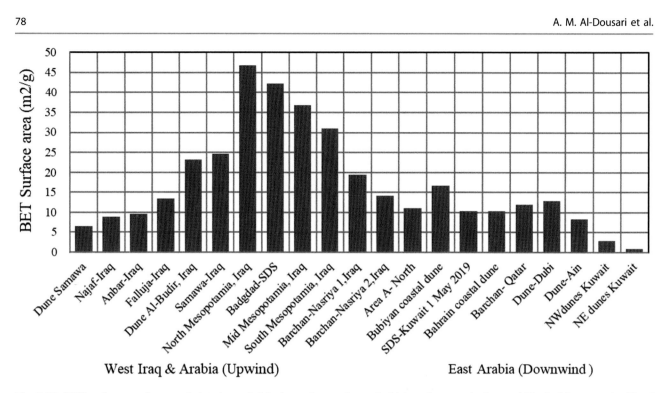

Fig. 3.20 BET surface area from upwind to downwind in the study area shows the high surface area in dunes within the Mesopotamian Flood Plain

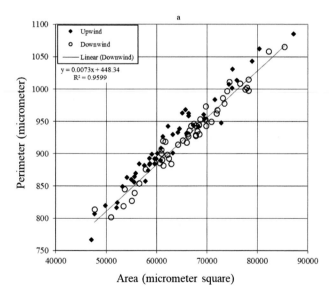

Fig. 3.21 Area (a) versus perimeter diagrams for upwind and downwind samples

- The total volume of dunes up to date is about 93,379,916 m^3.
- Bhaith dunes contain along with Um Eish and Jal Zur, the purest silica content in Kuwait.
- Dhubaiyah and Huwaimiliyah dunes contain the lowest silica content.

- Generally, the aeolian sand dunes in Kuwait contain low percentages of silica (about 84% on average) that can be of economic value for glass manufacturing.
- Most of the dunes are in the Huwaimiliyah field.
- The significant sand dune dynamics and migration rates varied from 121° to 142° except in Buhaith area: a dune displaced to the east (87°) by 2 m yr^{-1}.
- Dune migration rate had a negative correlation with the dune area ($r = -0.81$).

References

Abdullah, J. (1988). *Study of control measures of mobile barchan dunes in the Umm Al-Eish and west Jahra areas.* Kuwait Institute for Scientific Research, KISR 2580, Kuwait.

Abolkhair, Y. M. S. (1986). The statistical analysis of the sand grain size distribution of Al-Ubay-lah barchan dunes, Northwestern Ar-Rub-Alkhali desert. *Saudi Arabia. Geo Journal, 13*, 103–109. https://doi.org/10.1007/BF00212712.

Abu-Eid, R., El-Sayed, M., Salman, A. S. (1983). *Field and laboratory investigation of sand dune deposits in north-east and north-west of Kuwait.* Kuwait Institute for Scientific Research, Kuwait.

Al-Awadhi, J., & Cermak, E. (1995). Application of Mathematical Modelling of Sand Transport to Kuwait Environment. *Kuwait Journal of Science, 22*, 183–197.

Al-Bakri, D., Kittaneh, W., & Shublaq, W. (1988a). Sedimentological characteristics of the surficial deposits of the Jal Az-Zor area. *Kuwait, Sedimentary Geology, 59*, 295–306.

Al-Bakri, D., Rejehrt, M., Al-Sulaimi, J., & Kittaneh W. (1988b). *Assessment of the sand and gravel resources of Kuwait: Geology and geomorphology, Final Report*, Vol.3, KISR 2830, Kuwait.

Al-Dousari, A. M. (1998). *Textural characteristics and mineralogy of free dunes of the Al-Huwamiliyah–Al-Atraf zone in Kuwait* (Doctoral dissertation, MS thesis, Kuwait University, Kuwait)..

Al-Dousari, A. M., Ahmed, M. O. D. I., Al-Senafy, M., & Al-Mutairi, M. (2008). Characteristics of nabkhas in relation to dominant perennial plant species in Kuwait. *Kuwait Journal of Science and Engineering, 35*(1A), 129.

Al-Dousari, A. M., & Pye, K. (2005). Mapping and monitoring of dunes in northwestern Kuwait. *Kuwait Journal of Science and Engineering, 32*(2), 119.

Al-Dousari, A., Pye K., Al-Hazza, A., Al-Shatti, F., Ahmed, M., Al-Dousari, N., & Rajab, M. (2020). Nanosize Inclusions as A Fingerprint for Aeolian Sediments. *Journal of Nanoparticle Research*. https://doi.org/10.1007/s11051-020-04825-7.

Al-Dabi, H., Koch, M., Al-Sarawi, M., & El-Baz, F. (1997). Evolution of sand dune patterns in space and time in north-western Kuwait using Landsat images. *Journal of Arid Environments, 36*, 15–24.

Alenezi, A., (2001). Morphodynamic and sedimentology of falling dunes northeast of Kuwait. Ph.D. Thesis, Royal Holloway, University of London, unpublished.

Al-Saud, M. (1986). Sand drift and its size characteristics in Ad-Dahna desert on the Riyadh-Dammam highway. Unpublished M.A. Thesis, Geogr. Dept, King Saud University, Riyadh, Saudi Arabia, 1986 (Arabic).

Al-Sheeb, A. I. (1998). Sand movement in the state of Qatar—The problem and solution. Assessment and monitoring of desert ecosystemsIn S. A. S. Omar, R. Misak, & D. Al-Ajmi (Eds.), *Sustainable development in arid zones* (Vol. 1, pp. 223–239). Balkema.

Anthonsen, K., Clemmensen, L., & Jensen, J. (1996). Evolution of a dune from crescentic to parabolic form in response to short-term climatic changes: R¥bjerg Mile, Skagen Odde. *Denmark. Geomorphology, 17*, 63–77. https://doi.org/10.1016/0169-555X(95)00091-I.

Bagnold, R. A. (1941). *Physics of blown sand and desert dunes* (p. 205). Methuen.

Beadnell. H. (1910). The sand-dunes of the Libyan desert. Their Origin, form, and rate of movement, considered in relation to the geological and meteorological conditions of the region. *The Geographical Journal, 35*(4), 379–392.

Benaafi, M., & Abdullatif, O. (2015). Sedimentological, mineralogical, and geochemical characterization of sand dunes in Saudi Arabia. *Arabian Journal of Geosciences, 8*, 11073–11092. https://doi.org/10.1007/s12517-015-1970-9.

Binda, P. L. (1983). On skewness of some eolian sands from Saudi Arabia. In M. E. Brookfield & T. S. Ahlbrandt (Eds.), *Developments in sedimentology* (Vol. 38, pp. 27–39). Elsevier.

Blumberg, D. (1998). Remote Sensing of Desert Dune Forms by Polarimetric Synthetic Aperture Radar (SAR). *Remote Sensing of Environment, 65*(2), 204–216. https://doi.org/10.1016/S0034-4257(98)00028-5.

Bowler, J. M. (1973). Clay Dunes: Their occurrence. formation and environmental significance. *Earth-Science Reviews, 9*(4), 315–338. https://doi.org/10.1016/0012-8252(73)90001-9.

Breed, C. S., Grolier, M., McCauley, J. (1979a). Morphology and distribution of common 'sand' dunes on Mars: Comparison with the Earth *84*(14), 8183–8204. https://doi.org/10.1029/JB084iB14p08183.

Breed, C. S., & Grow, T. (1979b). Morphology and distribution of dunes in sand seas observed by remote sensing. In E. D. McKee (Ed.), *A study of global sand seas*, United States Geological Survey Professional Paper 1052, pp. 253–304.

Breed, C. S., et al. (1979b). Regional studies of sand seas using Landsat (ERTS) imagery. In E. D. McKee (Ed.), *A study of global sand seas*, United States Geological Survey Professional Paper 1052, pp. 305–398.

Bull, P. A. (1981). Environmental reconstruction by scanning electron microscopy. *Progress in Physical Geography, 5*(3), 368–397.

Bullard, J. E., Thomas, D. S. G., Livingstone, I., & Wiggs, G. S. F. (1997). Dunefield Activity and Interactions with Climatic Variability in the Southwest Kalahari Desert. *Earth Surface Processes and Landforms, 22*(2), 165–174. https://doi.org/10.1002/(SICI)1096-9837(199702)22:2%3c165::AID-ESP687%3e3.0.CO;2-9.

Chaundhri, R. S., & Khan, H. M. (1981). Textural parameters of desert sediments—Thar desert (India). *Sedimentary Geology, 28*(1), 43–62. https://doi.org/10.1016/0037-0738(81)90033-6.

Cooke, R., Warren, A., & Goudie, A. (1993). *Desert geomorphology*. UCL press.

Culver, S. J., Bull, P. A., Campbell, S., Shakesby, A., & Whalley, W. B. (1983). Environmental discrimination based on grains surface textures: A statistical investigation. *Sedimentology, 30*, 129–136.

Dougill, A. J., & Thomas, A. D. (2002). Nebkha dunes in the Molopo Basin, South Africa and Botswana: Formation controls and their validity as indicators of soil degradation. *Journal of Arid Environments, 50*(3), 413–428.

Dougrameji, J. (1984). The physical and mineralogical characteristics of some sand dunes in Iraq. *Presented at the first Arabian conference for dunes stabilization and desertification control*, University of Baghdad: Iraq (in Arabic). 14–22 October.

Elsayed, Z., & El-Sayed. (1999). Geoarchaeology and Hydrogeology of Deir El-Hagar Playa, Dakhla. *Bulletin Society Bulletin Society Geography*. Egypte Tome LXXII, V. 72, 81–90.

El-Baz, F. (1986). On the reddening of quartz grains in dune sand. Physics of desertification. Springer, Dordrecht, pp. 191–209. https://doi.org/10.1007/978-94-009-4388-9_14.

Embabi, N. S., & Ashour, M. M. (1993). Barchan dunes in Qatar. *Journal of Arid Environments., 25*(1), 49–69. https://doi.org/10.1006/jare.1993.1042.

Foda, M. A., Khalaf, F. I., Gharib, I.M., Al-Hashash, M. Z., & Al-Kadi, A. S. (1984). Assessment of sand encroachment and erodability problems in Kuwait. Technical report. Kuwait Institute for Scientific Research, KISR 1297, Kuwait.

Folk, R. (1971). Longitudinal dunes of the northwestern edge of the Simpson desert, northern territory, Australia, 1. *Geomorphology and grain size relationships. Sedimentology, 16*(1), 5–54. https://doi.org/10.1111/j.1365-3091.1971.tb00217.x.

Fryberger, S. G., AL-SARI, A. M., Clisham, T. J., Rizvi, S., & Al-Hinai, K. (1984). Wind sedimentation in the Jafurah sand sea, Saudi Arabia. *Sedimentology, 31*(3), 413-431.

Fryberger, S. G., & Schenk, C. (1981). *Wind sedimentation tunnel experiments on the origins of Aeolian strata sedimentology, 28*(6), 805–821.

Fuchs, F., Gattinger, T. E., & Holzer, H. F. (1968). *Explanatory text to the synoptic geologic map of Kuwait: A surface geology of Kuwait and the Neutral Zone*. Geological Survey of Austria.

Gharib, I., Foda, M. A., Al-Hashash, M., & Marzouk, F. (1985). *A study of control measures of mobile sand problems in Kuwait air bases*. Kuwait Institute for Scientific Research, KISR 1696, Kuwait.

Greeley, R., & Iversen, J. (1985). *Wind as a geological process* (p. 33). Academic press.

Goudie, A. S., & Middleton, N. J. (2001). Saharan dust storms: Nature and consequences. *Earth-Science Reviews, 56*, 179–204. https://doi.org/10.1016/S0012-8252(01)00067-8.

Howard, A., Morton, J., Hack, M., & Pierce, D. (1978). Sand transport model of barchan dune equilibrium. *Sedimentology, 25*, 307–338.

Hunter, R. (1977). Basic types of stratification in small eolian dunes. *Sedimentology, 24*(3), 361–387. https://doi.org/10.1111/j.1365-3091.1977.tb00128.x.

Jensen, J. (2015). *Introductory digital image processing: A remote sensing perspective* (4th edition). Pearson.

Khalaf, F. I. (1989). Textural characteristics and genesis of the aeolian sediments in Kuwait desert. *Sedimentology, 36*, 253–271.

Khalaf, F. I., Misak, R., & Al-Dousari, A. M. (1995). Sedimentological and morphological characteristics of some nabkha deposits in the northern coastal plain of Kuwait, Arabia. *Journal of Arid Environments, 29*, 267–292.

Khalaf, F. I., Gharib, I., & Al-Hashash, M. Z. (1984). Types and characteristics of recent surface deposits of Kuwait. *Journal of Arid Environments, 7*, 9–33.

Khalaf, F. I., Kadib, L., Gharib, I., Al-Hashash, M. Z., Al-Saleh, S., & Al-Kadi, A. (1980). *Dust fallout in Kuwait*. Kuwait Institute For Scientific Research, KISR/PPI 108/EES-RF-8016. Kuwait.

Khalaf, F., & Al-Ajmi, D. (1993). Aeolian processes and sand encroachment problems in Kuwait. *Geomorphology, 6*, 111–134. https://doi.org/10.1016/0169-555X(93)90042-Z.

Krinsley, D., & Doornkamp, J. (1973). *Atlas of quartz sand surface textures*. Cambridge University Press.

Lancaster, N. (1989). Star dunes. *Progress in Physical Geography: Earth and Environment, 13*(1), 67–91. https://doi.org/10.1177/030913338901300105.

Lancaster, N. (1981). Grain size characteristics of namib desert linear dunes. *Sedimentology, 28*, 115–122.

Livingstone, I., Bullard, J. E., Wiggs, G. F. S., & Thomas, D. S. T. (1999). Grain-size variation on dunes in the Southwest Kalahari, Southern Africa. *Journal of Sedimentary Research, 69*(3), 546–552. https://doi.org/10.2110/jsr.69.546.

Mainguet M. (1984). A classification of dunes based on aeolian dynamics and the sand budget. In F. El-Baz (Eds.), *Deserts and arid lands. Remote Sensing of Earth Resources and Environment*, Vol. 1. Springer, Dordrecht. https://doi.org/10.1007/978-94-009-6080-0_2.

Misak, R., Zaghloul, M., & Ahmed, M. (1996). New approach to the classification of aeolian landforms in Kuwait. *Presented at the international conference on desert development in Arab Gulf Countries*, Kuwait. 23–26 March.

Misak, R., & Kawarteng, A. (2000). Recent environmental changes in the surface geological features of Kuwait: An effective tool for environmental management. *5th international conference on the geology of the Arab world (GAW-5)*, p. 109.

McKee, E. D. (1979). Introduction to study of global sand seas. In E. D. Mckee (Ed.), *A Study of Global Sand Seas*. United States Geological Survey, Professional Paper 1052, pp. 3–19.

Nayfeh, A. (1990). *Geotechnical Characteristics of the Dune Sands for Kuwait*. Thesis, Kuwait University (Unpublished), Kuwait.

Neuman, C. M., Lancaster, N., & Nickling, W. G. (1997). Relations between dune morphology, air flow, and sediment flux on reversing dunes, Silver Peak. *Nevada. Sedimentology, 44*(6), 1103–1111. https://doi.org/10.1046/J.1365-3091.1997.D01-61.X.

Nielsen, O., Dalsgaard, K., Halgreen, C., Kuhlman, H., Møller, J. T., & Schou, G. (1982). Variation in particle size distribution over a small dune. *Sedimentology, 29*(1), 53–65. https://doi.org/10.1111/j.1365-3091.1982.tb01708.x.

Nickling, W., & Wolfe, S. A. (1994). The morphology and origin of nabkhas, Region of Mopti, Mali, West Africa. *Journal of Arid Environments, 28*, 13–30.

Omar, S., El-Bagouri, I., Anwar, M., Khalaf, F., Al-hashash, M., & Nassef, A. (1989). *Measures to control mobile sand in Kuwait*. Kuwait Institute for Scientific Research, Technical report no. KISR2760, Kuwait.

Paisley, E., Lancaster, N., Gaddis, L., & Greeley, R. (1991). Discrimination of active and inactive sand from remote sensing: Kelso dunes, Mojave desert. *California. Remote Sensing of Environment, 37*(3), 153–166. https://doi.org/10.1016/0034-4257(91)90078-K.

Phillip, G. (1968). Mineralogy of the recent sediments of Tigris and Euphrates Rivers and some of the older detrital deposits. J. Sediment. Petrol. 38. 35–44. Power, M.C. 1953. A new roundness scale for sedimentary particles. *Sedimentary Petrology, 23*, 117–119.

Planet. (2021). Available at https://www.planet.com/.

Powers, M. C. (1953). A new roundness scale for sedimentary particles. *Journal of Sedimentary Petrology, 23*, 117–119.

Pye, K., & Tsoar, H. (1990). *Aeolian sand and sand dunes*. Unwin Hyman, London. https://lib.ugent.be/catalog/rug01:001642518.

Ramakrishna, Y., Kar, A., Rao, A., & Singh, R. (1994). Micro-climate and mobility of barchan dune in the thar desert. *Annals of Arid Zone, 33*(3), 203–214.

Satellite Imaging Corporation. (2021). Available at https://www.satimagingcorp.com/.

Skocek, V., & Saadallah, A. A. (1972). Grain size distribution, carbonate content and heavy minerals in aeolian sands, southern desert, Iraq. *Sedimentary Geology* pp. 29–46.

Shehata, W., Bader, T., Irtem, O., Ali, A., Abdallah, M., Aftab, S. (1992). Rate and mode of Barchan Dunes advance in the central part of the Jafurah sand sea. *Journal of Arid Environments, 23*(1), 1–17. https://doi.org/10.1016/S0140-1963(18)30537-8.

Thomas, D. S. G. (1986). Dune pattern statistics applied to the Kalahari Dune Desert. *Southern Africa. Zeitschrift Für Geomorphologie, 30*(2), 231–242. https://doi.org/10.1127/zfg/30/1986/231.

Thomas, D. S. G. (1988a). Analysis of linear dune sediment-form relationships in the Kalahari dune desert. *Earth Surface Processes and Landforms, 13*(6), 545–553. https://doi.org/10.1002/esp.3290130608.

Thomas, D. S. G. (1988b). Arid geomorphology. *Progress in Physical Geography: Earth and Environment, 12*(2), 595–606. https://doi.org/10.1177/030913333900140020.4.

Thomas, D. S. G. (1989). *Aeolian sand deposits*, 232–261.

Thomas, D. S. G. (1988c). Desert dune activity: Concepts and significance. *Journal of Arid Environments, 22*(1), 31–38. https://doi.org/10.1016/S0140-1963(18)30654-2.

Tsoar, H., & Yaalon, D. (1983a). Deflection of sand movement on a sinuous longitudinal (seif) dune: use of fluorescent dye as tracer. *Sedimentary Geology, 36*(1), 25–39.

Tsoar, H. (1983b). Wind tunnel modeling of echo and climbing dunes. *Eolian sediments and processes*, pp.247–259.

Tsoar, H. (1986). Two-dimensional analysis of dune profile and the effect of grain size on sand dune morphology. *Physics of desertification*, pp. 94–108.

Vincent, P., & Lancaster, N. (1985). Some Saudi Arabian dune sand: A note on the use of response diagram. *Zeitschrift Fur Geomorphology, 29*, 117–122.

Warsi, W. E. K. (1990). Gravity field of Kuwait and its relevance to major geological structures. AAPG Bull. 1610–1622 (American Association of Petroleum Geologists; United States, Medium: X; Size).

Watson, A. (1990). The control of blowing sand and mobile desert dunes. Techniques for desert reclamation, pp. 35–85 ref.164.

Watson, A. (1985). The control of wind blown sand and moving dunes: A review of the methods of sand control in deserts, with observations from Saudi Arabia. *Quarterly Journal of Engineering*

Geology and Hydrogeology, 18, 237–252. https://doi.org/10.1144/GSL.QJEG.1985.018.03.05.

White, B., & Tsoar, H. (1998). Slope effect on saltation over a climbing sand dune. *Geomorphology, 22*(2), 159–180.

Wilson, I. G. (1973). *Ergs. Sedimentary Geology, 10*, 77–106. https://doi.org/10.1016/0037-0738(73)90001-8.

Zimbelman, J., & Williams, S. (2002). Geochemical indicators of separate sources for eolian sands in the eastern Mojave Desert, California, and western Arizona. *Geological Society of America Bulletin, 114*(4), 490–496.

Marine Geology of Kuwait

4

Qusaie Ebrahim Karam

Abstract

Kuwait marine geology can be directly or indirectly influenced by various environmental, anthropogenic, industrial, and geomorphological factors continuously altering its nature and affecting its numerous compartments like fauna/flora, water chemistry, physical oceanography, etc. In this chapter, we attempted to describe salient characteristics about Kuwait onshore, and offshore marine geology from historical and modern perspectives. It highlights unique and dynamic ecological and geological factors that are continuously influencing its nature concerning local and transboundary environmental impacts in addition to the significantly booming oil industry that is persistently transforming Kuwait's marine geology.

4.1 Introduction

The Arabian Gulf as a geological entity is unique in its existence bounded by variable geomorphological terrains which are 990 km long, and its width ranges from 56 to 338 km, with a total area of 226,000 km². Located between Iran and the northeastern Arabian Peninsula, the Gulf is considered one of the hottest sea basins on planet earth (Primavera et al., 2018). This chapter is dedicated to highlighting the major marine geological, and geo-environmental features of Kuwait; however, it is worth linking Kuwait as an integral part of the Arabian Gulf marine compartment to the totality of the Gulf. The importance of the Arabian Gulf resides in its economic, strategic, social, and environmental characteristics. The Gulf as a body of water is a valuable source of drinking water through the process of seawater desalination, food commodities like fish, shrimp, etc., navigation route for oil tankers and merchants shipping. Also, it provides recreation, and tourism activities, cooling water for power plants, and a source for offshore oil and minerals (Al-Yamani et al., 2004a, b).

Sea surface temperature can reach >35 °C in summer (Spalding et al., 2007), and <20 °C in winter as surface sea temperatures (SSTs) have increased by 0.4 °C since the 1980s which is documented to be double the global average of SSTs (Al-Rashidi et al., 2009). Due to this extreme summer temperature, the Gulf can experience high evaporation rates leading to water loss, and elevated water salinity levels (>40 PSU) (Reyolds, 1993). The bulk of seawater that reaches the Arabian Gulf originates from the high evaporation processes that generate halocline forces affecting eventually the water circulation patterns in the Gulf. By entering through the Strait of Hormuz, seawater travels to the northwest adjacent to the Iranian coastline circulating near the Kuwait/Saudi borders as it returns south to discharge from the Gulf (Reyolds, 1993).

The environmental conditions rendering the Arabian Gulf an extreme marine region stem from its natural isolation as a geological basin characterized by shallow depths (mean depth 35 m along the Arabian coast and 60 m along the Iranian coast) and limited freshwater input (Persur & Seilbold, 1973). Those high evaporation levels lead to the formation of salt flats (locally named sabkhas), and salt marshes which are considered an ecologically important habitat housing algal mats (Burt et al., 2014). Also, tidal mudflats which are a significant niche for algal mats constitute an important part of the north Arabian Gulf, specifically in Kuwait Bay (Al-Zidan et al., 2006).

Over the decades, the Arabian Gulf have experienced an increased, and rapid urban development along its coasts which resulted in multiple artificial structures like piers, breakwaters, and seawalls in significant cities along the coastline (Burt et al., 2013; Khan, 2007; Price et al., 1993). Other man-made structures that have been integrated at

Q. E. Karam (✉)
Crises Management & Decision Support Program, Environmental
& Life Sciences Research Center, Kuwait Institute for Scientific
Research, Kuwait City, Kuwait
e-mail: qkaram@kisr.edu.kw

© The Author(s) 2023
A. el-aziz K. Abd el-aal et al. (eds.), *The Geology of Kuwait*, Regional Geology Reviews,
https://doi.org/10.1007/978-3-031-16727-0_4

several regions in the gulf are oil and gas facilities that along with other infrastructures house diverse marine habitats and communities for shrimps, fish, birds, etc. (Torquato et al., 2017). In Kuwait, artificial lagoons, Khairan recreational, and residential development projects, also support a high percentage of intertidal biodiversity which is of a sandy and rocky nature (Jones & Nithyanandan, 2013). It's worth mentioning that approximately 40% of the Gulf's coastline has been urbanely developed, modified, and restructured over the past 50 years to support the growing human population, for multiple purposes like tourism, residential projects, and industrial facilities (Naser, 2014). Consequently, numerous dredging, land reclamation, and coastline infilling activities have occurred along the Gulf's coastline continuously modifying it, causing serious environmental stress on its natural marine ecosystems (Sheppard, 2016). Those coastal modifications can further result in hydrological and biophysical alterations of the coastal habitats leading to a severe decline in marine species biodiversity, and abundance (Lokier, 2013).

4.2 Kuwait Marine Environment

Kuwait as an integral state member of the Gulf Cooperation Council (GCC) and the Arabian Gulf environmentally and contribute greatly to the Gulf's marine ecosystem. Its marine bottom sediment is a unique type of sediment representing various textural classes and characteristics. Over the years the marine sediments of Kuwait were the subject of many studies like Purser (1973), Mohammed and Shamlan (1977), Khalaf and Ala (1980), Al-Bakri et al. (1984).

Kuwait's marine environment is much influenced by numerous factors; one of many is the arid climate which is a major characteristic of the entire country. Other factors which also influence the sedimentological and oceanographic traits of the region are low rainfall, high evaporation rate, and high atmospheric temperature (Khalaf et al., 1984). Also, the "Shamal", a prevailing dry northwesterly wind, plays an integral role in contributing much of the terrigenous material to the offshore marine environment through the frequent dust storms associated with this wind in particularly the months of June and July (Khalaf & Al-Hashash, 1983). That aeolian sediments transport contributes much to the total sediment budget to Kuwait marine area (Al-Abdul Razzaq et al., 1982).

4.2.1 Kuwait Bay

One of the prominent features of Kuwait territorial waters is Kuwait Bay which is a unique embayment that hosts various ecological, geological, and anthropogenic compartments. It's

a northwestern semi-enclosed water body extending westward from the Arabian Gulf with a maximum depth of 20 m reaching less than 30 m at Ras Al-Ardh area (Dames & Moore, 1983; Khalaf et al., 1984). In its northern sector, a submerged estuarine flat exists, and a steep shelf-slope extends to the south; both features border the central Kuwait Bay (Al-Matrouk & Karam, 2007). The Bay accommodates multiple industrial, tourists, and aesthetic activities, and as a result of the continuous urban development in the past decade; it has exerted environmental stress on the marine ecosystem. Such anthropogenic activities are Al-Subiya, Doha east power plants, Shuwaikh port, Doha west desalination, and power plants which can contribute to elevated seawater temperature as a result of the warm water discharge from desalination power plants. Also, organic and inorganic compounds have been introduced into Kuwait Bay as a result of wastewater effluent discharges. In addition, urban and industrial discharges are frequent events from ports and other industrial activities which are severely toxic to resident biota in the Bay (Al-Muziani et al., 1991; Al-Muziani & Jacob, 1997).

Heavy minerals in Kuwait Bay are characterized by a unique combination of minerals that share a large resemblance to the sediment of Tigris—Euphrates region in Iraq. Khalaf et al., 1982 have revealed that Kuwait sediment size classes are of polygenetic origins, and they constitute two main classes. The first class is autochthonous material-derived erosion of ancient submerged sediment, sub-recent coastal sediments, and as a result of degradation of marine biota shells. The second class was observed to be allochthonous material which has originated from onshore desert sediment transported by northwestern winds. A mixture of hornblende, mica, and pyroxenes make up the bulk of the heavy mineral content of Kuwait Bay sediments. Other minerals which originate from recent surface deposits and dust fallout are amphiboles, dolomites, epidotes, pyroxenes, zircon, tourmaline, and garnet (Ali, 1976). Also, Fig. 4.1a and b depicts main sediment textural classes, and types of sediments common in Kuwait Bay.

4.3 Offshore Geology of Kuwait

The offshore area of Kuwait is characterized by a depositional environment of a low-energy couple with little sediment transport potential (Khalaf et al., 1984). The offshore marine area is situated in the northwestern corner of the Arabian Gulf which was exposed to nature elements around 10,000 yr BP (Fairbridge, 1961). Historically, the northern regions of Kuwait's offshore marine environment are a part of the ancient delta which is submerged in the northern part of the Arabian Gulf recently. The bottom submarine topography of the marine area was formed as a result of a

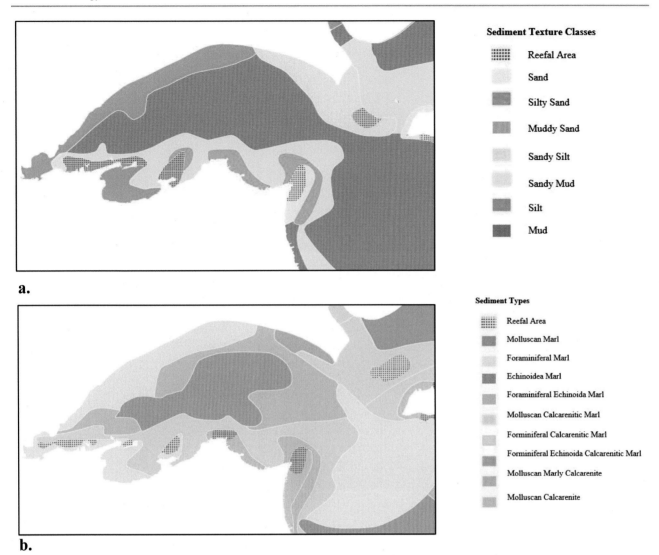

a.

b.

Fig. 4.1 **a**. Main sediment textural classes and **b**. types of sediments common in Kuwait Bay

geomorphic and tectonic process in the late Pliocene–Pleistocene (Kassler, 1973). Also, as it's geologically different than other parts of the Gulf, it's categorized as a shallow depth region that can reach a maximum depth of 30 m in the SE direction. Being diverse in its formation geology, the bottom seafloor can be categorized into five distinctive physiographic divisions like Kuwaiti Bay trough, submerged estuarine distributary channel and bar system submerged estuarine flat, shelf-slope, and the islands which are distributed along with the entire Kuwait offshore marine environment (Khalaf et al., 1984). The Tigris and Euphrates River systems play a pivotal role in supplying Shatt Al-Arab and eventually the Kuwait marine area with sediments. Aeolian materials constitute a significant percentage of sediment cores collected from different locations in the Arabian Gulf (Foda et al., 1985). More specifically, bottom sediments sampled from the northern regions of the Arabian

Gulf have aeolian grains extending 70 km offshore (Emery, 1956). Foda et al. (1985), have also concluded that the contribution of dust particles from dust clouds in the northern Arabian Gulf to sea surface sediment load can be considered as a sediment transport diffusion issue. The authors have also estimated that sedimentation rates of dust origins were 0.8 mm/year which is many orders of magnitude higher than world records of dust fallouts on earth seas (Fig. 4.2).

4.4 Effects of Water Circulation on Sediment Transport

In a study by Khalaf et al. (1984), the textural classes of Kuwait offshore marine environment are characterized by mud and muddy sediments, with relatively narrow belts of

Fig. 4.2 A map showing the **a**. trajectory of dust storms over the northern Arabian Gulf including Kuwait territorial waters which eventually contribute to marine sediment particle load (*Source* Al-Ghadban, 2004 after Vingoradov et al., 1973), **b**. image by Terra MODIS satellite, and **c**. image by Aqua satellite

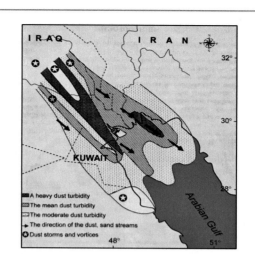

a.

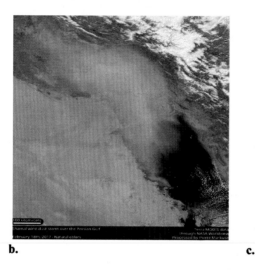

b.

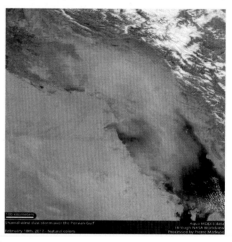

c.

coarse-grained sediment like sand, muddy sand, and silty sand present in the nearshore shallow environments in Kuwait Bay and the southern coastline. Moreover, the study classified Kuwait's offshore marine area as either a low-energy zone or a moderate-to-high energy zone. The low-energy zone comprises most of the marine study area, however the moderate-to-high energy zone covers the sand and sandy deposits. The energy classification scheme is initially influenced by weak tidal currents which are governed by surface currents and waves generated by either locally named "Kaus", a southeasterly prevailing wind or "Shamal", a northwesterly wind. The coarse-grained sandy deposits in the high-energy zones are primarily composed of organic biogenic material and rock rubbles, and they originated from beach rocks, reef flats, and coastal ridges. The pattern of sediment movements to the Kuwaiti shoreline is perpendicular to the coast with net sediment transport in a parallel fashion to the shoreline (Al-Yamani et al., 2004a, b). Offshore energy zones control the types of marine sediments, i.e., muddy sediments prevail where wave and current energy are considered low. Also, sand and sandy deposits prevail where energy in the area is moderate to high.

Water circulation heavily controls marine sediment transport, and anticlockwise water circulation dominates the western side of the Arabian Gulf including Kuwait's marine area. Based on that observation, three distinctive coastal regions were classified for sediment movement along the coastline which is from the north: a. Khor Subiya, b. Kuwait Bay (KB), and c. southern coast (Lardner et al., 1993; Reyolds, 1993). The controlling tidal current/wave factors are responsible for the high concentration of suspended sediments in Khor Subiya coupled with high sedimentation transport through the channel (Al-Bakri et al., 1985). This high concentration of transported sediment made the amount of total suspended solids (TSS) around Bubyan and Failaka Islands, and in Khor Subiya higher in the magnitude of 10 than TSS in offshore marine areas. The southern region is characterized by a wintery sediment transport with an accretion-dominated mechanism, while other southern regions of headlands and reaches are subjected to erosional effects. As for Kuwait Bay, it is controlled by a modest net sediment transport, and the low-energy zone of KB experiences deposition of suspended sediments (Al-Yamani et al., 2004a, b). Alosairi et al. (2011a, b) have indicated that water residence time in the Arabian Gulf was more than 3 years along the Arabian coast, and that wind/tide energies are responsible factors for seawater dispersion along the coast.

Kuwait coastline was a subject of numerous urban development projects which induced geomorphological alterations to natural marine habitats. And one of the examples of such various anthropogenic activities which led to coastline developments was new waterfronts, accommodations, tourist attractions, etc. Those projects focused on land reclamation, filling/excavations near coastal and off-shore areas and dredging processes have caused substantial transport of suspended sediments into coastal waters, consequently increasing turbidity levels in the water column. Also, an increase in sediment sorption processes along with nutrient levels was evidently observed (Alosairi et al., 2011a, b). Al-Ghadban et al. (2002) have indicated that as nutrients are transferred via suspended sediments, it assists in enriching the Kuwait Bay marine ecosystem transforming it into a rich nursery ground for fish and shrimps. The amounts of sediment particles being deposited over Kuwait terrestrial and marine areas were estimated according to Kuwait Environment Public Authority to be 55 tons/km^2 which contributes heavily to the total sediment load in sea-water. Another significant source of sediment particles to seawater compartment is Shatt Al-Arab, which heavily contributes to total sediment concentration in the north Arabian Gulf (Saad & Al-Azmi, 2002; Al-Ghadban & El-Sammak, 2005). Moreover, Al-Ghadban (2004) has estimated that approximately 29 tons of sediments are loaded in Kuwait Bay specifically on the top layer (1 m) of surface waters which can eventually be deposited by wind and tide forces. This finding confirms the conclusion that Kuwait Bay, as a geomorphological feature of Kuwait, can receive significant amounts of suspended sediments from offshore regions as well as from Shatt Al-Arab.

4.5 Effects of Draining of the Iraqi Marshes

Transboundary environmental activities in one geographical region can result in tremendous adverse effects on indigenous biota and marine ecosystems in another region. Northern territorial waters of Kuwait have experienced alterations in sediment characteristics; hence, there is an increase in sediment deposition in the past since the construction of embanked canal termed "The Third River" (Fig. 4.3). The purpose of constructing the Third River in 1990 is to divert the Euphrates River waters before reaching the marches region (locally named Al-Ahwar) in Iraq through Shatt Al-Basra to Khor Al-Zubair to solve the increasing salinity issue in farmlands (Al-Hilli et al., 2009).

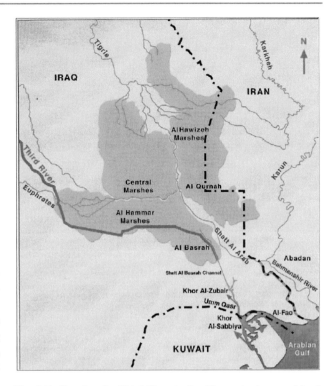

Fig. 4.3 Showing the Third River in the Mesopotamian marshlands (*Source* Al-Yamani et al., 2004a, b)

Al-Ghadban et al. (1999) have confirmed that water depths in the northern region were shallower in 1998 than in previous years like 1956, 1984. Also, grain size analysis of sediment samples collected from the area revealed that finer sediments were observed in 1998 than what was reported in 1982. Overall, sedimentological assessment of the study area demonstrated a north–south sediment transport gradient from the Shatt Al-Arab region. All of the above observations indicate the sensitivity of Kuwait's marine ecosystem to external factors like the drainage of the Iraqi marshes and the construction of the Third River.

4.6 Grain Size and Textural Classes

The classification of Kuwait's marine sediment is based on multiple schemes such as sediment particle sizes, sediment composition, or the combination of both categories. Seven textural classes comprise surface offshore marine sediments which are described in Table 4.1 and Fig. 4.4. Also, while muddy sediment covers a large percentage of Kuwait's area, there is another important fraction that constitutes marine sediment which is then represented by sand, muddy sand, and silty sand (Al-Yamani et al., 2004a, b). In the nearshore and intertidal areas, coarse-grained sediments can be found, in Ras Ajouzah (near Kuwait Tower) in the northern regions

Table 4.1 Textural classes of surface sediments of Kuwait offshore marine area (adopted from Khalaf et al., 1984)

Type	Grain size (mm)	Composition
Sand	0.063–2.0	95% sand 3.6% silt 1.4% clay
Silty sand		
Muddy sand		
Sandy silt		
Sandy mud		
Silt		
Mud	Finer than 0.063	54% silt 43% clay 3% sand

Fig. 4.4 Textural classes of Kuwait bottom seabed (*Source* Al-Yamani et al., 2004a, b)

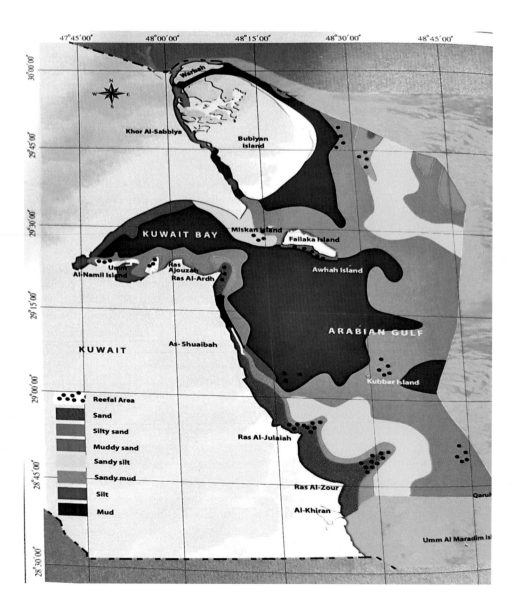

Table 4.2 Biolithofacies sediment categories of Kuwait's marine sediment (adopted from Al-Yamani et al., 2004a, b)

	Biolithofacies type	Weight % mud (<0.063 mm)
1	Molluscan marl	>90
2	Foraminiferal marl	>90
3	Echinoid marl	>90
4	Foraminiferal echinoid marl	90
5	Mollascan calcarenitic marl	50–90
6	Foraminiferal calcarenitic marl	50–90
7	Foraminiferal echinoid calcarenitic marl	50–90
8	Mollascan marly calcarenitic	10–50
9	Mollascan calcarenitic	<10

to the south where Khairan is located. Moreover, the marine areas around Kuwaiti islands can be characterized as having the same type of sediments as Failaka, and the southern islands. Therefore, we conclude that areas comprised of coarse-grained sediments are of a limited distribution range and can be associated with shallow rocky bottoms.

With a close focus on the sediment texture classification of Kuwait Bay (KB), it was observed that two categories comprised its sediment classes. (1) Allochthonous sediment materials, which originate from disintegrated onshore desert sediments as a result of force action of the prevailing northwesterly winds. (2) Autochthonous sediment materials originate from the fragmentation of fresh marine fauna shells, and the result of the erosional process of submerged ancient coastal sediment (Khalaf et al., 1982).

The granulometric and textural variation and distribution of Kuwait Bay's marine sediments are largely driven by the following environmental factors: nature of coastline sediment, ecological surroundings of the bay, and kinetic and hydrodynamic energy of the water compartment. Moreover, the dust storm's born fallout adds greatly to the KB sediment budget. Kuwait Bay is mostly covered with muddy sediments and is composed of mudflats in the northern section and central channel, and tidal flats in Suliabikhat Bay, a prominent geological part of KB composed of mud. These muddy sediments are principally consisting of silty clay and clayey silt types of sediments, while the southern, eastern, and western flats of KB are characterized by sandy sediments (Khalaf et al., 1982).

4.7 The Biogenic Nature of Coarse Sediment Fractions

Since the sand of Kuwait's marine areas constitutes a considerable fraction of the coarse sediments, the biogenic components also play a role in the formation of this sand category. Biogenic sediments are the fractions derived from a biological entities such as living organisms most commonly mollusks which are composed of gastropods (snails), echinoids (sea urchins, sea cucumbers), foraminifera, bivalves, ostracodes, and sponge spicules. In addition to other petrographic materials like feldspar, quartz, and other rock fragments, 90% of the biogenic material comes from the three distinctive faunal groups which are echinoids, mollusks, and foraminifera which eventually make the coarse-grained fraction of marine sediments. Based on this biogenic material and carbonate content, nine biolithofacies groups of Kuwait's marine sediment have been categorized as shown in Table 4.2. The abundant sediments characterizing Kuwaiti beaches, coastal reef flats originate from the fragments of cemented calcareous sandstone, and the resultant from beach rocks, while quartz predominates the northern part of the marine environment as it is concentrated in fine and very fine sand fractions which constitute 10–45% of the marine sediments (Al-Yamani et al., 2004a, b).

Biogenic sediment types cover Kuwait Bay, which originates from quartz and shell fragments (constitute 60% of total sand fractions), in addition to feldspar, pellets, and oolites which exist in lower amounts in bad sediments. The following biological entities comprise the biogenic coarse-grained fraction of KB like gastropods, pelecypods (bivalves), echinoids spines and plates, ostracodes, and foraminifera. It's been observed that those grains underwent severe micritization process as cyanobacteria (blue-green algae or filamentous endolithic algae) within the sediment which have bored into the shells of marine organisms that constitute the biogenic sediments. As a result, micrite has formed and precipitated in the bores, eventually the micritization process led to the breakdown of coarse sand grains to form fine silt and sand grains (Bathurst, 1966).

4.8 Kuwaiti Islands

Kuwait territorial waters contain nine islands that are unique in their biodiversity along with their geological and geomorphological features. For the State of Kuwait, those

Table 4.3 Classification of Kuwaiti Islands according to their location in Kuwait Territorial Waters

Onshore islands	Offshore islands
Warba	Kubbar
Bubyan	Qaruh
Miskan	Um Al-Maradim
Failaka	
Auha	
Um Al-Namil	
Akkaz[a]	

[a] During construction of Shuwaikh Port in Kuwait Bay, Akkaz Island was connected to Shuwaikh Port, permanently closing the tidal channel that used to connect it to the mainland

islands are significant and have a strategic value. They are classified into two distinctive categories according to their physiographical characteristics: (1) onshore islands located in the northern region of the Kuwait Sea and (2) offshore islands which are located in southern waters (Table 4.3; Fig. 4.5). Bubyan, Failaka, Warba, and Um Alnamil Islands are considered the largest of all of the nine islands in Kuwait territorial waters. Miskan and Awhah Islands, relatively smalls islands, are characterized as having a mesotidal regime and considered part of the Failaka Island geological structure and occupy the northern region of Kuwait Sea which are situated in a shallow platform (<5 m below sea level). While the southern islands like Kubbar, Qaruh, and Um Almaradem have a microtidal regime situated on top of a platform of aggrading perched reef structure descending deeply for 20–30 m depths. Alternating beach rock platforms (4–46%) cover the island's beaches along with beach-ridge sets, bioclastic material constitutes the island's beach sediments, consisting of very coarse-to-medium sand (Buynevich et al., 2020). Also, other geomorphological features of Kuwaiti Islands are cuspate spits, thinly vegetated beach ridges and salients in addition to coral reef platforms surround the southern islands. In general, the islands are characterized by having a minor elliptical shape (<0.6 value), while Kubbar and Qaruh Islands further in the southern region can be considered almost round in shape. Since those islands have a low elevation, they can be susceptible to sea-level rise; along with their geomorphological features like hard beach rock, sandy beaches will experience erosion–deposition patterns.

4.9 Offshore Petroleum Geology of Kuwait

4.9.1 Historical Overview

The history of Kuwait's offshore oil exploration goes back to the year 1961 by conducting a threefold analog seismic survey in collaboration between Kuwait Oil Company (KCO) and Shell International Exploration and Production B.V. This endeavor culminated in multiple offshore exploration activities from the 1960s to 1980s but none of the drilled wells were intended for commercial purposes.

The geological structure of Kuwait's offshore area is characterized by a gentle regional dip to the northeast, and all of the offshore wells have been explored at a very low relief geological structure. This monocline nature of offshore structures created stratigraphic trapping which has caused the highly potential hydrocarbon entrapment within the Middle Cretaceous reservoirs. And based on that observation, a high-resolution sequence-stratigraphic framework was conducted which has generated quantitative biostratigraphical data (Fig. 4.6). Those data were used to perform a detailed paleoenvironmental and chronostratigraphic calibration of this data series (Al-Fares et al., 1998).

The first drilled wells revealed 720 barrels of oil per day (BOPD) of 38°–40° API oil quality from Lower Cretaceous Ratawi limestones, but unfortunately, production declined after due to low amounts of oil discovered (Al-Fares et al., 1998). However, in the 1990s, a more in-depth investigation was carried out by conducting a quantitative biostratigraphical study for the Lower to Middle Cretaceous interval. From the 11 drilled wells, 500 samples (cuttings) were analyzed and revealed multiple fossil microfaunal distributions of foraminifera, nanoplanktons, and ostracodes (Robertson Research International, 1996; Varol Research, 1996; Lacustrine Basin Research, 1996).

In 2019, the KOC has signed an offshore drilling service contract with renowned drilling expert Haliburton Co. The drilling operation will include six high-pressure/high-temperature exploration wells on two jack-up rigs in offshore Kuwait territorial waters. Planned drilling operations were supposed to start in mid-2020, but the coronavirus pandemic that struck the world delayed the projects. The forecasting goals of Kuwait Petroleum Corporation (KPC) are to increase the production capacity to 4 million B/D by 2020.

Fig. 4.5 A map showing
Kuwait's islands

Fig. 4.6 Map of Kuwait offshore
area depicting well locations and
categories based upon oil
exploration operations (*Source*
Al-Fares et al., 1998)

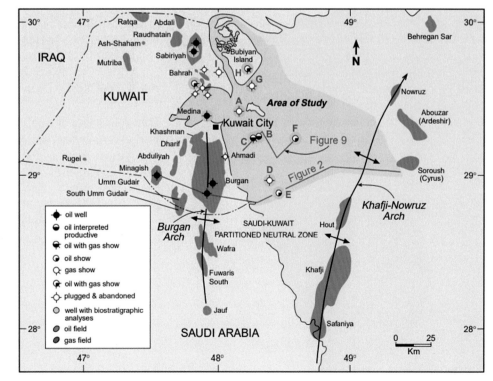

4.9.2 Oil Pollution in Kuwait Territorial Waters

One of the busiest regions in the world for shipping oil as a commercial commodity is the Arabian Gulf which receives large numbers of oil tankers and ships to load crude oil in addition to other goods (Karam, 2011). Two-thirds of the world's crude oil reserves are in the Arabian Gulf countries which produce about 25% of the world's oil. Among this percentage, the State of Kuwait has 11% of that estimate (1.7 million bbl/d). Also, 27% of the total gross oil imports for the Organization of Economic Co-Operation and Development (OECD) is the gross oil imports of the Arabian Gulf countries for 2002 which is averaged 10.6 million barrels per day (bbl./d) (www.eia.doe.gov, 2001) (Fig. 4.7).

There is always a chance of oil spillage if crude oil is transported via the seas to other energy-demanding countries around the world. As a result, approximately 25,000 metric tons of crude oil are spilled per year into the Arabian Gulf from multiple sources like offshore drilling operations and pipelines. Also, 16, 800 metric tons of crude oil find a route to the Arabian Gulf from natural seeps, coastal refineries, and municipal and non-refining industrial wastes (Hayes & Gundlach, 1977).

It is worth mentioning that as there is an accidental oil spillage in Kuwait's marine environment, there is a natural oil seepage. A study by Al-Sammak et al. (2005) has documented two sites on sea floor in the marine area between Qaruh and Umm Al-Maradem Offshore Islands where oil is naturally seeping to the surface. And natural oil footprint analysis has confirmed that the seeping oil is not caused by leaks or spills from foreign oil tankers, rather its Kuwait crude oil. Total petroleum hydrocarbons (TPH) analysis of collected sediment samples from the seepage area (22 km^2) revealed that TPH concentrations were 200–500 µg/g.

Kuwait's economy primarily depends on petroleum exports which led to active shipping and oil tanker traffic in the Arabian Gulf water frequently transporting crude oil to other countries as an energy commodity. Those activities made oil spills more common in the region whether it was due to accidental spillage incidents due to offshore oil platforms, oil tankers, and leakages from ships or pipeline networks, offshore drilling, and oil exploration operations (RMSI, 2020). All of the aforementioned activities led to the firm belief that oil pollution is to be considered a significant environmental concern in the Arabian Gulf.

Historically, Kuwait's marine environment suffered an environmental catastrophe due to political/war conflicts between neighboring countries. The 1991 Iraqi invasion of Kuwait resulted in a devastating marine environment from the spillage of 6–10 million barrels of oil from export terminals and oil tankers (Thorhaug, 1992; Al-Ghadban et al., 1996). The resultant oil spills during the Gulf War have depleted 2% of Kuwait's oil reserves (RMSI, 2020) which led to Kuwait losing about $60 million/day in revenues which sums up to $22 billion in a year.

However, the impact of oil pollution on marine ecosystems of Kuwait including marine resources like fish is not well understood. Toxicological work by Karam (2011) and Karam et al. (2019, 2021 IN PRESS) have demonstrated that a water-accommodated fraction of Kuwait crude oil mixed with seawater can result in toxic effects such as mortality in marine fish species native of Kuwait marine areas like sea bream and orange-spotted grouper.

4.9.3 Oil Spills Data Acquisition

Data concerning oil spills for onshore and offshore Kuwait marine environment are collected by two scientific entities like Kuwait Institute for Scientific Research (KISR) and Kuwait Environment Public Authority (KEPA). Unfortunately, the available data cannot differentiate between natural and accidental oil spill events in the region.

One of the methods used to monitor, document, and collect data on oil spill events is using satellite remote sensing image systems and geographic information system (GIS). Those tools enable research and regulatory authorities to assess and map the magnitude and extent of oil spills. Also, it enables the processing of satellite imagery data to estimate different parameters of the oil spill event like the volume of the oil spill and the depth of oil over water (0.87 µm). To process Sentinel-1 image data, SNAP software (European Space Agency Software) was utilized heavily to identify oil spill hotspots and their frequency in Kuwait's marine region. Figure 4.8 demonstrates the methodology adopted by KEPA, KISR, and RMSI consulting companies to document oil spill events.

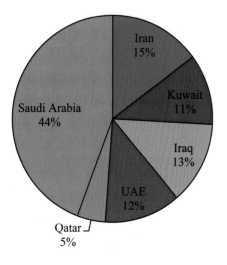

Fig. 4.7 Net oil exports from the Persian Gulf for 1999 (million barrels per day–MBPD) *Note* Bahrain net oil export was 0.02 MBPD

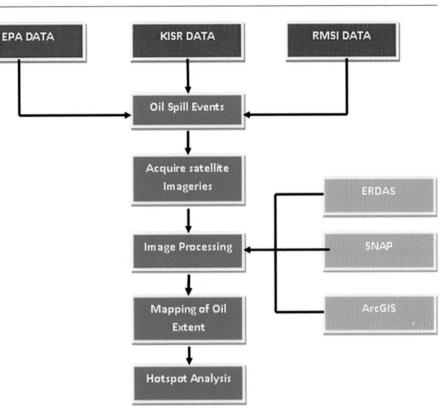

Fig. 4.8 Flowchart of oil spill events documentation adopted by KEPA, KISR, and RMSI

Sentinel-1 and Landsat 1–5 MSS (80 m) (RMSI, 2020) were proved to be a reliable instruments to monitor not only oil spills in the Kuwait sea area, but also for various physicochemical parameters like sea surface temperature, chlorophyll-a concentrations, etc. In addition to multiple water quality parameters that assist in assessing the adverse impact of oil spills on the marine ecosystem. The Sentinel-1 satellite permitted the acquisition of multiple oil spill data concerning the following sites in Kuwait Territorial Waters and in the neighboring regions like Bubyan, Failaka, and Qaru Islands' sea area, Ahmadi, and Al-Basra Terminal in Iraq. While the Landsat 1–5 MSS satellite has acquired data for offshore marine areas. Also, GIS was used heavily for oil spill identification and classification in SAR (Synthetic Aperture Radar) imagery system where it can detect differences in oil and water signatures. The second stage is to process the images through SNAP software to determine the extent of the oil spill and classify oil and water in the vicinity of the spillage zone. It was observed from the analyzed data that most of the offshore oil spill events have occurred from an approximate distance of about 100–150 km from Kuwait City (Fig. 4.9).

Hotpot analysis of all 39 offshore oil spill events documented by remote sensing and GIS has demonstrated that most of the spill events have occurred in the southeastern region of Kuwait Territorial Waters. This result can be explained by the fact that the southern region is where all the oil industry facilities are located, in addition to where most transport ships and oil tankers traffic exist, making this area a hotspot for frequent oil spill events (RMSI, 2020).

4.9.4 Oil Spills Response Strategy

The Kuwait Regional Convention (23rd April 1978) was formulated to serve as a central agreement for the protection of the marine environment. Regional Cooperation in Combating Pollution by Oil and other Harmful Substances in Cases of Emergency in the GCC area was the main objective of this convention to monitor and control marine pollution that originates from various sources. Also, cooperation between states of the region should focus on the environmental management of water bodies in case of pollution scenarios (MEMAC, 2016).

The Marine Emergency Mutual Aid Centre (MEMAC) has prepared a manual for the Regional Organization for the Protection of Marine Environment (ROPME) state members (Fig. 4.10) which contains information and instructions for joining marine pollution operations. Also, it is considered as a guideline for pollution incident management, strategy, and policymaking.

In the case of oil pollution in the ROPME sea area, precautionary preventative measures have to be taken to avoid pollution by hazardous and noxious chemicals like oil

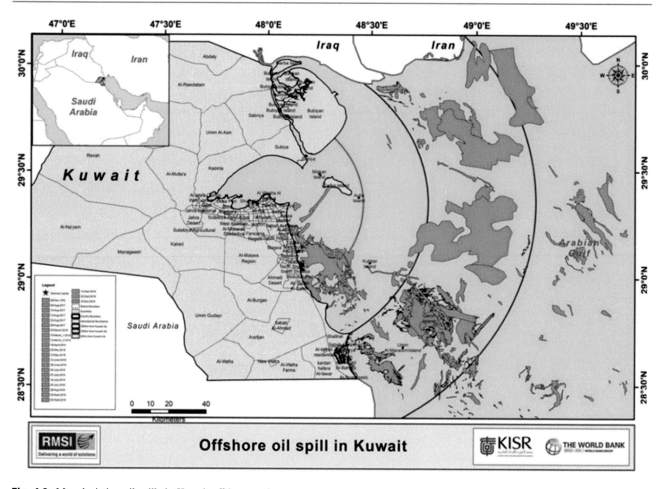

Fig. 4.9 Map depicting oil spills in Kuwait offshore marine area

in offshore marine areas (ROPME, 2011). Therefore, in 2000, the International Maritime Organization (IMO) took on the responsibility to adopt the Protocol on Preparedness, Response and Cooperation to Pollution Incidents by Hazardous and Noxious Substances (OPRC-HNS Protocol) in ROPME Sea Area (RSA). The primary goal of this protocol is for RSA state members to cooperate and provide mutual assistance in preparing and combating chemical pollution incidences that threaten the integrity of RSA's marine resources and ecosystem.

4.10   Conclusion

In this chapter, we have highlighted the salient features of Kuwait's onshore and offshore marine geology. There are numerus natural and anthropogenic factors that continuously govern the marine geology of Kuwait forever interacting with

such dynamic system. Increasing coastal rapid urban development projects, elevated oil exploration operations, and accidental oil spillages from oil tankers all threaten to change the main features of Kuwait's marine geological compartment. Therefore, altering its composition as finer muddy sediments are incoming from transboundary regions like Iraq, and from newly developed coastal dredging projects. Those onshore modifications will ultimately alter the biophysical and biological characteristics of coastal habitats, thus affecting marine species biodiversity and abundance. By understating all anthropogenic and ecological factors governing Kuwait's marine geology, environmental mitigation and rehabilitation will be possible in the future. Moreover, in-depth research of marine sediment transport patterns from coastal land reclamation/dredging projects, and from transboundary regions like Khor Zubair, Shatt Al-Arab in Iraq will assist in assessing how geological aspects of the marine environment will respond to such external and internal factors.

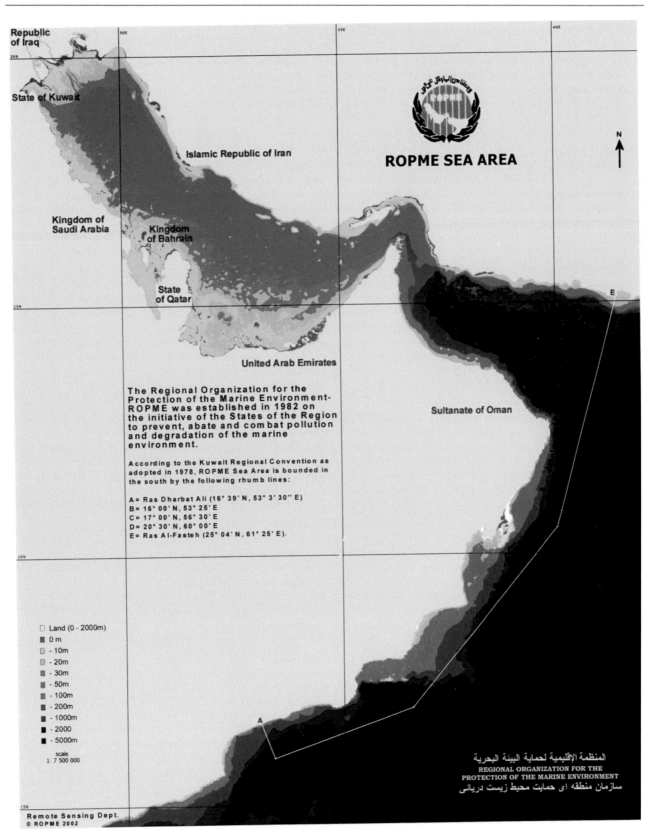

Fig. 4.10 A map depicting the ROPME sea area with state members

References

Al-Bakri, D., Khalaf, F., & A1-Ghadban, A. N. (1984). Mineralogy, genesis and sources of surficial sediments in the Kuwait environment, northern Arabian Gulf. *Journal of Sedimentary Petrology, 54*, 1266–1279.

Al-Bakri, D., Foda, M., Behbehani, M., Khalaf, F., Shublaq, W., El-Sayed, M., Al-Sheikh, Z., Kittaneh, W., Khuraibit, A., & Al-Kadi, A. (1985). The environmental assessment of the intertidal zone of Kuwait. Kuwait Institute for Scientific Research Report No. 1687, Kuwait.

Al-Abdul Razzaq, S., Khalaf, F., Al-Bakri, D. & Shublaq, W. (1982). Marine sedimentology and benthic ecology of kuwait marine environment. Kuwait Institute for Scientific Research, Kuwait, KISR 694.456 pp.

Albanai, J. A. (2019). *A GIS science simulation for the expected sea level rise scenarios on Failaka Island in the State of Kuwait* (1st ed.). Kuwait centre for research and studies.

Al-Fares, A. A., Bouman, M., & Jeans, P. (1998). A new look at the Middle to Lower Cretaceous stratigraphy, offshore Kuwait. *GeoArabia, 3*(4), 543–560. https://doi.org/10.2113/geoarabia0304543

Al-Ghadban, A. N., Massoud, M. S., & Abdali, F. (1996). Bottom sediments of the Arabian Gulf: I. Sedimentological characteristics. *Kuwait Journal of Science and Engineering, 23*, 71–88.

Al-Ghadban, A. N., Saeed, T., Al-Dousari, A. M., Al-Shemmari, H., & Al-Mutairi, M. (1999). Preliminary assessment of the impact of draining of Iraqi marshes on Kuwait's northern marine environment. parti. physical manipulation. *Water Science and Technology, 40*(7), 75–87.

Al-Ghadban, A. N., Al-Majed, N., & Al-Muzaini, S. (2002). The state of marine pollution in Kuwait: Northern Arabian Gulf. *Technology, 8*(1–2), 7–26.

Al-Ghadban, A. N., & El-Sammak, A. (2005). Sources, distribution and composition of the suspended sediments, Kuwait Bay, Northern Arabian Gulf. *Journal of Arid Environments, 60*(4), 647–661.

Al-Ghadban, A. N. (2004). Assessment of suspended sediment in Kuwait Bay using Landsat and SPOT images. *Kuwait Journal of Science and Engineering, 31*(2), 155.

Al-Hilli, M., Warner, B., Asada, T., & Douabul, A. (2009). An assessment of vegetation and environmental controls in the 1970s of the Mesopotamian Wetlands of Southern Iraq. *Wetlands Ecology and Management, 17*(3), 207–223.

Al-Matrouk, K., & Karam, Q. (2007). Bottom sediment characteristics in Kuwait's marine environment. Kuwait Institute for Scientific Research. Report Number KISR 8586.

Al-Muziani, S., Samhan, O., & Hamouda, N. F. (1991). Sewage related impact on Kuwait's marine environment: A case study. *Water Science and Technology, 32*, 181–199.

Al-Muziani and P.G., Jacob,. (1997). Trace metals in the near shore sediment of the Shuaiba Industrial area of Kuwait from August 1993 to June 1994. *UAE Journal, 9*(1), 1–10.

Alosairi, Y., Imberger, J., & Falconer, R. F. (2011a). Mixing and flushing in the Persian Gulf (Arabian Gulf). *Journal of Geophysical Research: Oceans, 116*, no. C3.

Alosairi, Y., Al Enezi, E., Falconer, R. A., & Imberger, J. (2011b). Modelling phosphorus sorption processes in Kuwait Bay: Effects of sediment grain size. In *Proceedings of the 34th World Congress of the International Association for Hydro-Environment Research and Engineering: 33rd Hydrology and Water Resources Symposium and 10th Conference on Hydraulics in Water Engineering* (p. 3175). Engineers Australia.

Al-Rashidi, T. B., Al-Gimly, H. I., Amos, C. L., & Rekha, K. A. (2009). Sea surface temperature trends in Kuwait Bay, Arabian Gulf. *Natural Hazards, 50*, 73–82.

Al-Sammak, A., Govao, B., Al-Din, S., Al-Wadi, M., Al-Matrouk, K., Al-Anzi, A., Karam, Q., & Al-Rushaid, R. (2005). Characterization of natiral oil seepage and its effects on the marine ecosystem and the coastal zone management of Umm Al-Maradem and Qaru Islands in the State of Kuwait. Kuwait Institute for Scientific Research. Report Number KISR 7663.

Al-Yamani, F. Y., Bishop, J., Ramadhan, E., Al-Husaini, M., & Al-Ghadban, A. (2004a). *Oceanographic atlas of Kuwait's waters.* Kuwait Institute for Scientific Research Press.

Al-Yamani, F. Y., Bishop, J., Ramadhan, E., Al-Husaini, M., & Al-Ghadban, A. N. (2004b). Oceanographic atlas of Kuwait's waters.

Al-Zidan, A., Kennedy, H., Jones, D., & Al-Mohanna, S. (2006). Role of microbial mats in Suliabikhat Bay (Kuwait) mudflat food webs: Evidence from δ ^{13}C analysis. *Marine Ecology Progress Series, 308*, 27–36.

Ali, A. J. (1976). Heavy mineral provinces of the Recent sediments of the Euphrates-Tigris basin. *Journal of the Geological Society of Iraq, 10*, 33–46.

Bathurst, R. G. C. (1966). Boring algae, micrite envelopes and lithification of molluscan biosparites. *Geological Journal, 5*, 15–32.

Burt, J. A., Al-Khalifa, K., Khalaf, E., AlSuwaikh, B., & Abdulwahab, A. (2013). The continuing decline of coral reefs in Bahrain. *Marine Pollution Bulletin, 27*, 357–363.

Burt, J., Van Laviern, H., & Feary, D. (2014). Persian Gulf reefs: An important assess for climate science in urgent need for protection. *Ocean Challenges, 20*, 49–56.

Buynevich, I. V., FitzGerald, D. M., Al-Zamel, A. Z., & Al-Sarawi, M. A. (2020). Comparative morphosedimentary framework of small subtropical islands offshore Kuwait. *Marine Geology, 429*, 106294.

Dames & Moore. (1983). Aquatic biology investigation: Studies of the Subyia area, Kuwait Bay and development of electrical networks. Vol. 1. Government of Kuwait, Ministry of Electricity and Water, Contract N. MEW/CP/PGP-1113–80/81, Kuwait.

EIA (Energy Information Administration). (2001). U.S. Energy Information Administration–Independent Statistics and Analysis. 1000 independence Ave., SW, Washington, DC 20585. www.eia.doe.gov.

El-Gindy, A., & Hegazi, M. (1996). Atlas on hydrographic conditions in the Arabian Gulf and the upper layer of the Gulf of Oman. University of Qatar, p. 170

Emery, K. O. (1956). Sediments and water of Persian Gulf. *AAPG Bulletin, 40*(10), 2354–2383.

Fairbridge, R.W. (1961). Eustatic changes in sealevel. In: Athrens ei a/ A. (Eds.), Physics and Chemistryo/the Earth, 4. Oxford: Pergamon Press, p 185

Foda, M. A., Khalaf, F. I., & Al-Kadi, A. S. (1985). Estimation of dust fallout rates in the northern Arabian Gulf. *Sedimentology, 32*(4), 595–603.

Hayes, M.O., & Gundlach, E.R. (1977). Oil pollution in the Arabian Gulf: A preliminary survey. Technical Report No. 1-GOP, Research Planning Institute Inc., Study carried out for: Kuwait Engineering Operation and Management Co. (K.S.C.), Kuwait.

Jones, D. A., & Nithyanandan, M. (2013). Recruitment of marine biota into hard and soft artificially created subtidal habitats in Sabah Al-Ahmed Sea City, Kuwait. *Marine Pollution Bulletin, 72*, 351–356.

Khalaf, F., Al-Bakri, D., & Al-Ghadban, A. (1984). Sedimentological characteristics of the surficial sediments of the Kuwaiti marine environment, northern Arabian Gulf. *Sedimentology, 31*(4), 531–545.

Khalaf, F. I., Al-Ghadban, A., Al-Saleh, S., & Al-Omran, L. (1982). Sedimentology and mineralogy of Kuwait Bay bottom sediments Kuwait—Arabian Gulf. *Marine Geology, 46*(1–2), 71–99. https://doi.org/10.1016/0025-3227(82)90152-9

Khalaf, F. I., & Al-Hashash, M. (1983). Aeolian sedimentation in the north-western part of the Arabian Gulf. *Journal of Arid Environments, 6*, 60–64.

Khalaf, F. I., & Ala, M. (1980). Mineralogy of the recent intertidal muddy sediments of Kuwait, Arabian Gulf. *Marine Geology, 35*, 331–342.

Khan, N. Y. (2007). Multiple stressors and ecosystem-based management in the Gulf. *Aquatic Ecosystem, Health, and Management, 10*, 259–267.

Karam Q. E. (2011). Toxicity of Kuwait crude oil and dispersed oil on selected marine fish species of Kuwait [PhD dissertation]. Newcastle University.

Karam, Q., Ali, M., Subrahmanyam, M. N. V., Al-Abdul Elah, K., Bentley, M., & Beg, M. U. (2019). A comparative study on the effect of dispersed and undispersed Kuwait crude oil on egg hatching and larval survival of Epinephelus coioides. *Journal of Environmental Biology, 40*, 192–199

Karam Q., Annabi-Trabelsi, N., Ali, M., Al-Abdul Elah, K., Beg, M. U., & Bentley, M. (2021). The response of Sobaity Sea bream *Sparidentex hasta* Larvae to the Toxicity of Dispersed and Undispersed Oil. *Polish Journal of Environmental Studies*. IN PRESS.

Kassler, P. (1973). The structural and geomorphic evolution of the Persian Gulf. In: B. H. Purser (Eds.), The Persian Gu/f, pp. I 1–32. Berlin: Springer-Verlag

Lardner, R. W., Al-Rabeh, A. H., Gunay, N., Hossain, M., Reynolds, R. M., & Lehr, W. J. (1993). Computation of the residual flow in the Gulf using the Mt Mitchell data and the KFUPM/RI hydrodynamical models. *Marine Pollution Bulletin, 27*, 61–70.

Lacustrine Basin Research. (1996). Environmental Interpretation of Ostracod Faunas from Wells Onshore/ Offshore Kuwait. Kuwait Joint Study Team Internal Report.

Lokier, S. W. (2013). Coastal sabkha preservation in the Arabian Gulf. *Geoheritage, 5*, 11–22.

MEMAC. (2016). Marine oil pollution manual-ROPME Sea Area.

Mohammed, M. A., & Shamlan, A. (1977). Organic matter content in Kuwait Bay sediments as an index of pollution. *Journal of the University of Kuwait (science), 5*, 170–185.

Naser, H. A. (2014). Marine ecosystem diversity in the Arabian Gulf: Threats and conservation. In O. Grillo (Ed.), *Biodiversity-The dynamic balance of the planet* (pp. 297–327). Rijeka. In Tech Open Access Publisher.

Price, A., Sheppard, C., & Roberts, C. (1993). The gulf: Its biological settings. *Marine Pollution Bulletin, 27*, 9–15.

Purser, B. H. (1973). *The Persian Gulf: Holocene carbonate sedimentation and diagenesis in a shallow epicontinental sea* (p. 471). Springer.

Persur, B. H., & Seilbold, E. (1973). The principal environmental factors influencing Holocene sedimentation, and diagensisin the Persian Gulf. In B. H. Persur (Ed.), *The Persian Gulf: Holocene carbonate sedimentation and diagenesis in a shallow Epicontinental Sea* (pp. 1–9). Springer.

Primavera, J. H., Friess, D. A., Van Lavieren, H., Lee, S. Y., & Sheppard, C. (2018). World seas: an environmental evaluation: Volume II: The Indian Ocean to the Pacific.

Reyolds, R. M. (1993). Physical oceanography of the Gulf, Strait of Hormuz, and the Gulf of Oman-results from Mt Mitchell expedition. *Marine Pollution Bulletin, 27*, 35–59.

RMSI. (2020). *Development of national hazard profile for the state of Kuwait and identification of existing data and gaps for exposure and vulnerability assessment report*. Prepared by RMSI Private Limited.

Robertson Research International. (1996). Mid and Lower Cretaceous, Coastal/Offshore Kuwait; Lithostratigraphy, Biostratigraphy and Palaeoenvironments of 7 wells. Kuwait Joint Study Team Internal Report.

ROPME. (2011). ROPME SEA AREA Regional Contingency Plan to Combat Pollution of the Sea By Hazardous and Noxious Substances (HNS) (ChemPlan) Report.

Saad, H. R., & Al-Azmi, D. (2002). Radioactivity concentrations in sediments and their correlation from space: Some new perspectives. *Applied Radiation and Isotopes, 56*, 991–997.

Sheppard, C. (2016). Coral reefs in the Gulf are mostly dead now, but can we do anything about it? *Marine Pollution Bulletin, 105*, 593–598.

Spalding, M. D., Fox, H. E., Allen, G. R., Davson, N., Ferdana, Z. A., Finlayson, M., et al. (2007). Marine ecoregions of the world: A bioregionalization of coastal and sheld area. *BioScience, 57*, 573–583.

Thorhaug, A. (1992). The environmental future of Kuwait. In *Proceedings of the International Symposium on the Environmental and Health Impacts of the Kuwait Oil Fires*. University of Birmingham, Edgbaston, England (pp. 69–72).

Torquato, F., Jensen, H. M., Range, P., Bach, S. S., Ben-Hamadou, R., Sigsgaard, E. E., et al. (2017). Vertical zonation and functional diversity of fish assemblages revealed by ROV videos and oil platform in the Gulf. *Journal of Fish Biology, 91*, 947–967.

Varol Research. (1996). Report on the stratigraphical implications of Nannoplankton analyses on eleven wells. KJST Internal Report.

Vinogradov, B., Grigorev, A., & Lipatov, V. (1973). The structures of dust storms according to information from television pictures from the ITOS-1 artificial earth satellite (Structure of dust storms determined from ITOS-1 television pictures), Image obtained over Iraq and the Gulf of Persia, Institute for Aerospace Methodology, Moscow, USSR.

www.beatona.net. (2013).

www.ierre-markuse.net. (2017).

Structures and Tectonics of Kuwait

5

Mohammad Naqi and Aimen Amer

Abstract

Despite the surface geology of Kuwait appears to be scarce and most of the country is covered with Quaternary deposits except for a few outcrops of Oligo-Miocene to Pleistocene age, the subsurface geology of Kuwait is quite unique and astonishing. The discovery of hydrocarbon in Kuwait at the beginning of the last century helped geologists to better understand the structural geology of Kuwait especially by utilizing geophysical methods such as potential field methods (e.g., gravity and magnetic) and seismic reflection. Being part of the Arabian Peninsula, the structural geology of Kuwait shares many of the Arabian Peninsula structural trends. The dominant N-S trending structures of the Arabian Plate are manifested in the Kuwait Arch which is one of the major structures of the country where many of the oil and gas oil fields are associated with. Other dominant structural trends of the Arabian Plate such as NE-SW and NW–SE are resembled in Kuwait as Jal Az-Zor and Dibdibah Trough, respectively. Paleo- and in-situ stress analysis is an important subject for oil and gas exploration, and many studies have been commissioned to better understand them in most of the Kuwaiti fields. The present-day in-situ stress in Kuwait is oriented NE-SW resembling the current tectonic setting of the region due to the collision of the Arabian Plate with the Eurasia Plate since the Oligocene. This chapter will present a thorough review of the previous studies discussing the surface and subsurface structural geology of Kuwait.

5.1 Introduction

Kuwait is located in the north-eastern region of the Arabian Peninsula (Fig. 5.1).

Geologically, Kuwait is suited as part of the Arabian Plate which was part of the Proterozoic Gondwana Supercontinent through most of the Phanerozoic. Therefore, it shares the geological history and the tectono-stratigraphy of the Arabian Peninsula.

Even though the surface geology of Kuwait looks scarce and most of the country is covered by Quaternary deposits except for a few outcrops of the Oligo-Miocene to Pleistocene age, the subsurface geology of Kuwait is quite unique and astonishing. Interests of the geology of Kuwait started in the early 20th century (around 1914) during the exploration of oil in the region and after the discovery of oil in Iran in 1908 (Milton, 1967). Few geological reconnaissances have been prepared to understand the geology in Kuwait by the British and Americans. In 1914 S. L. James, who held a British Admiralty commission, recommended shallow wells to be drilled based on oil seepage (Milton, 1967). Oil seepages were described in two locations near Bahrah—north of Kuwait Bay—and Burgan area—south of Kuwait. Further investigations of Kuwait were made in 1917 by S. L. James and G. W. Halse; in 1926, by B. K. N. Wyllie and A. G. H. Mayhew; in 1931, by T. Dewhurst; and in 1931–32, by P. T. Cox (Milton, 1967). Also, based on the surface geological survey, a gentle anticline in the Miocene Formations at the Bahrah area was identified (Fox, 1956). This preliminary surface geology survey provided limited surface evidence of structures; thus, it was decided to drill a test hole in the Bahrah area, north of Kuwait Bay in 1936 along with the geophysical survey. The geophysical survey included gravity, magnetic and seismic surveys (Boots & McKee, 1946).

After the successful discovery of commercial oil in 1936, more geological studies were a must to improve the understanding of the subsurface geology of Kuwait especially by

M. Naqi (✉)
Department of Earth and Environmental Sciences, Kuwait University, P.O. Box 5969 13060 Safat, Kuwait
e-mail: m.naqi@ku.edu.kw

A. Amer
Schlumberger, East Ahmadi, Block No. 6 - Building 193, 194, P.O. Box 9056 61001 Ahmadi, Kuwait

A. el-aziz K. Abd el-aal et al. (eds.), *The Geology of Kuwait*, Regional Geology Reviews,
https://doi.org/10.1007/978-3-031-16727-0_5

Fig. 5.1 Digital Elevation map of the Arabian Plate with the main plate boundaries. The Arabian Plate is bounded from the northeast by collision margin of Zagros Fold and Thrust Belt. Rift margin to the west and southwest, Red Sea and Gulf of Aden, respectively. Strike-slip margin of the Dead Sea to the northwest (Modified after Sharland et al., 2001)

utilizing more sophisticated geophysical methods such as gravity, magnetic and reflection seismology. This has revealed the complicated geological history of Kuwait and resulted in major oil and gas discoveries over the past several decades which made Kuwait one of the major oil exporters in the world.

This chapter will review and summarize the regional tectonics settings of the Arabian Plate and the structural elements of the State of Kuwait. Discussing the development and evolution of the major surface and subsurface structural elements of the country through the geological history. Geological evidence reveals that the structural geology of Kuwait and the major structural trends were shaped since the Precambrian and many of the subsequent structural pulses reactivated those structures. Concluding with shedding the light on some geomechanical aspects; present-day in-situ stress and paleostress. These studies have grown in the past decade and shown their significance in order to improve oil recovery and to minimize risks related to Hydrocarbon exploration and production.

5.2 Geological Settings

The Arabian plate is bounded by different structural regimes (Figs. 5.1 and 5.2); from the north and northeast margins by a compressional tectonic setting forming the Bitlis Suture and eastward by the Zagros Fold and Thrust Belt and to the east with the Makran subduction zone, where the subduction occurs in the Gulf of Oman; the Owen Fracture Zone and Gulf of Aden rifting to the south and southwest, and to the west the Red Sea rift system and the Gulf of Aqaba and the Dead Sea transform fault to the northwest. The Arabian Plate has been a coherent and potentially stable block since the

Fig. 5.2 Arabian Plate tectonic boundaries and kinematics (after Amer & Al-Hajeri, 2020)

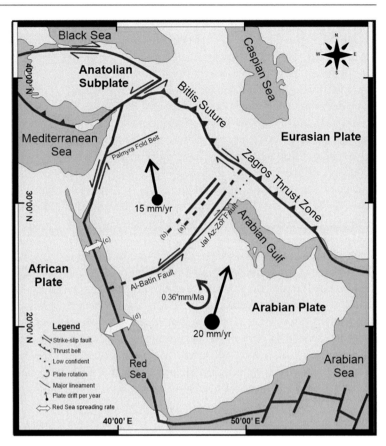

end of the Precambrian. The entire Middle East region was an integral part of the African Plate throughout most of the Phanerozoic. The separation of the Arabian Plate from the African Plate took place by the rise of the Afar Region triple junction in the Cenozoic (25–30 Ma ago). Since that time, the plate has rotated anticlockwise and drifted north, currently at a rate of 2-3 cm/year (Fig. 5.2) (Bird, 2002).

As part of the Arabian Plate during most of the Phanerozoic time, Kuwait's structures and tectonics capture the complex geological history of the region. The manifestation of the geological record, in both surface and subsurface, represents several long-term periods of tectonic deformation intermitted with periods of relative quiescence.

The Arabian Plate formed during the late Precambrian by suturing of several basic volcanic and plutonic terranes (Haq & Al-Qahtani, 2005; Stern & Johnson, 2010). The main structural elements were formed during the formation of the Arabian Shield trending northeast and north-northwest basement sutures (Al-Husseini, 2000; Ziegler, 2001).

The Najd strike-slip system took place during the late Precambrian time, which is observed from Egypt to Oman, and might have a continuation into the Indian Plate (Al-Husseini, 1988, 2000). The Najd strike-slip system deformation was accompanied by the evolution of rift basins where the Hormuz and Ara salt were deposited. The deposition of thick Precambrian Hormuz salt might play a role in

the architecture of some of the oil field structures in Kuwait (Fox, 1956; Murris, 1980; Singh, 2012).

Another significant structural pulses that widely influenced the Arabian Plate and resembled the structural geology of Kuwait was the so-called Hercynian Orogeny, which took place during the Late Devonian to Permian in North America and Europe. However, the timing of this Orogeny in the Arabian Plate is more restricted to Late Devonian through Late Carboniferous forming multiple episodes of compression and faulting (Haq & Al-Qahtani, 2005; Stern & Johnson, 2010). This deformation pulse formed a major unconformity in the subsurface of the Arabian Plate (Haq & Al-Qahtani, 2005) and reactivated many of the north-south trending structures that were initially formed during the Precambrian assembly of the Arabian Plate. In Burgan and Umm Gudair area, in Kuwait, the Permian Khuff Formation unconformably overlies altered basic igneous rocks and older clastic of presumably of Proterozoic age, while in the north of Kuwait Arch Khuff overlies Permo-Carboniferous Unayzah Formation (Tanoli et al., 2015). This observation along with regional analogues and Gravity-Magnetic data can be interpreted that the proto-Burgan structure was reactivated during the Hercynian Orogeny (Tanoli et al., 2015).

From the thickness variation of the Triassic Sudair and Lower Jilh Formations, where both formations are thinner over the Kuwait Arch and thicker on both sides, Dibdibba

Trough to the west and eastward in the offshore trough. This variation indicates that the Kuwait arch remained relatively higher during that time. However, during Late Triassic to Late Jurassic, little tectonic activity took place on the Arabian plate during this time and the Kuwait Arch was mainly inactive during this period (Tanoli et al., 2015).

A compressive regime was developed in the Arabian plate during the Late Cretaceous because of the convergence between the Asian and Arabian Plates which resulted in the obduction of the Ophiolite sequence of the Neo-Tethyan oceanic crust in the northeastern margin of the Arabian Plate (Haq & Al-Qahtani, 2005; Sharland et al., 2001). This compressive regime is called Alpine-1 Orogeny. Uplift and erosion took place over many of the reactivated structural highs due to the compressive stress and were recorded as Pre-Aruma Unconformity. This Unconformity is well demonstrated at the Burgan and Magwa structures where onlaps are clearly observed, on seismic data, on the eastern and western sides of the structures. The Alpine-1 structural event is captured in Kuwait by the accentuation of earlier structures in a more NW-SE trend. Many of the older structural highs such as Burgan, Minagish, Umm Gudair, Raudhatain and Sabriyah were strongly reactivated (Tanoli et al., 2015). The majority of these faults ceased at the Late Cretaceous Tayarat Formation.

The Alpine-2 Orogeny took place in Late Eocene to the present day. The Arabian Plate begun to collide with Asia and the closure of the Neo-Tethyan Ocean was complete by the Late Oligocene. The development of the Zagros Fold and Thrust Belt is directly associated with this Orogeny (Richard et al., 2014) and caused uplift and erosion in Kuwait developing into the top Dammam Unconformity. Most of the major structures, in Kuwait, show truncation and gentle arching over them. This compressive tectonic event led to the inversion of many normal faults and amplified anticlines at the Jurassic section and the development of long wave anticline in the shallower successions (Tanoli et al., 2015).

5.3 Basement Structural Configuration and Trends

The Arabian Plate is divided into the Arabian Shield to the west and the Arabian Platform to the east (Fig. 5.3). Basement structural trends vary across the Arabian Peninsula. In the west where the Arabian shield is present, basement structures such as folds, faults, shear zones and dykes can be well identified, and their orientation can be easily determined by direct observation. However, to the east and north of the Arabian shield, where Kuwait is located, the basement is covered, and structural trends can only be inferred from geophysical methods e.g., magnetic, gravity, and seismic data, and as a result, are less well known.

The Arabian Plate is characterized based on basement structural trends into three main basement structural domains (Stern & Johnson, 2010) (Fig. 5.3). Stern and Johnson (2010) divided the domains a follow: (1) The Arabian Shield and the south-central part of the Arabian platform which is characterized by northerly and northwesterly structural trends; (2) a north-central domain marked by northerly structural trends and a broad northeasterly trend; and (3) northeasterly dominated structural trends in the eastern domain. Based on this classification, Kuwait is located in the north-central domain where the basement structures are dominated by northerly and northeasterly trends such as Ahmadi ridge, Kuwait Arch (Khurais-Burgan anticline) and Al-Batin Arch.

The Arabian Plate crustal thickness varies. The thickest part is under the Arabian Platform and thins toward the west near the Red Sea. Several studies have been conducted to measure the crustal thickness of the Arabian Plate (Stern & Johnson, 2010), finding that the Arabian Plate crustal thickness reaches about 22 km near the Red Sea and 53 km in the eastern part of the Arabian Plate. Pasyanos et al. (2007) measured the crustal thickness beneath Kuwait by using surface waves and receiver-function data and concluded that 8 km sediments overlaying 37 km thick crust. On the other hand, Midzi (2005) utilized receiver function techniques to teleseismic arrivals at six seismic stations and found out the crust was about 42.6±1.8 km thick beneath Kuwait. Sediment thickness in Kuwait is between 7 and 8 km based on seismic methods (Pasyanos et al., 2007, Laske & Masters, 1997).

5.4 Structural Architecture and Elements of Kuwait

5.4.1 Kuwait Arch

One of the major structural expressions manifested in Kuwait is the N-S trending Kuwait Arch (Fig. 5.4) representing the northern end of the giant Khurais-Burgan Anticline, a structure that is underlain by faulted Precambrian basement blocks. The Khurais-Burgan structure is about 500 km long, and tens of kilometers wide (Al-Husseini, 2004). This structure along with the other N-S trending structures (e.g., Summan Platform, Hawtah and En Nala Ghwar structures) is bounded by an ancestral fracture system of N-S trending faults that formed in the late-Protereozoic Amar Collision between about 640 and 620 Ma (Al-Husseini, 2000) and most likely rejuvenated during the mid-Carboniferous compressional structural pulse (Al-Husseini, 2004).

Kuwait Arch has a relief of about 100 meters (m) south of Kuwait. This structural high was first captured by Warsi (1990) via gravity survey. His Bouguer anomaly map

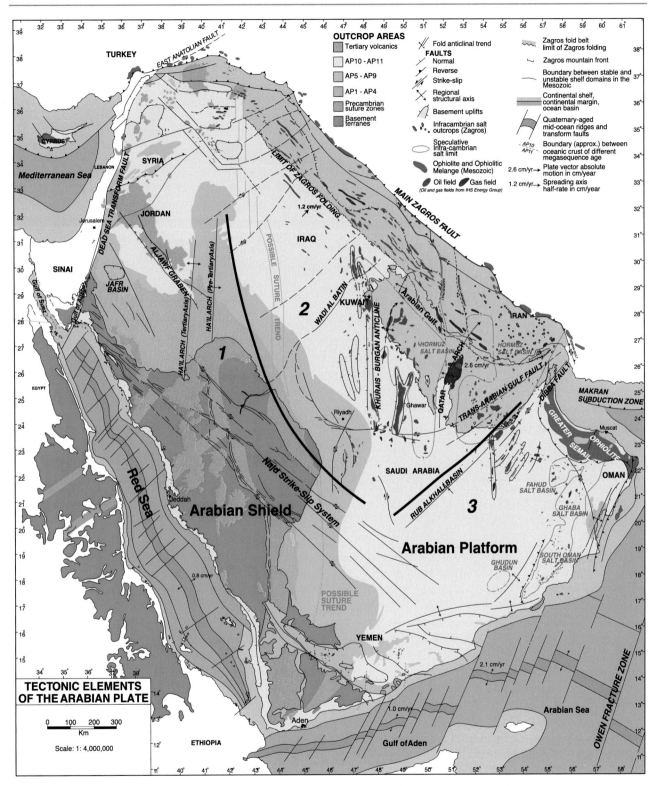

Fig. 5.3 Tectonic elements of the Arabian plates showing the main structural trends. Also, showing three structural domains (after Stern & Johnson, 2010)

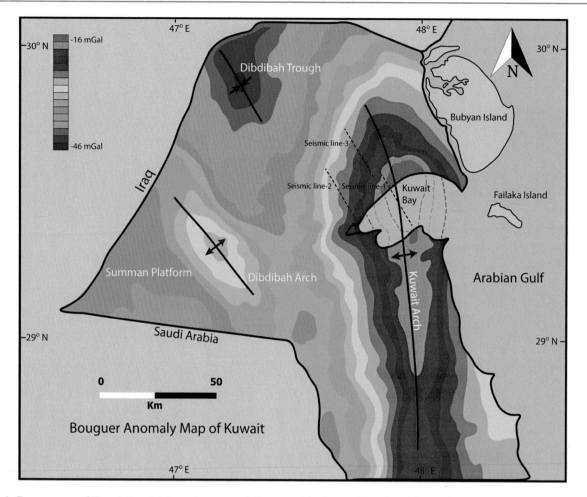

Fig. 5.4 Bouguer map of Kuwait showing the major structural elements of the State of Kuwait trending N-S and NNW-SSE (after Warsi, 1990). Dashed lines are location of seismic lines in Fig. 5.7

(Fig. 5.4) shows that the crestal area of the arch is located within Kuwait Bay. While subsurface seismic and well data indicate that the center of the Arch is located to the south in the Kuwait-Saudi Partitioned Neutral Zone (PNZ) (Carmen, 1996). The geometric configuration of Kuwait Arch resembles asymmetric where its western flank dips about 2°–3° and the eastern flank shows a steeper dip of about 10°. Many of Kuwait oil fields are associated with Kuwait Arch e.g., Burgan, Magwa, Ahmadi, Bahrah Sabriyah, Raudhatain, and Wafra oil fields (Fig. 5.5).

Amer and Alhajeri's (2020) study shows that the thickness of the middle Eocene Dammam Formation is relatively constant suggesting the structural quiescence of Kuwait Arch during its deposition. However, the drastic decrease in thickness of the overlain Kuwait Formation indicates the reactivation of the Kuwait Arch during Late Eocene to

Pliocene. This structural reactivation could be associated with the Alpine-2 structural pulse in the Arabian Peninsula region (Richard et al., 2014).

5.4.2 Dibdibba Trough

Dibdibba Trough is a major structural element in the state of Kuwait (Fig. 5.4). It is located to the west of Kuwait Arch and east and northeast of Summan Platform and it might be related to the Najd fault system. It is trending NNW-SSE. It is estimated that the maximum sedimentary thickness reaches up to 28,000 feet in the central part of Dibdibba Trough (Singh et al., 2011). Along the center of the Trough, West Kuwait High is located. This High comprises of Minagish, Kra Al-Maru and Khabrat Ali structural highs that

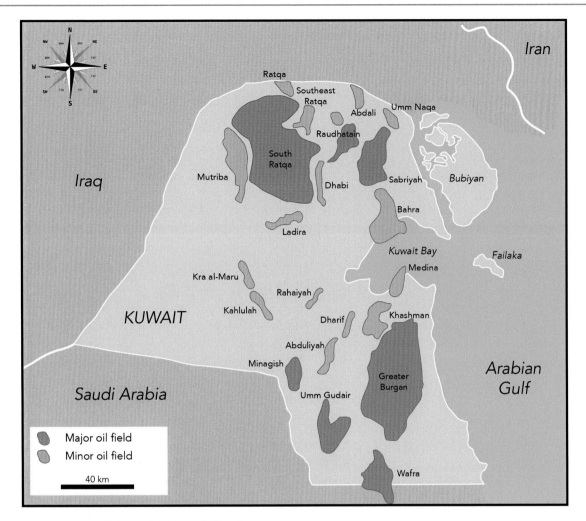

Fig. 5.5 Major and minor oil fields of the State ok Kuwait

trending NNW. These structures were likely formed and modified by reactivation of basement faults and by the halokinesis of the Infracambrian Hormuz salt during major tectonic episodes (Tanoli et al., 2015).

5.4.3 Summan Platform

This structure located west of Kuwait Arch (Fig. 5.4) and extends to the south into the northeastern part of Saudi Arabia where is the Summan Platform is bounded by N-S trending faults following the Late Proterozoic structural grain. These faults were reactivated during the mid-Carboniferous tectonic pulse (Al-Husseini, 2004). It is located on the western side of Kuwait and is characterized by a gradual rising basement to the west.

5.4.4 Wadi Al Batin

Wadi Al Batin is located west of Kuwait forming a natural boundary between the State of Kuwait and Iraq (Fig. 5.6). The wadi exhibits a linear, steep-sided valley about 5–10 km wide and trends N35 °E with a relief of about 60 m. It can be tracible for 700 km on satellite imagery from Saudi Arabia and extends to the south of Iraq (Al-Sarawi, 1980). It might represent the northern continuation of Wadi Ar-Rimah in Saudi Arabia. Parts of the Wadi are covered by Quaternary and Tertiary gravels of igneous and metamorphic rock that have been transported by large streams from the Saudi Arabian and Syrian deserts during Pleistocene pluvial episodes (Holm, 1960). From Seismic and water well drawdown data Al-Sarawi (1980) identified three en-echelon faults trending northeast with a throw of about 25 m to the

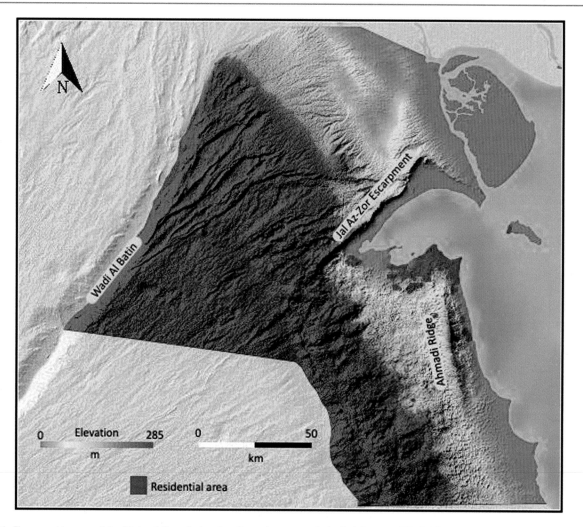

Fig. 5.6 Topographic map of the State of Kuwait showing the main geomorphological features; Wadi Al Batin, Ahmadi Ridge and Jal Az-Zor escarpment

southwest. He assigned the age of the faults to be Late Eocene to Miocene, although minor movement may have continued during Pliocene. The trend of these faults is similar to the major basement faults system in Saudi Arabia (e.g., Wadi Ar-Rimah). Hancock et al. (1981) proposed a 20 km of sinistral offset along a strike-slip fault, positioned along the continuation of Wadi Al Batin in Saudi Arabia, based on outcrop contacts of Cretaceous and Paleogene and Paleogene and Neogene strata. In addition, a few small faults cutting the Neogene rocks of the wadi was identified from the high pass, filter band 7, Landsat image 179 (path)/40 (row) from the Saudi Arabia mosaic (Hancock et al., 1981).

5.4.5 Jal Az Zor

The Jal Az-Zor escarpment (Fig. 5.6) represents the most prominent surface rock exposure in Kuwait. It extends for about 60 km from the north of Al-Jahra to the Al-Sabiyah

area, trending NE-SE parallel to the northern shore of Kuwait Bay, forming a cliff that reaches about 36m high (Amer & Al-Hajeri, 2020). Several studies investigated the origin of the Jal Az-Zor escarpment (Salman, 1979, Al-Sarawi, 1982; Carmen, 1996; Amer & Al-Hajeri, 2020) proposing a range of scenarios. Al-Sarawi (1982) proposed based on the lithostratigraphic correlation between several wells drilled around the escarpment, suggesting that origin of the Jal Az-Zor escarpment to be the retreat of a major fault lineament located offshore in the center of Kuwait Bay during the Neogene through erosion to its current position. Bou-Rabee and Kleinkopf (1994) based on free air and Bouguer gravity anomaly proposed that the Jal Az-Zor escarpment is a surface expression of a basement strike-slip fault with minor displacement. Based on NW-SE 2D seismic section acquired in 1960s, Al-Anzi (1995) proposed the occurrence of a series of wrench faults with no major throw close the Jal Az-Zor escarpment without surface expression. However, this interpretation was based on notably poor-quality seismic data

(Amer & Al-Hajeri, 2020). Carman (1996) also placed the Jal Az-Zor fault in the center of Kuwait Bay.

Amer and Al-Hajeri (2020) from reprocessed seismic data were able to clearly identify faults and their associated structures beneath the Jan Az-Zor escarpment. From 3 seismic lines across the escarpment (Fig. 5.7), they interpreted the existence of detachment folds, recumbent folds, thrust faults, and fault propagation folds, forming a complex duplex system within the upper Dammam Formation below Jal Az-Zor (Fig. 5.8). They suggested that the present-day relief of Jal Az-Zor is a manifestation of fault kinematics along with complex folding as a result of increased basal friction towards the south. They also indicated based on 2D structural restoration and balancing along with GPS velocities and vectors for the Arabian Plate that the layers under Jal Az-Zor escarpment accommodated about 6.25 km of shortening that occurred within the last 1 Ma.

5.5 Kuwait Tectonic Stresses

Having discussed the various structural elements over Kuwait, in this section in-situ stress will be discussed and its relation to Kuwait's surface and subsurface geology. The following section will be divided into two parts, the first will discuss the present day in-situ stress that effect Kuwait today. The second part will discuss the paleostresses or historic stresses that resulted in the formation of the fold and fault systems in the past but still observed today.

5.5.1 Present day In-Situ Stress

For decades in-situ stress measurements have been collected over the different hydrocarbon fields in Kuwait, using different techniques such as borehole images, caliper logs, and sonic logs. The borehole image logs are arguably the most effective measurement that can provide details on the stress direction by recognizing drilling induced features. Over electrical borehole image logs these features can range between borehole breakouts and drilling induced fractures (Amer & Alexander, 2005; Cesaro et al., 2000). Over vertical wells the borehole breakouts would indicate the minimum horizontal stress direction (Shmin), whereas the drilling induced fractures will indicate the maximum horizontal stress direction (SHmax) (von Winterfeld et al., 2005). Caliper and sonic logs, on the other hand, can only detect one of the in-situ horizontal stresses. Caliper logs for instance can only detect borehole breakouts, reflecting the minimum horizontal stresses (Plumb & Hickman, 1985), and sonic logs through dipole-shear anisotropy processing can detect the fast shear direction indicating the maximum

horizontal stress direction (Esmersoy et al., 1994, 1995). The overburden stresses or vertical stresses (Sv) is the last component of the 3 main principal stresses, and it generally determines the stress regime (Fig. 5.9). In Kuwait this stress exhibits a vertical direction and is calculated by the integration of the rock densities with depth, hence it increases as we go deeper into the earth's crust. The horizontal stresses magnitudes, on the other hand, are calculated using the poroelastic equation:

$$S_{hmin} = \frac{v}{1-v}(S_v - aP_p)$$
$$+ \frac{1-2v}{1-v}aP_p + vE_{ps}e_{Hmax} + E_{ps}e_{Hmin}$$

$$S_{Hmax} = \frac{v}{1-v}(S_v - aP_p)$$
$$+ \frac{1-2v}{1-v}aP_p + E_{ps}e_{Hmax} + vE_{ps}e_{Hmin}$$

where plane-strain Young's modulus (E_{ps}); Poisson's ratio (v); vertical stresses (S_v); Biot's coefficient (a); considering tectonic strains (e_{Hmax} and e_{Hmin}); pore pressure (P_p).

Young's modulus and Poisson's ratio can be either measured in the lab or calculated from openhole logs. The Biot's coefficient is best measured in the lab, but in most cases, it is equal to 1. The pore pressure is best measured using downhole pressure measuring tools or estimated using Eaton's method. Using these techniques, the stress magnitudes have been calculated over Kuwait and it shows a general strike-slip stress regime (Al-Kandary et al., 2012; Perumalla et al., 2014). However over specific zones along the stratigraphy of Kuwait this stress regime can change to normal or reverse depending on the proximity to the major fault plane and mechanical layering that can alter the stress regime and direction (Amer et al., 2013) (Fig. 5.10).

Several geomechanical studies have been commissioned by Kuwait Oil Company (KOC) to study the in-situ horizontal stresses over Kuwait (Al-Kandary et al., 2012; Amer et al., 2013; Carman, 1996; Dey et al., 2009; Perumalla et al., 2014; Richard et al., 2014).

Carman (1996) studied the structural elements of onshore Kuwait and identified the in-situ stress over the Cretaceous successes. Though he identified the stress over Minagish, Umm Gudair, Burgan, Magwa, Sabriyah, and Raudhatain fields (Fig. 5.5), due to its complexity and notable variation it was suggested that more work is required to explain the local variations (Fig. 5.11).

Al-Kandary et al., (2012) focused on the northern region of Kuwait and covered Abdali, Bahrah, Dhabi, Mutriba, Raudhatain, Ratqa, Sabriyah, and Umm Niqa fields. By focusing on the Jurassic and Cretaceous sections, the authors were able to identify discrepancies between the two sections. This was found to be a result of the stress decoupling as a

Fig. 5.7 Interpreted seismic sections across Jal Az-Zor escarpment showing the different structure in the subsurface (after Amer & Al-Hajeri, 2020). Location of the seismic lines are shown in Fig. 5.4

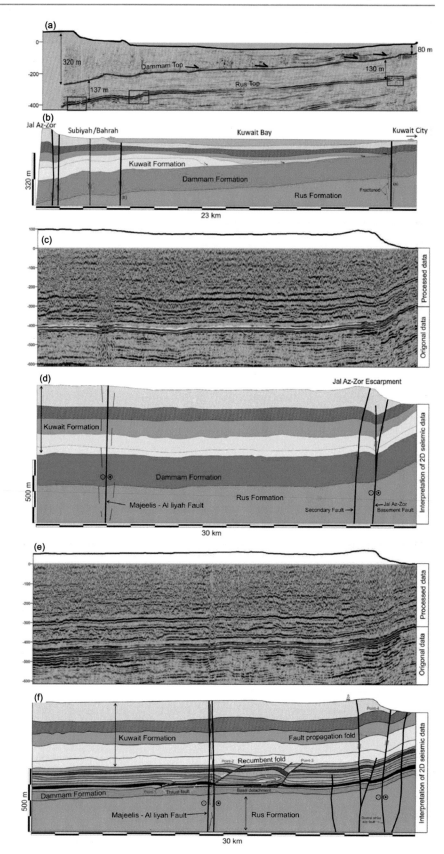

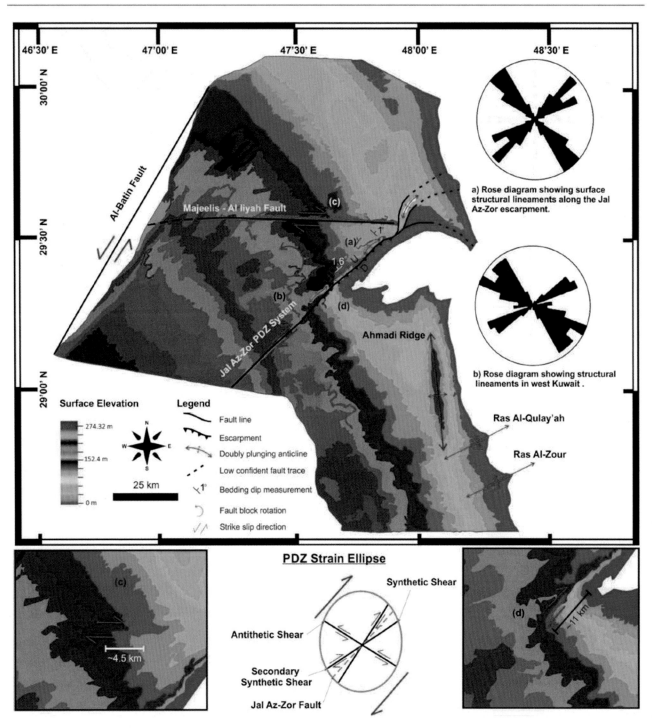

Fig. 5.8 Surface topographic contour map of Kuwait showing the interpreted structures from surface expressions (after Amer & Al-Hajeri, 2020)

result of the Gotnia Formation salt. Amer et al., (2013, 2014) confirmed this observation and documented a total of 1.8 km offset between the structural crests observed at the Jurassic Najmah Formation and the Gotnia Formation tops in the SW direction suggesting detachment at the base of Gotnia Formation associated with thrust faults (Fig. 5.12). Amer et al (2013) for the first time indicated the reserves stress regime

component that is intermittently observed especially over the Jurassic section, helping in forming a better understanding of Kuwait's stress regime.

Perumalla et al. (2014) followed up on the work performed by Al-Kandary et al. (2012) and extended their quest to understand the stress regime for both Jurassic and Cretaceous sections for all Kuwait using a wider range of stress

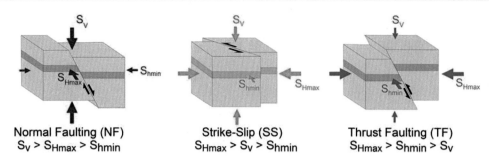

Fig. 5.9 The main principal stress regimes

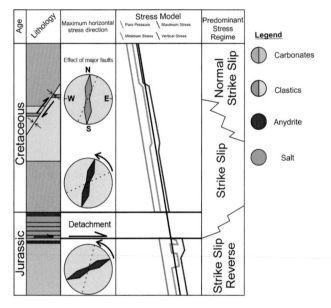

Fig. 5.10 Schematic representation of the Jurassic and Cretaceous section of Kuwait and the different maximum in-situ stress general change in direction in areas with significant detachment along Gotnia Formation (dominantly composed of salt). The illustration also shows the potential of stress rotation around major faults. The stress model shows the evolution of stress with depth and the predominant stress regime over each interval

measurements such as caliper logs and borehole images (Fig. 5.11). In general, they found that the vertical stresses (Sv) for the Cretaceous section is on average 19.5 ppg and for the Jurassic section it is 21 ppg. Such increase in the vertical stresses is expected as depth increase. The minimum horizontal stress (Shmin) magnitude on average for the Cretaceous section is 16 ppg and for the Jurassic is 18 ppg. The increase can be attributed to the increasing pore pressure gradient with depth. And finally, the maximum horizontal stress (SHmax) magnitude for the Cretaceous is on average 20.5 ppg and for the Jurassic 24.3 ppg. Such increase is generally a result of the continues NE drift of the Arabian Plate towards the Eurasia plate combined with rotation (Amer & Al-Hajeri, 2020) (Fig. 5.2).

5.5.2 Kuwait Paleostresses

The paleostresses that effected the region of Kuwait during the Jurassic and Cretaceous are governed by transtensional tectonics (Alpine-1), whereas the mid-to-late Tertiary Kuwait is dominated with transpression (Alpine-2) (Amer & Al-Hajeri, 2020; Filbrandt et al., 2006; Richard et al., 2014; Sharland et al., 2001; Ziegler, 2001).

Based on the work of Richard et al. (2014) over Bahrah, Dhabi, Raudhatain, Sabriyah, and Umm Niqqa fields during the Jurassic the overall stress regime was characterized by an extensional stress regime setting associated with a N-S maximum horizonal stress direction (Fig. 5.13). This evolved to a transtensional stress regime during the Cretaceous that was associated with a NW-SE maximum horizontal stress direction (Al-Kandary et al., 2012; Perumalla et al., 2014; Richard et al., 2014; Verma et al., 2007). The final stress rotation event in Kuwait occurred during the Tertiary. These times were associated with a transpressional stress regime associated with the Oligo-Miocene Zagros Orogeny and exhibits a NE-SW maximum stress direction, which is consistent with the present day in-situ stress (Richard et al., 2014).

Abu-Hebail et al. (2018) attempted to model the paleotectonic stress history of the fault systems over Kra Al-Maru and Riksah structures, west Kuwait. This system area is characterized by a principal displacement zone that exhibits a NE-SW direction (Amer et al., 2013, 2014). The technique is based on a new generation paleostress inversion geotechnical based method (Maerten, 2010). The paleotectonic stress history inversion simulated over 1000 possible scenarios and resulted in a stress inversion plot that showed the structures analyzed where formed under a NE-SW stress orientation. The stress regime is characterized by normal and strike slip tectonics (transpressional stress regime). By combining the understanding of the geological stress history and paleotectonic stress history the timing of faulting or fracturing can be better assisted, which in this case is during the Tertiary (Fig. 5.14).

Fig. 5.11 Kuwait map showing major surface structural elements along with the Jurassic and Cretaceous present day horizontal in-situ stresses

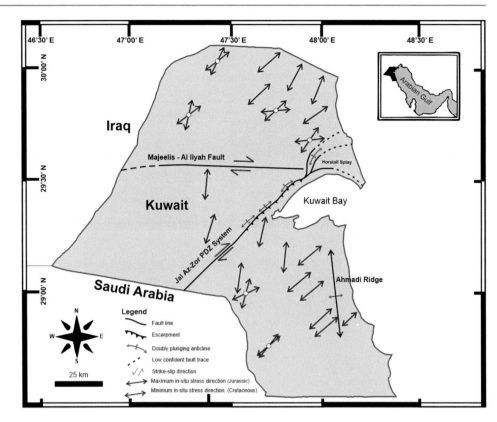

5.6 Conclusions

Geologically, Kuwait is suited as part of the Arabian Plate which was part of the Proterozoic Gondwana Supercontinent through most of the Phanerozoic. Therefore, it shares the geological history and the tectono-stratigraphy of the Arabian Peninsula. The surface geology of Kuwait looks scarce and most of the country is covered by Quaternary deposits except for few outcrops of Oligo-Miocene to Pleistocene age however, the subsurface geology of Kuwait is quite unique and astonishing. Kuwait went through multiple structural episodes.

The oldest recorded structural pulse was during the formation of the Arabian Plate in the Precambrian. This pulse generated NE and N-NW trending structures (Najd Fault System). Another important structural pulse is the so-called Hercynian Orogeny, which took place during the Late Devonian to Late Carboniferous. This compressional event has reactivated many of the older structures that trend N-S. During the Jurassic and Late Cretaceous a transtensional tectonics event took place (Alpine-1), whereas the mid-to-late Tertiary Kuwait is dominated with transpression (Alpine-2).

One of the major structural expressions manifested in Kuwait is the N-S trending Kuwait Arch, representing the northern end of the giant Khurais-Burgan Anticline, a structure that is underlain by faulted Precambrian basement blocks. Many of Kuwait oil fields are associated with Kuwait Arch e.g., Burgan, Magwa, Ahmadi, Bahrah Sabriyah, Raudhatain, and Wafra oil fields.

Dibdibba Trough is a major structural element in the state of Kuwait, trending NNW-SSE. It is located to the west of Kuwait Arch and east and northeast of Summan Platform and it might be related to the Najd fault system.

The Summan Platform structure located west of Kuwait Arch and extends to the south into the northeastern part of Saudi Arabia where is the Summan Platform is bounded by N-S trending faults following the Late Proterozoic structural grain.

Another major structural element in Kuwait is Wadi Al Batin which is located west of Kuwait forming a natural boundary between the State of Kuwait and Iraq trending N35°E. From Seismic and water well drawdown data Al-Sarawi (1980) identified three en-echelon faults trending northeast with a throw of about 25 m to the southwest.

The Jal Az-Zor escarpment represents the most prominent surface rock exposure in Kuwait. Based on a recent study by Amer and Al-Hajeri (2020), they were able to clearly

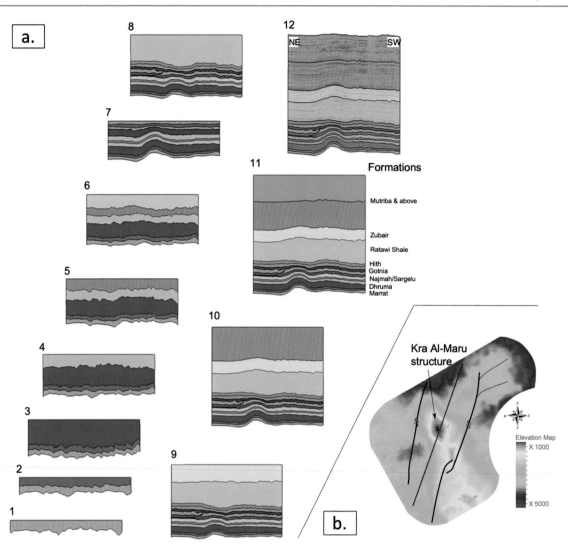

Fig. 5.12 **a** Structural restoration of Kra Al-Maru field showing shift in the structural crest between Najmah and Gotnia formations, along with the thrust faulting at the base of Gotnia salts. **b** Elevation map showing the location of the section in red crossing Kra Al-Maru structure (see Fig. 5.5 for the location of the Kra Al-Maru structure), (Amer et al., 2014)

identify faults and their associated structures beneath Jan Az-Zor escarpment. They interpreted the existence of detachment folds, recumbent folds, thrust faults, and fault propagation folds, forming a complex duplex system within the upper Dammam Formation below Jal Az-Zor.

For decades in-situ stress measurements have been collected over the different hydrocarbon fields in Kuwait, using different techniques such as borehole images, caliper logs, and sonic logs. The present day in-situ stress in Kuwait is oriented NE-SW resembling the current tectonic setting of the region due to the collision of the Arabian Plate with the Eurasia Plate since the Oligocene. The stress magnitudes have been calculated over Kuwait and it shows a general strike-slip stress regime. However over specific zones along the stratigraphy of Kuwait this stress regime can change to normal or reverse depending on the proximity to the major fault plane and mechanical layering that can alter the stress regime and direction (Amer et al., 2013).

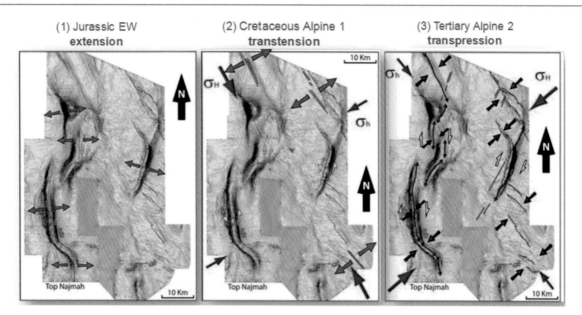

(1) Jurassic EW extension

(2) Cretaceous Alpine 1 transtension

(3) Tertiary Alpine 2 transpression

Fig. 5.13 Variation of the paleostress of Kuwait through time. (1) Extensional phase of deformation during the Jurassic with N-S maximum horizontal stress. (2) Transtensional phase of deformation during Cretaceous with NW–SE maximum horizontal stress. (3) Transpressional phase of deformation during the Tertiary with NE-SW maximum horizontal stress (after Richard et al., 2014)

Fig. 5.14 3D model for Kra Al-Maru and Riksah structures showing the major faults and stress inversion plot for the area. Note the red area represents the stress direction and regime, where a normal and strike-slip components is represented

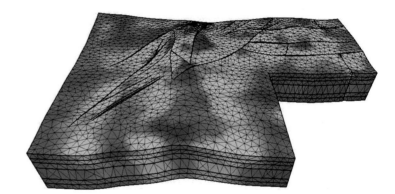

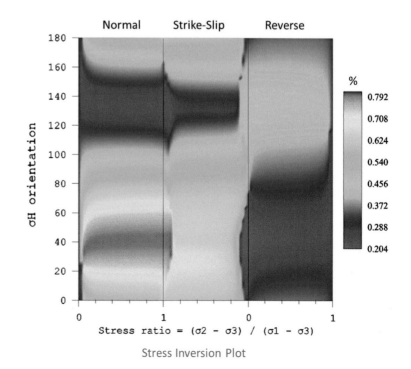

References

Abu-Hebail, H., Al-Wadi, M., Al-Otaibi, T. M., Amer, A. (2018). Modeling fracture networks during exploration stages; A case study from West Kuwait. In *GEO 2018 the 13th Middle East Geosciences Conference and Exhibition March 5–8, 2018, Manama, Bahrain Search and Discovery* Article #90319 (p. 1).

Al-Anzi, M. (1995). Stratigraphy and structure of the Bahrah field, Kuwait. In M. I. Al-Husseini (Ed.), *Middle East Petroleum Geosciences, GEO'94*. Gulf PetroLink, Bahrain (Vol. 1, pp. 53–64).

Al-Husseini, M. I. (1988). The Arabian Infracambrian extensional system. *Tectonophysics, 148*, 93–103.

Al-Husseini, M. I. (2000). Origin of the Arabian plate structures: Amar Collision and Najd Rift. *GeoArabia, 5*(4), 527–542.

Al-Husseini, M. I. (2004). Pre-unayzah unconformity, Saudi Arabia. In M. I. Al-Husseini (Ed.), *Carboniferous, Permian and Early Triassic Arabian Stratigraphy* (pp. 15–59). GeoArabia Special Publication 3, Gulf PetroLink, Bahrain.

Al-Sarawi, A. M. (1980). Tertiary faulting beneath wadi Al Batin (Kuwait). *GSA Bulletin, 91*, 610–618.

Al-Sarawi, A. M. (1982). Origin of the Jal Al Zor escarpment. *Journal of the University of Kuwait (science), 9*, 151–162.

Al-Kandary, A., Al-Fares, A., Mulyono, R., et al. (2012). Regional In-situ stress in Northern Kuwait-Implications for the oil Industry. Society of Petroleum Engineers Implications for the oil Industry. Society of Petroleum Engineers.

Amer A, Alexander SJ (2005) Interpreting the Depositional Facies of Upper Gharif Fluvial

Amer, A., Al-Ammar, H., Sajer, A., et al. (2013). Principal displacement zone evaluation over the Jurassic section of Kra Al-Maru and Riksah Areas, NW Kuwait. In *AAPG GTWExploring and Producing Fractured Reservoirs in the Middle East Conference and Workshop, Jordan on* (pp. 22–24).

Amer, A., Salem, H., Adwani, T. A., et al. (2014). Structural restoration of the Jurassic section at Kra Al-Maru and Riksah structures, West Kuwait. In *AAPG International Conference & Exhibition, Istanbul, Turkey*

Amer, A., & Al-Hajeri, M. (2019). Strontium isotope radiometric dating reveals the late Eocene and Oligocene successions in northern Kuwait. *Arabian Journal of Geosciences, 12*, 288. https://doi.org/10.1007/s12517-019-4455-4. (Springer).

Amer, A., & Al-Hajeri, M. (2020). The Jal Az-Zor escarpment as a product of complex duplex folding and strike-slip tectonics; A new study in Kuwait, northeastern Arabian Peninsula. *Journal of Structural Geology*. https://doi.org/10.1016/j.jsg.2020.104024

Bird, P. (2002). An updated digital model of plate boundaries. *Geochemistry, Geophysics, Geosystems, 4*(3), 1027. https://doi.org/10.1029/2001GC000252

Boots, P. H., & McKee, A. H. (1946). Geophysical operations in Kuwait. *Geophysics, 11*, 164–177.

Bou-Rabee, F., & Kleinkopf, D. (1994). Crustal structure of Kuwait: Constraints from gravity anomalies (Open-File Report No. 94–210). US Department of the Interior, US Geological Survey.

Carman, G. J. (1996). Structural elements of onshore Kuwait. *Gulf PetroLink, Bahrain, GeoArabia, 1*, 239–266.

Cesaro, M., Gonfalini, M., Cheung, P., Etchecopar, A. (2000). Shaping up to stress in the Apennines: Schlumberger Well Evaluation Conference Italy.

Dey, A. K., Singh, S. K., Banik, N., et al. (2009). An integrated analysis for the re-assessment of hydrocarbon potential of a low prospect area: A case study on Jurassic Marrat reservoir of Burgan Structure in South East Kuwait. In *International Petroleum Technology Conference*

Esmersoy, C., Koster, K., Williams, M., et al. (1994). Dipole shear anisotropy logging. Society of Exploration Geophysicists.

Esmersoy, C., Kane, M., Boyd, A., Denoo, S. (1995). Fracture and stress evaluation using DipoleShear anisotropy logs. Society of Petrophysicists and Well-Log Analysts.

Filbrandt, J. B., Al-Dhahab, S., Al-Habsy, A., et al. (2006). Kinematic interpretation and structural evolution of North Oman, Block 6, since the Late Cretaceous and implications for timing of hydrocarbon migration into Cretaceous reservoirs. *GeoArabia, 11*, 97–140.

Fox, A. F. (1956). Oil occurrences in Kuwait. In *Symposium sobre yacimientos de petroleo y gas, XX Congress Geologico Internacional, II Mexico* (pp. 131–158).

Hancock, P. L., Al-Khatieb, S. O., & Al-Kadhi, A. (1981). Structural and photogeological evidence for the boundaries to an East Arabian Block. *Geological Magazine, 118*(5), 533–538.

Haq, B. U., & Al-Qahtani, M. (2005). Phanerozoic cycles of sea-level change on the Arabian Platform. *GeoArabia, 10*(2), 2005.

Heidbach, O., Rajabi, M., Reiter, K., et al. (2016) World stress map database release 2016.

Holm, D. A. (1960). Desert geomorphology in the Arabian Peninsula. *Science, 132*, 13691379.

Laske, G., & Masters, G. (1997). A global digital map of sediment thickness. *Eos, Transactions of the American Geophysical Union, 78*, F483.

Maerten, F. (2010). Geomechanics to solve geological structure issues: forward, inverse and restoration modeling. PhD Thesis.

Midzi, V. (2005). The receiver structure beneath the Kuwait National Seismic Network. *MESF Cyber Journal Earth Science, 3*, 1–21.

Milton, D. I. (1967). Geology of the Arabian Peninsula Kuwait. United States Geological Society, Professional Paper 560-F (p. 7).

Murris, R. J. (1980). Middle East: Stratigraphic evolution and oil habitat. *American Association of Petroleum Geologists Bulletin, 64*(5), 597–618.

Nelson, P. H. (1968). Wafra field Kuwait-Saudi Arabia neutral zone.

Pasyanos, M. E., Tkalcic, H., Gök, R., Al-Enzi, A., & Rodgers, A. J. (2007). Seismic structure of Kuwait. *Geophysical Journal International, 170*, 299–312.

Perumalla, S., Al-Fares, A., Husain, R., et al. (2014). Regional in-situ stress mapping: An initiative for exploration & development of deep gas reservoirs in Kuwait. In *IPTC 2014: International Petroleum Technology Conference*.

Plumb, R. A., & Hickman, S. H. (1985). Stress-induced borehole elongation: A comparison between the four-arm dipmeter and the borehole televiewer in the Auburn geothermal well. *Journal of Geophysical Research: Solid Earth, 90*, 5513–5521.

Richard, P., Bazalgette, L., Kidambi, V. K., et al. (2014). New structural evolution model for the North Kuwait carbonate fields and its implication for fracture characterization and modelling. In *IPTC 2014: International Petroleum Technology Conference*.

Salman, A. M. S. (1979). Geology of the Jal Az-Zor-Al-Liyah Area, Kuwait. Unpublished MSc Thesis, Kuwait University (p. 129).

Singh, P., Husain, R., Al-Kandary, A., Al-Fares, A. (2011). Basement configuration and its impact on Permian–Triassic Prospectivity in Kuwait, The Permo–Triassic Sequence of the Arabian Plate Abstracts of the EAGE's Third Arabian Plate Geology Workshop, Part II.

Singh, P. (2012). Role of Eocambrian Hormuz salt in evolution of structures in Kuwait-An integrated approach. In *European Association of Geoscientists & Engineers Conference*.

Stern, R. J., & Johnson, P. (2010). Continental lithosphere of the Arabian Plate: A geologic, petrologic, and geophysical synthesis. *Earth-Science Reviews, 101*, 29–67. https://doi.org/10.1016/j.earscirev.2010.01.002

Sharland, P. R., Archer, R., Casey, D. M., et al. (2001). Arabian plate sequence stratigraphy. GeoArabia, Spec Publ 2, Gulf PetroLink. Oriental Press, Manama, Bahrain (p. 371).

Tanoli, S. K., Husain, R., & Al-Khamiss, A. (2015). Geological handbook of Kuwait. Kuwait Oil Company.

Verma, N. K., Al-Medhadi, F., Franquet, J. A., Maddock, R. H., Dakshinamurthy, N., Al-Mayyas, E. M. (2007). Critically stressed fracture analysis in naturally fractured carbonate reservoir-A case study in West Kuwait. Society of Petroleum Engineers.

Von Winterfeld, C., Babajan, S., Amer, A., & Marsden, R. (2005). Geomechanics analyses of the crestal region of an Omani Gas Field. In *International Petroleum Technology Conference*.

Warsi, W. E. K. (1990). Gravity field of Kuwait and its relevance to major geological structures. *American Association of Petroleum Geologists Bulletin, 74*(10), 1610–1622.

Ziegler, M. A. (2001). Late Permian to Holocene paleofacies evolution of the Arabian Plate and its hydrocarbon occurrences.

Petroleum Geology of Kuwait

6

Mohammad Naqi, Ohood Alsalem, Suad Qabazard, and Fowzia Abdullah

Abstract

Kuwait has proven conventional oil reserves of about 100 billion barrels which makes it one of the major oil-producing countries worldwide. Most of this reserve is found in Cretaceous and Jurassic with minor quantities in the Paleogene sedimentary successions. Most hydrocarbon production comes from the siliciclastic Burgan Formation which is the most important reservoir in Kuwait. The Jurassic and Lower Cretaceous exhibit good quality source rocks that charged most of the hydrocarbon reservoirs in Kuwait and entered the oil window in Late Cretaceous to Eocene. Most of the hydrocarbon is trapped in very gentle four-way closure structures that are related to the deep-seated fault system of the Arabian Peninsula such as Khurais-Burgan Anticline. Hydrocarbon reservoirs in Kuwait are sealed and capped mainly by shale rocks and to a less extent by evaporites. In the last 15 years, Kuwait Oil Company (KOC) displayed interest in commercially exploiting unconventional hydrocarbon reserves and started laying significant emphasis on the exploration and development of unconventional resources. The aim of this work is to summarize the different petroleum systems of Kuwait including the Paleozoic, Mesozoic, and Cenozoic systems.

6.1 Introduction

After several preliminary surface geological reconnaissances commissioned by the British and Americans early last century, the first oil well spudded in Bahrah, north of Kuwait, in 1936. This well was in the vicinity of oil seepage and close to the center of a gentle anticline that was mapped north of Kuwait Bay. However, this well didn't show huge potential. Then, another location was chosen to drill a new well Burgan-1 (BG-001) at the Burgan oil field and it produced oil from the Cretaceous sedimentary succession. Since the first major oil discovery in 1938 in the Burgan field, Kuwait has become one of the main oil-producing countries worldwide with proven crude oil reserves of about 100 billion barrels (BP Statistical Review of World Energy 2021).

The State of Kuwait is located in the easternmost upper part of the Arabian Peninsula. The tectonic, subsidence histories, and the diagenetic processes that prevailed, on the eastern part of the Arabian Peninsula, are the main drivers for exceptional geological conditions that resulted in the prolific hydrocarbon reserves. These geological conditions are manifested by the widespread geographic distribution of rich, mature source rocks, high-quality reservoir rocks, and highly efficient seal rocks. Along with the formation of large structural traps during or subsequent to oil and gas generation. These geological conditions helped in the formation of multiple petroleum systems in Kuwait, contributing to its enormous oil and gas fields (Fig. 6.1) (Alsharhan and Nairn, 1997). Although multiple petroleum systems of different ages are discovered in Kuwait, only Jurassic, Cretaceous, and Tertiary petroleum systems where the majority of the production is coming from, while deep stratigraphic sections are underexplored yet.

The Paleozoic formations such as Unayzah and Khuff formations are very important hydrocarbon reservoirs in many parts of the Arabian Peninsula (e.g., Laboun, 1987a, b; McGillivray & Husseini, 1992; Al-Jallal, 1995; Alsharhan and Nairn, 1997; Wender et al., 1998; Konert et al., 2001). The Lower Silurian organic-rich "hot shale" of Qusaiba Formation is believed to be the source rock of these Permian reservoirs (Mahmoud et al., 1992; Abu-Ali et al., 1999; Parmijit et al., 2010). However in Kuwait, the Paleozoic section is very deep and remains an exploratory target for decades due to the hostile drilling environment. This deep section contains mainly continental clastic deposits with intercalated shallow shelf carbonates with a maximum

M. Naqi (✉) · O. Alsalem · S. Qabazard · F. Abdullah
Department of Earth and Environmental Sciences, Kuwait University, P.O. Box 5969 13060 Safat, Kuwait
e-mail: m.naqi@ku.edu.kw

© The Author(s) 2023
A. el-aziz K. Abd el-aal et al. (eds.), *The Geology of Kuwait*, Regional Geology Reviews,
https://doi.org/10.1007/978-3-031-16727-0_6

Fig. 6.1 Major and minor oil fields (onshore) map of Kuwait

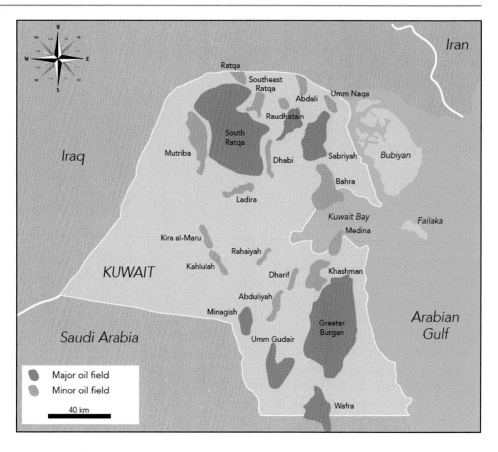

thickness of 2000 m (6562 ft) (Alsharhan et al., 2014). Although the primary depositional characteristics of the Paleozoic clastics such as coarse-grained sandstones, good sorting, and low matrix content suggest they had potentially good reservoir quality (e.g., Abdullah et al., 2017; Tanoli et al., 2008), the reservoir potential of the deep Paleozoic section is very poor (Strohmenger et al., 2003). Recent studies are focusing on the evaluation of the Paleozoic source rock, yet the reservoir quality of that deep section is still under exploration. Therefore, we are excluding the Paleozoic reservoirs and seal rocks from this work as more studies are needed to be conducted on that deep section.

The aim of this work is to summarize the different petroleum systems of Kuwait including the Paleozoic, Mesozoic, and Cenozoic (Fig. 6.2). Data is collected from several published literature.

Here, we will discuss the elements of the Kuwait petroleum system from older to younger.

6.2 Source Rocks

6.2.1 Paleozoic Source Rocks

6.2.1.1 Qusaiba Formation

The Lower Silurian organic-rich mudstone of Qusaiba Formation of the Qalibah Group is accepted to be the source rock for most Permian reservoirs in the Arabian Peninsula (Mahmoud et al., 1992; Abu-Ali et al., 1999; Parmijit et al., 2010), and it is also considered as one of the effective Paleozoic source rocks in Kuwait (Al-Khamiss et al., 2012; Alsharhan et al., 2014). The formation unconformably overlies the siliciclastic equivalent deposits of Saq Formation and unconformably overlies the thick siliciclastic sediments of the Tabuk Formation, with a hiatus of Ludlow age (Fig. 6.2).

The productive hot shale unit of the lower part of Qusaiba Formation was deposited within intra-shelf anoxic

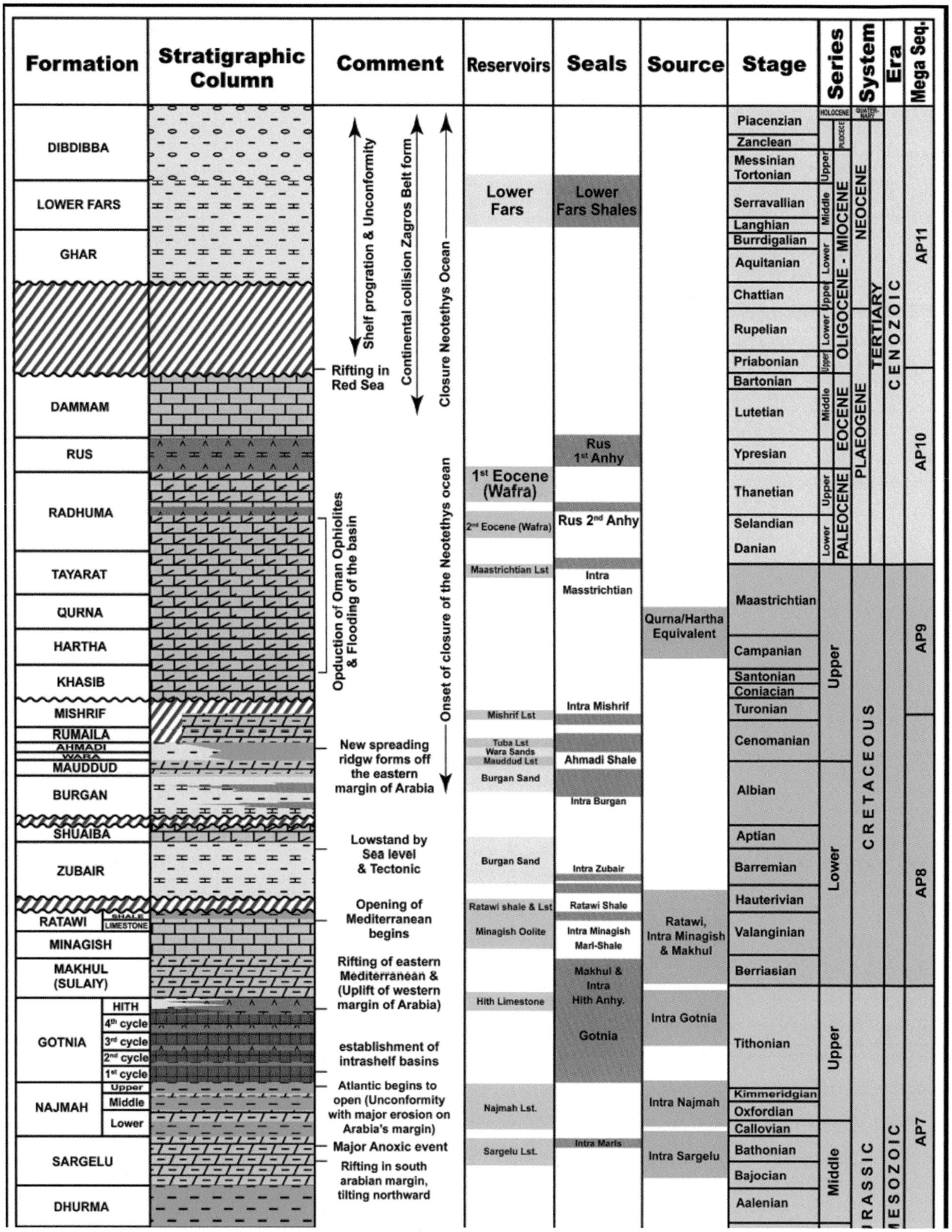

Formation	Stratigraphic Column	Comment	Reservoirs	Seals	Source	Stage	Series	System	Era	Mega Seq.
DIBDIBBA		Shelf progration & Unconformity / Continental collision Zagros Belt form / Closure Neotethys Ocean				Piacenzian	HOLOCENE / PLIOCENE Upper	QUATER-NARY / NEOCENE	CENOZOIC / TERTIARY	AP11
						Zanclean				
						Messinian				
LOWER FARS			Lower Fars	Lower Fars Shales		Tortonian				
						Serravallian	Middle			
						Langhian		OLIGOCENE - MIOCENE		
GHAR						Burrdigalian	Lower			
						Aquitanian				
						Chattian	Upper			
		Rifting in Red Sea				Rupelian	Lower			
						Priabonian	Upper			
DAMMAM						Bartonian		PLAEOGENE		AP10
						Lutetian	Middle	EOCENE		
RUS			1st Eocene (Wafra)	Rus 1st Anhy		Ypresian				
RADHUMA		Opduction of Oman Ophiolites & Flooding of the basin / Onset of closure of the Neotethys ocean				Thanetian	Upper	PALEOCENE		
			2nd Eocene (Wafra)	Rus 2nd Anhy		Selandian	Lower			
						Danian				
TAYARAT			Maastrichtian Lst	Intra Masstrichtian		Maastrichtian	Upper		CRETACEOUS	AP9
QURNA					Qurna/Hartha Equivalent				MESOZOIC	
HARTHA						Campanian				
KHASIB						Santonian				
MISHRIF			Mishrif Lst	Intra Mishrif		Coniacian				
RUMAILA						Turonian				
AHMADI / WARA		New spreading ridgw forms off the eastern margin of Arabia	Tuba Lst / Wara Sands / Mauddud Lst	Ahmadi Shale		Cenomanian				
MAUDDUD			Burgan Sand							
BURGAN				Intra Burgan		Albian	Lower			
SHUAIBA						Aptian				
ZUBAIR		Lowstand by Sea level & Tectonic	Burgan Sand	Intra Zubair		Barremian				AP8
						Hauterivian				
RATAWI SHALE / LIMESTONE		Opening of Mediterranean begins	Ratawi shale & Lst	Ratawi Shale	Ratawi, Intra Minagish & Makhul	Valanginian				
MINAGISH			Minagish Oolite	Intra Minagish Marl-Shale						
MAKHUL (SULAIY)		Rifting of eastern Mediterranean & (Uplift of western margin of Arabia)		Makhul & Intra		Berriasian				
HITH			Hith Limestone	Hith Anhy.			Upper			AP7
GOTNIA 4th cycle / 3rd cycle / 2nd cycle / 1st cycle		establishment of intrashelf basins		Gotnia	Intra Gotnia	Tithonian		JURASSIC		
NAJMAH Upper / Middle / Lower		Atlantic begins to open (Unconformity with major erosion on Arabia's margin)	Najmah Lst.	Intra Najmah		Kimmeridgian				
						Oxfordian				
SARGELU		Major Anoxic event / Rifting in south arabian margin, tilting northward	Sargelu Lst.	Intra Marls	Intra Sargelu	Callovian	Middle			
						Bathonian				
DHURMA						Bajocian				
						Aalenian				

Fig. 6.2 Generalized stratigraphic column of Kuwait along with petroleum system elements and tectonic events (Alsharhan et al., 2014)

M. Naqi et al.

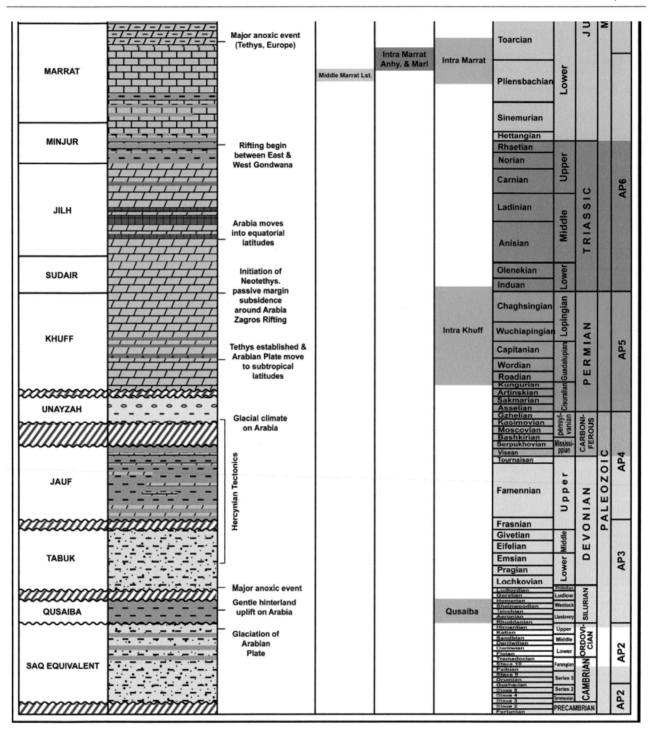

Fig. 6.2 (continued)

depressions during early Silurian in response to major transgression following deglaciation that was observed as maximum flooding surfaces over most of the Gondwana passive margin shelf (Sharland et al., 2001; Inan et al., 2016). Few studies have conducted the burial and thermal history modeling of the Qusaiba shales for testing different burial scenarios to estimate present-day gas potentials. These studies suggest that the formation has reached the oil-generation window around 170 Ma and the gas maturation window around 53 Ma. It is believed that it has expelled all of its potential, and it is overmatured in the present (Al-Khamiss et al., 2012; Alsharhan et al., 2014).

6.2.1.2 Unaizah Formation

The Carboniferous–lower Permian Unayzah Formation and its stratigraphically equivalent deposits are widespread over most of the Greater Arabian Basin (Tanoli et al., 2008). In northern Kuwait, the Unayzah Formation unconformably overlies a thick carbonate succession of the Hormuz Group and is unconformably overlain by a thick carbonate sequence of the Basal Khuff Clastics (Fig. 6.2) (Tanoli et al., 2008). The thickness of the formation varies in value, and it tends to increase westward reaching a maximum depth of ~6334 m (20,780 ft.) while decreasing gradually toward the NE and SE and reaching the minimum depth of 5197 m (17,049 ft.) (Fig. 6.3).

The formation is composed of fine sandstones, siltstones, and claystone cycles with thin beds of argillaceous limestone as shown in Fig. 6.4 (Abdullah et al., 2017). The upper part of the formation has porosity values ranging from 2 to 15%, with water saturation ranging from 65 to 100% reflecting the presence of hydrocarbon concentrations (Abdullah et al., 2017). Based on the neutron and density logs, relatively high gas saturation is recorded at the lower part of the formation in the Raudhatain field, north Kuwait (Fig. 6.5) (Abdullah et al., 2017).

Due to its great hydrocarbon-bearing potential, the formation has great exploration significance in the region (Wender et al., 1998). A recent organic geochemical study conducted on the Unayzah Formation in Kuwait shows a total organic carbon ranging from 0.8 to 3.7 wt. % of a mixture of algal-marine and terrestrial organic matter with

the highest TOC values being in the north fields (Abdullah et al., 2021). The thermal maturity data shows that there is a range from oil to overmature with elevated R_o values ranging from 1.85 to 2.57% (Abdullah et al., 2021).

6.2.1.3 Khuff Formation

The Permian–Early Triassic Khuff Formation in Kuwait occurs at greater depth in comparison to adjoining regions where it reaches more than 4572 m (15,000 ft.) deep (Strohmenger et al., 2003), and it is fully penetrated only in the wells located over the Kuwait Arch (Husain et al., 2011). The formation depth varies in values, and it tends to increase westward reaching a maximum depth of ~6271.6 m (20,576 ft.) while decreasing gradually toward the NE and SE and reaching the minimum depth of ~4727.8 m (15,511 ft.) as shown in Fig. 6.6. The formation unconformably overlies Unayzah Formation and underlines the Lower–Middle Triassic Sudair Formation (Fig. 6.2).

The formation is divided into the Lower Khuff of Middle Permian and the Upper Khuff of the Late Permian at its base to Early Triassic at its top, and they are separated by the Median Anhydrite (Husain et al., 2011). The formation is composed mainly of dolomite and anhydrite interbeds with few limestone and shale deposited in evaporitic low-energy inner to middle ramp setting (Abdullah et al., 2017; Husain et al., 2011) (Figs. 6.7 and 6.8). The formation has low porosity of 2% and low permeability that decreases with clay and anhydrite content and no remarkable hydrocarbon saturation (Abdullah et al., 2017).

A recent organic geochemical study conducted on the Khuff Formation in Kuwait shows a total organic carbon ranges from 0.6 to 2.8 wt.% of a mixture of algal-marine and terrestrial organic matter with the highest TOC values being in the north fields (Abdullah et al., 2021). The thermal maturity of Khuff Formations shows similar values of Unayzah Formation with a range from oil to overmature with elevated R_o values ranging from 1.85 to 2.57% (Abdullah et al., 2021).

6.2.2 Mesozoic Source Rocks

The petroleum system of Kuwait contains excellent source, reservoir, and seal rocks of the Mesozoic carbonate and clastic sequences. Detailed organic geochemical investigations of crude oil and source rocks and extensive reservoir characterization and basin modeling indicated that hydrocarbons trapped in the Lower Cretaceous sandstone and carbonate reservoirs were generated from organic matter-rich Upper Jurassic to Lower Cretaceous carbonate source rocks (Abdullah, 2001; Abdullah et al., 2005; Alsharhan et al., 2014).

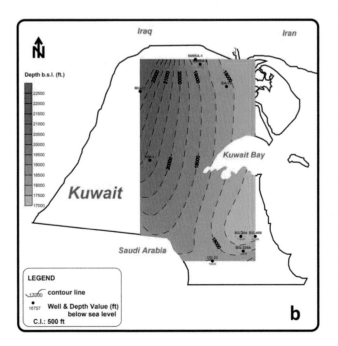

Fig. 6.3 Depth contour maps on the top of the Unayzah Formation (Abdullah et al., 2017)

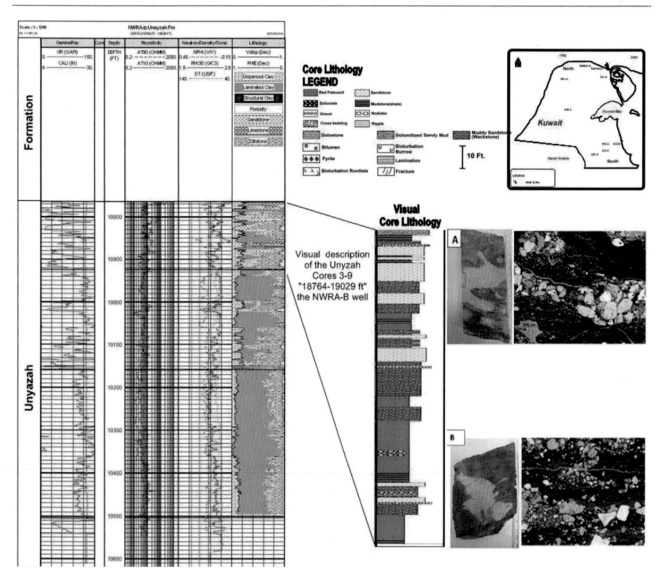

Fig. 6.4 Correlation of the well logging interpreted lithologies with the visual core lithology of the Unayzah Formation in Raudhatain field, north Kuwait (Abdullah et al., 2017)

The type and maturity level of organic matter in the Makhul (Sulaiy) and Minagish Formations in NW Kuwait indicate that they are the most probable Cretaceous source rocks and were probably responsible for generating a proportion of the oil which has accumulated in Kuwait oil fields. Moreover, the Middle Jurassic Najmah-Sargelu are significant contributors to Kuwait's Jurassic and Cretaceous hydrocarbon accumulations (Abdullah, 2001; Al-Bahar et al., 2019).

6.2.2.1 Triassic Source Rocks

The Triassic section in Kuwait comprises the Upper Khuff, Sudair, Jilh, and Minjur formations. The section is under-explored due its depth and poor-quality seismic data. However, a few exploration wells in west Kuwait produced small quantities of condensates and gas in two oil fields

(Mutriba and Kra Al-Maru) (Husain et al., 2009). The hydrocarbon shows were from a layer locally named Kra Al-Maru found between underlying Sudair and overlying Jilh formations. The lower clayey carbonate section of Kra Al-Maru is believed to be the source rock. It has TOC of 3–4% and Hydrogen Index of 300–500. It reached the oil window in Middle Cretaceous and the wet gas window during Late Cretaceous. Andriany and Al-Khamiss (2011) based on a crude oil assessment concluded that the source rock is a carbonate of Permo-Triassic age that deposited in an anoxic marine depositional environment.

6.2.2.2 Jurassic Source Rocks

The Jurassic sequence of Kuwait which constitutes one of the most important petroleum systems of the Arabian Gulf

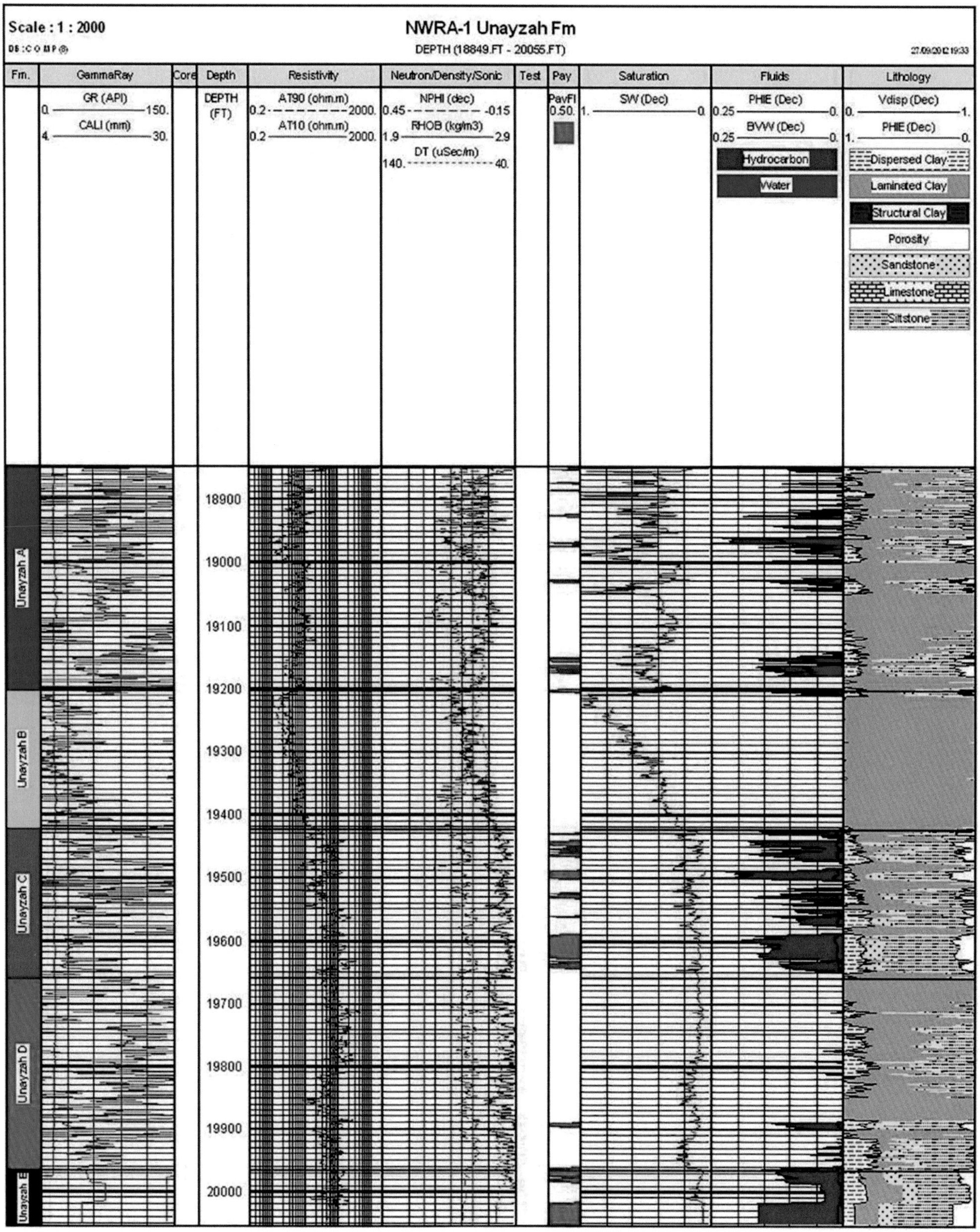

Fig. 6.5 Lithosaturation crossplot of the Unayzah Formation in Raudhatain field, north Kuwait (Abdullah et al., 2017)

Fig. 6.6 Depth contour maps on the top of the Khuff Formation (Abdullah et al., 2017)

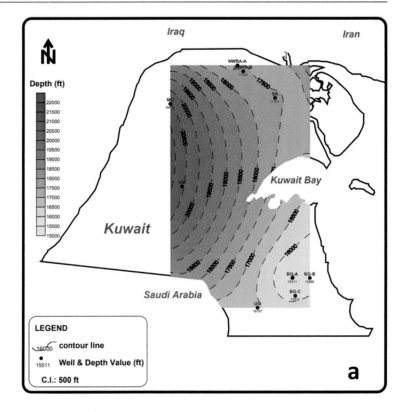

Region is composed of Lower Jurassic Marrat Formation, Middle Jurassic Dhruma, Sargelu, and Najmah Formations, and Upper Jurassic Gotnia and Hith Formations. In the south and southwest of Kuwait, the thickest Jurassic rocks are found and gradually thinned toward the north (Hawie et al., 2021). Organic geochemical analysis of the Jurassic source rocks Marrat, Dhruma, Sargelu, and Najmah Formations which are mainly limestones and calcareous shales shows the significant potential for oil generation in Kuwait (Abdullah, 2001).

Marrat Formation

The Marrat Formation was first defined in the Minagish oil field in the west of Kuwait at a depth between 13,515 ft (4120 m) and 15,350 ft (4679 m), with an average of 2000 ft (610 m). The Marrat Formation has been classified into the Lower Marrat (lime mudstone and dolomite, interbedded with anhydrite and shale), Middle Marrat (limestone), and Upper Marrat (limestone interbedded with shale and dolomite and calcareous shale) members—mainly—based on the presence of argillaceous sediments in the Lower and Upper members. The Marrat Formation conformably overlies the Triassic–Lower Jurassic Minjur Formation in Kuwait (Al Wazzan, 2021). Total organic carbon (TOC%) content in Marrat Formation from several fields in Kuwait show a range of 0.35–5.5%, indicating good source potential. In particular, the dark-colored, fine-grained sediments contain kerogen type II and type III with higher maturity levels as indicated

by the kerogen elemental analysis and vitrinite reflectance (0.8–0.9 R_o) of bitumen (Fig. 6.9) (Alsharhan et al., 2014).

Dhruma Formation

Dhruma Formation, overlying Marrat and underlying Sargelu Formation, is composed of calcareous shale interbedded with limestone. Analysis of source potential for samples collected from the Burgan and Minagish fields shows ranges of TOC from 0.30 to 2.50% wt., with higher concentrations in laminated shales. The type of kerogen is amorphous marine kerogen types II to III with the abundance of some oxidized and biodegraded terrestrial particles (spores and woody particles) which as a result lowered their level of maturity (Fig. 6.9). Organic matter maturity level was interpreted as the very first stage of oil generation as indicated by ranges of 0.4–0.5 Ro of vitrinite reflectance (Alsharhan et al., 2014).

Sargelu Formation

Sargelu Formation consisting of Lower and Upper carbonate members with overlying Najmah fractured limestones comprises important hydrocarbon sources for the producing reservoir in the south and east Kuwait oil fields (Al-Enezi, 2019). Organic geochemical results of samples taken from southern oil fields show fair to good source potential of Sargelu Formation with TOC ranges from 0.40 to 5.5% wt and averages about 2.0%. The elemental analysis

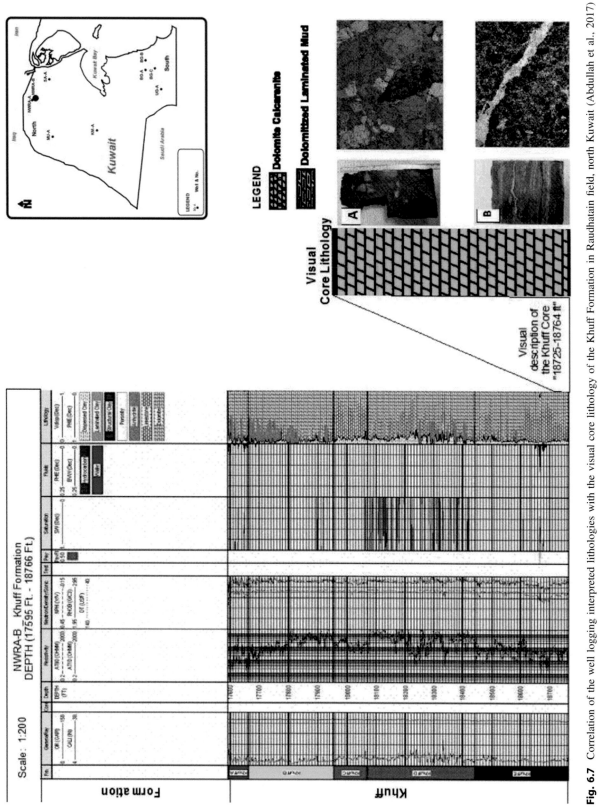

Fig. 6.7 Correlation of the well logging interpreted lithologies with the visual core lithology of the Khuff Formation in Raudhatain field, north Kuwait (Abdullah et al., 2017)

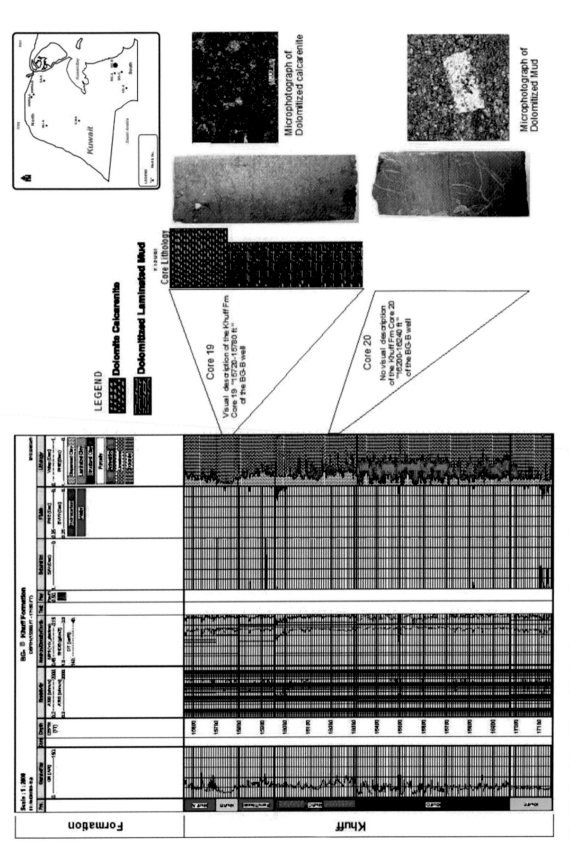

Fig. 6.8 Correlation of the well logging interpreted lithologies with the visual core lithology of the Khuff Formation in Burgan field, south Kuwait (Abdullah et al., 2017)

Fig. 6.9 Kerogen type and thermal maturation of the Jurassic source rocks in Kuwait fields (After Alsharhan et al., 2014)

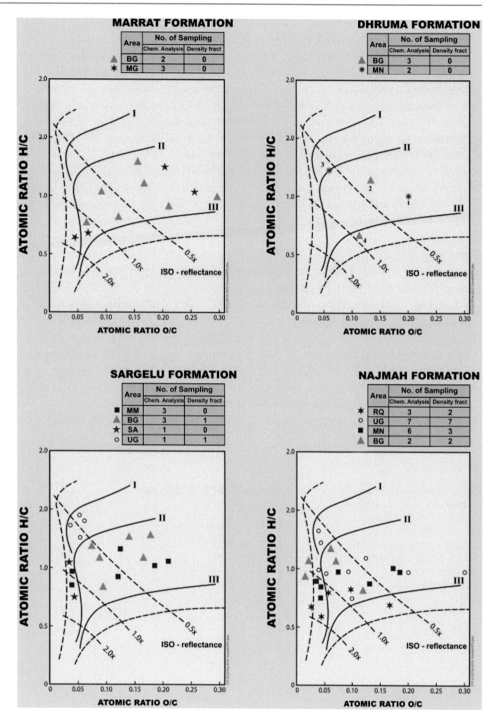

of kerogen indicates type II or types II–III amorphous marine organic matter as well as terrestrial oxidized woody materials that lowered their source potential (Fig. 6.9). Sargelu carbonates are at the onset of oil generation in both Minagish and Umm Ghudair fields as indicated by the vitrinite reflectance (0.6 Ro) and the kerogen dark color. The lowest kerogen maturity is shown by samples from the Burgan field because Sargelu Formation is found at the shallowest depths of 3215–3165 m (10,548–

10,384 ft), concluding the early maturity stage of organic matter (Alsharhan et al., 2014).

Najmah Formation

The contribution of Jurassic Najmah-Sargelu Formations to the petroleum system in Kuwait and NE of the Arabian Gulf region has been significantly discussed by many authors (Aqrawi & Badics, 2015; Al-Murakhi, 2019; Grader et al., 2019). Najmah Formation, composed of organic-rich bituminous shales and carbonate, is considered as the prolific major source rock for the main Jurassic Najmah, Sargelu, and Middle Marrat reservoirs. The excellent source beds of Najmah with TOC of 7% on average have generated light oil and condensate from North Kuwait and medium oil from west and south Kuwait (Al-Enezi et al., 2018). Organic richness is concentrated in the laminated Najmah carbonates at the Ratqa field in north Kuwait where the level of maturity is the highest and kerogen is mainly of type II amorphous sapropelic (Fig. 6.9). In contrast, the level of maturity and quality of kerogen decrease in the southern oil field (Minagish and Umm Ghudair) as a result of biodegradation (Alsharhan et al., 2014).

Cretaceous Source Rocks

The source rock characteristics of the Cretaceous formations in Kuwait have been investigated by many authors (Abdullah and Kinghorn, 1996; Abdullah et al., 1997, 2005; Alsharhan et al., 2014; Al Bahar et al., 2019) who concluded their significance as part of the petroleum system of Kuwait. The type of kerogen (type III) and its maturity state indicate that the Cretaceous Formations such as Ratawi, Mauddud, and Ahmadi may have generated oil and gas at a very low maturity level. Oil may also originate from deeper source rocks or may have migrated laterally from other areas in the Arabian Gulf Basin.

Makhul (Sulaiy) Formation

Lower Cretaceous Makhul or Sulaiy Formation is a potential source rock in the Mesopotamian Foredeep (Iraq and Kuwait) and in the Zagros Fold Belt Basins (Iran). Source rock potential of Sulaiy Formation in Kuwait and central and southern Iraq with its basal Garau equivalent in western Iran is well documented (Alsharhan et al., 2014; Aqrawi & Badics, 2015). The formation consists of basinal bituminous shale intercalated with dark gray argillaceous limestone, in which TOC values range from 0.40 to 2.70%. Results of H/C and O/C kerogen elemental analysis of samples from the Raudhatain field show less maturation at shallower depths with the presence of type II kerogen and a higher level of maturation as depth increases where type I amorphous, dark-colored organic matter is present and corresponds to the onset of the oil-generation zone (Fig. 6.10).

Minagish Formation

The Makhul (Sulaiy) Formation is overlain by Minagish Formation which consists of three members: Lower Minagish Member Peloidal–bioclastic, occasionally dolomitic limestone, Middle Minagish Member of medium- to very coarse-grained oolitic-pelletal grainstone and Upper Minagish Member of massive, oolitic, fossiliferous limestone. The oolitic units show higher TOC ranges (from 1.30 to 1.95% wt.) than the Lower member that has a TOC range

Fig. 6.10 Kerogen elemental analysis plot of the Lower Cretaceous Makhul (Sulaiy) and Minagish Formations showing the type organic matter and level of maturity of the source rocks in Kuwait fields (after Alsharhan et al., 2014)

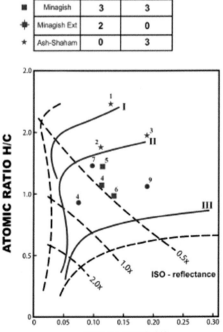
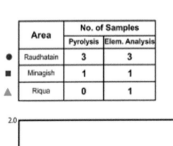

of 0.3–0.55% wt, indicating the more organic richness of the oolitic parts of the Minagish Formation. The quality of organic matter is better in the oolitic units than in the biodegraded non-oolitic member as indicated by the presence of very low dense, algal amorphous kerogen of marine origin (Fig. 6.10). The level of maturation of the ooilitc Minagish Members as indicated by the pyrolysis results for the hydrogen index (HI) and Tmax with an average of 550 mg HC/g to 355 mg HC/g and 435 °C, respectively, suggests the onset of the oil-generation stage (Abdullah et al., 1997) (Figs. 6.11 and 6.12).

Ratawi Formation

The Ratawi Formation consists of two members the Lower Ratawi Member, consisting of limestone thinly bedded with shale and the Upper Ratawi shale Member which is mainly calcareous shale with intercalated sandstone and siltstone. Organic geochemical analyses at the Raudhatain and Minagish fields show poor to fair TOC with an average of 0.55% wt. in the Lower Ratawi member and a range from 0.45 to 3.25% wt in the Upper Ratawi shale.

The type of organic matter is type II kerogen of marine origin as indicated by the abundance of foraminifera, dinoflagellates, and acritarchs. However, biodegradation and the presence of spores, cuticle and woody tissues, terrestrial organic components lowered the quality of this organic matter to type III kerogen. The overall level of maturity is immature or just entering the oil-generation stage as indicated by Tmax (430–435 °C), the Production Index (0.05–0.15), Thermal Alteration Index (TAI) of 2+ and 3–, and a vitrinite reflectance of 0.70% R_o results (Alsharhan et al., 2014).

Mauddud Formation

The Mauddud Formation is mainly composed of limestone interbedded with marl and fine, greenish-brown glauconitic sandstones, and contains abundant microfossils and some pyrite.

The Mauddud Formation which is considered as one of the major Cretaceous oil reservoirs varies in thickness between zero in the south and about 100 m (328 ft) in the north of Kuwait.

The average Total Organic Carbon (TOC) in the carbonate facies is 2.5 wt.% and the highest values (8.0 wt.%) are in the northern fields. The clastic intervals in the northern fields show higher total organic matter (1.3 wt.%) relative to the southern fields (0.6 wt.%). The total Production Index is higher in the carbonate (0.6) than in the clastic section (0.3). The kerogen type in the formation is immature types II–III and III, suggesting that oil accumulated in the reservoir might be largely related to migrated oil from mature deeply buried source rocks such as the Early Cretaceous Sulaiy Formation and the Upper Jurassic Najmah Formation (Abdullah et al., 2005) (Fig. 6.13).

6.3 Reservoir Rocks

6.3.1 Mesozoic Reservoir Rocks

There are many stratigraphic intervals of good-quality reservoirs in Kuwait mainly in the Mesozoic Era. The most famous reservoir is the siliciclastic reservoir of Burgan Formation which contribute to most of the hydrocarbon production in the state of Kuwait along with other several Cretaceous reservoirs which accounts for about 80% of hydrocarbon production in Kuwait (Alsharhan et al., 2014). The approximated production contribution of the Jurassic reservoirs (Marrat, Sargelu, and Najmah Formations) accounts for about 20% of the recoverable oil and gas reserves, while only 1% of recoverable oil and gas was found in the Miocene age (Alsharhan et al., 2014).

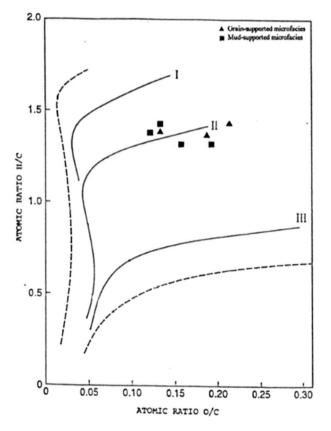

Fig. 6.11 Type of kerogen of the oolitic and non-oolitic samples in the Lower Cretaceous Minagish Formation (after Qabazard et al., 2000)

Fig. 6.12 Thin
section Photomicrographs
showing the oolitic
Middle Minagish limestones
(**a** and **b**) and the organic matter
isolated from the samples (**c** and
d) from the Minagish field.
a Typical oolitic grainstone (ppl,
5 mm), **b** A close view of
micritized ooids with a nucleus of
foraminifera (xpl, 1 mm),
c Amorphous marine organic
matter with sapropelic particles,
macerated from
the oolitic sample (ppl, 1 mm),
and **d** Marine organic matter
indicated by the foraminiferal test
lining (ppl, 1 mm) (after
Qabazard et al., 2000)

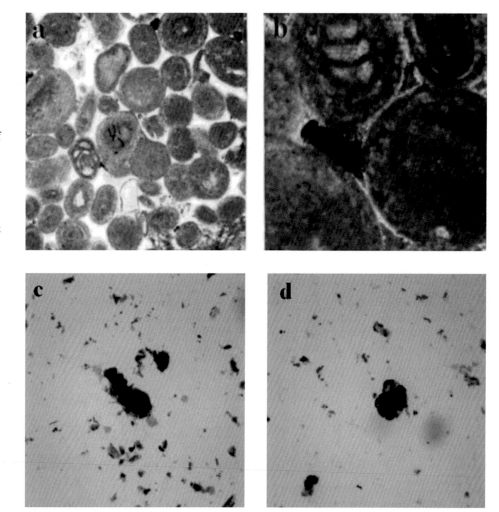

6.3.1.1 Triassic Reservoirs

Reservoirs in the Triassic section are very limited. In western Kuwait, the Kra Al-Maru unit which is underlain by Sudair and overlain by Jilh formation presents a potential reservoir with low conventional reservoir quality. This unit comprises cyclical packages of dolomudstones, laminated to bedded anhydrites and dolomitic shales (Husain, et al., 2009). Most matrix porosity is reduced by anhydrite, and matrix permeability values are less than 0.01 mD. The more brittle dolomudstone section is characterized by a fracture network and its related porosity range up to 3.3 mD.

6.3.1.2 Jurassic Reservoirs

Marrat Formation

The Lower Jurassic Marrat Formation is one of the most important carbonate reservoirs in north, west, and south Kuwait fields. Production from Marrat in the Greater Burgan field started in 1984 and has free gas production from the north fields (Raudhatain, Sabiriya, and Bahra) since 2008. The reservoir is classified as tight heterogenous with

natural fractures and an average permeability of about 17 md (Snasiri et al., 2015). Marrat Formation has a thickness of about 2400 ft and is subdivided into three members, Lower Marrat, Middle Marrat, and Upper Marrat. The main oil-producing unit is the Middle Marrat member which consists of dolomitized oolitic grainstone to packstone facies, representing high-energy inner-mid ramp shoals. The Lower Member of Marrat is an inner ramp dolostone and micritic limestone facies intercalated with anhydrite and the Upper Marrat Member is a transgressive complex of lime mudstones/wackestone deposited in a mid-outer ramp setting (Hawie et al., 2021). Typically, the Marrat reservoir quality was verified in the high-energy shoal packages and was enhanced by dolomitization (Chakrabarti et al., 2011).

6.3.1.3 Sargelu Formation

The significance of Najmah Sargelu carbonate self-sourced reservoirs has been widely investigated (Siddiqui et al., 2008; Nath et al., 2012). The Najmah Sargelu tight carbonates are considered as naturally fractured, organic-rich prolific reservoirs in west Kuwait (Al-Failakawi et al., 2018;

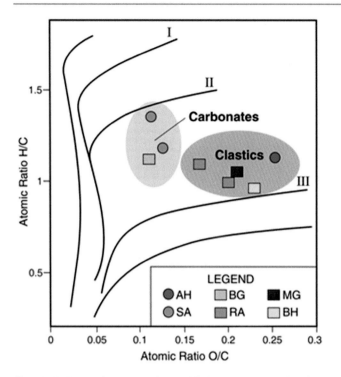

Fig. 6.13 Type of Kerogen in Mauddud Formation samples from Ahmadi (AH), Burgan (BG), Magwa (MG) southern fields and Sabiriya (SA), Raudhatain (RA), and Bahra (BH) northern fields in Kuwait (after Abdullah et al., 2005)

Al-Murakhi, 2019) as well as the north fields (Siddiqui et al., 2008; Al-Eidan et al., 2010). The Sargelu Formation overlies Dhruma calcareous shale formation which is an excellent cap rock for the Marrat Formation. The Sargelu Formation can be divided into a lower unit of mudstone and wackestone facies and an upper unit varying from wackestone to packstone facies with rare grainstone facies (Fig. 6.14). The shallowing upward sequence with the dominance of algal, skeletal, and peloidal packstones represents a typical shallow marine setting. The top of Sargelu Formation and the overlying Najmah are found to be similar to some extent in their composition of dark laminated mudstone to pelagic-pelecypod packstones, representing deeper facies accumulated in stagnant and restricted marine (Alsharhan et al., 2014).

Najmah Formation

The Upper Jurassic Najmah Formation is a black, oil-stained, naturally fractured, highly overpressured argillaceous mudstone and calcareous shale which was subdivided into two members: Upper Najmah limestone member and Lower organic-rich shale member (Rabie et al., 2014). The Najmah organic-rich shale is overlain by impermeable Salt and Anhydrite of the Gotnia and Hith Formations and underlain by the Middle Jurassic Sargelu Limestone and

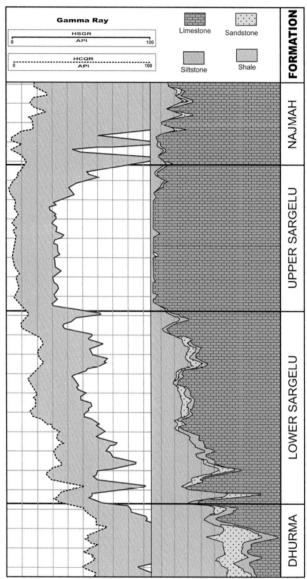

Fig. 6.14 Composite log of Dhruma, Sargelu, and Najmah Formations (after Alsharhan et al., 2014)

Dhruma Shale formations (Fig. 6.15). The Najmah Formation is a complex heterogenous tight carbonate reservoir in which porosity and permeability are essentially enhanced by high-density fracture and vuggy porosity (Nath et al., 2012, Al-Failakawi et al., 2018). The porosity percentage and distribution in the Najmah Formation are controlled by the lithofacies types and show good relation to permeability. The average porosity is about 2 to 10% mainly of fracture type and the permeability ranges 0.01–10 mD. Consequently, Najmah Formation is considered as a good commercial unconventional resource in Kuwait in spite of its low matrix porosity and ultra-low permeability (Nath et al., 2012).

Fig. 6.15 Composite Log of Najmah Formation (Rabie et al., 2014)

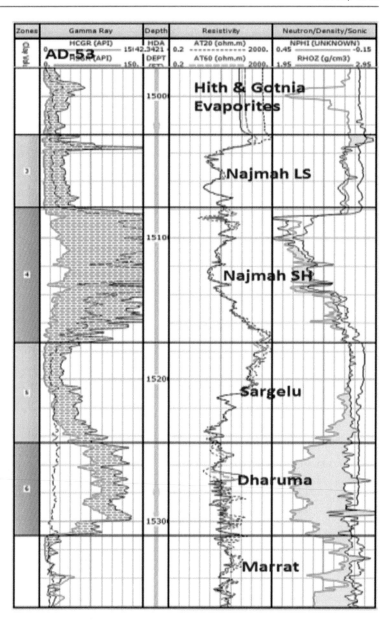

6.3.1.4 Cretaceous Reservoirs

Minagish Formation

The Minagish Formation is divided into 3 members where the top and bottom are relatively tight lime mudstone and wackestone separated by the middle bioclastic Oolitic grainstone member—Minagish Oolite (Nath et al., 2014). The latter is the primary reservoir in the Minagish Field and a significant contributor to Umm Gudair, South Um Gudair, and Greater Burgan field. The formation overlays the non-reservoir Makhul (Sulaiy) Formation and is underlying the Ratawi Limestone.

The lower Minagish Member is composed of Peloidal–bioclastic, occasionally dolomitic limestone with 22% porosity and nearly 500 md permeability. The middle

Minagish Member is composed of medium to very coarse - grained oolitic grainstone with occasional pellets, bioclastic leaching, and common development of vugs. The uppermost part of this member is mainly a well cemented, peloidal limestone and calcareous and argillaceous limestone.

The middle Minagish resembles a good quality reservoir, for example in Burgan field, porosity ranges 17–30% and permeability typically ranges 300–1000mD with average water saturation (S_w) of 14% (Nath et al., 2014). In the North of Kuwait, the Oolitic member is absent and commonly consists of fractured limestone with some fractures filled with calcite (Alsharhan et al., 2014). The upper Minagish Member is composed of massive, oolitic, fossiliferous limestone cemented with varying amounts of microcrystalline and sparry calcite (Alsharhan & Nairn, 1997).

A zone of heavy and dark, tarry oil (known as tar mat) is present near the oil–water contact in the Minagish field with thickness variation from 10 to 30 m. The continuity of this tar mat throughout the reservoir forms a seal isolating the reservoir from the underlying aquifer. Al-Ajmi et al. (2001) investigated the occurrence and the genesis of this tar mat and suggested that the formation of this tar mat is due to de-asphalting rather than gravity segregation.

Ratawi Formation

Ratawi Formation is considered as a minor reservoir in Kuwait while it is one of the major reservoirs in the Arabian Peninsula region. It is subdivided lithologically into Lower Ratawi and Upper Ratawi Members. The Lower Ratawi Limestone (Ratawi Oolite or Ratawi Limestone Member) consists mainly of limestone with thin shale laminations. It has a uniform thickness of about 90 m (295 ft) throughout Kuwait (Alsharhan et al., 2014). There are packstone and grainstone facies present in the Limestone Member with porosities up to 25% in southern and offshore Kuwait with good oil quality (28–46° API) (Al-Tendail et al., 2012).

The upper Ratawi member is also known as the Ratawi Shale member which is mainly composed of marine shale/claystone, interbedded by siltstone and fine-grained sandstone, and limestone streaks (Arasu et al., 2012). The origin of the sandstone is attributed to the uplift and erosion of the Arabian Shield (Al-Fares et al., 1998; Sharland et al., 2001). Thick coarser sandstones are encountered in the south of Kuwait, however, they are discontinuous in the north.

Major oil discoveries were found in the lower part of the Ratawi Limestone Member at Wafra and Umm Gudair fields since the 1950s. However, in the last 10 years, Kuwait Oil Company has shifted its exploration efforts to the Ratawi Shale Member where it tested oil in the Umm Gudair Field in the south, Rugae in the southwest, and Sabriyah, Raudhatain, and Abdali fields in north Kuwait (Arasu et al., 2012. And Al-Tendail et al., 2012). In north Kuwait, oil was produced from the sands and limestone streaks within Ratawi Shale Member where the hydrocarbon is entrapped stratigraphically.

A study was carried out in 2007 to evaluate the hydrocarbon potential of the Ratawi Limestone Member and consequently identified multiple factors which negatively affected the hydrocarbon potential of the Ratawi Limestone Member (Al-Tendail et al., 2012). Its diagenetic history records up to 11 periods of porosity reduction by calcite and other cements. Accordingly, it suggested that oil migration might have occurred during late diagenesis following the precipitation of the cement.

Burgan Formation

Burgan Formation (lower to middle Albian) is the most prolific oil-bearing sandstone reservoir throughout the Greater Burgan field (Burgan, Magwa, and Ahmadi fields) in the southeast and Raudhatain and Sabiriyah fields in the north of Kuwait (Fig. 6.1). It was named after Burgan hills in south Kuwait. The thickness of Burgan Formation varies from north to south, where in the north (Raudhatain and Sabiriyah fields) it is about 900 ft (275 m) and approximately 1250 ft (380 m) in the south (Greater Burgan field) (Bou-Rabee, 1996).

The formation is divided into Upper, Middle, and Lower Burgan. However, Kuwait Oil Company divides the formation into two members: Fourth and Third Sand. Generally, the formation is more sand prone to the south and southwest and increases in shale percentage toward the east and northeast of Kuwait.

It is composed of medium to coarse grained sandstone, well-sorted, rounded, grading upward into alternating fine-grained sandstone and siltstone (Al-Eidan et al., 2001).

Burgan Formation reservoir quality is highly related to the interpreted depositional environments (Strohmenger et al., 2006). The main reservoirs in Burgan Formation are fluvial and tidal deposits (Fig. 6.16). The best reservoir quality is associated with fluvial-dominated sandstones, where it has an average porosity of 25% and permeability of 1600 mD. The tidal-dominated sandstone displays moderately reservoir quality with average porosity and permeability values of 23% and 270 mD, respectively. The poorest reservoir quality presents in the marginal-marine deposits where the average porosity is about 19% and significantly lower permeability of about 10 mD.

Mauddud Formation

Mauddud Formation forms an important reservoir that extends over the Arabian Plate. It is characterized by rudist buildups with an improvement of reservoir quality by meteoric diagenesis. It was deposited in a broad ramp system ranging from the distal zone of clastic deltaic settings to the marine carbonate system to the outer ramp/basinal environment (Boix et al., 2014).

It is one of the main producing reservoirs in northern Kuwait with a minimal contribution in the south and southwest. Mauddud Formation thickness varies from north to south of Kuwait, where in the north of Kuwait, at Abdali, Raudhatain, and Sabriyah fields, it is approximately 140 (459 ft) and a few feet at Greater Burgan field (Bou-Rabee, 1996). Mauddud Formation can be divided into a lower part which is a mixed clastic and carbonate system, and an upper part which is carbonate-dominated (Fig. 6.17). Porosity

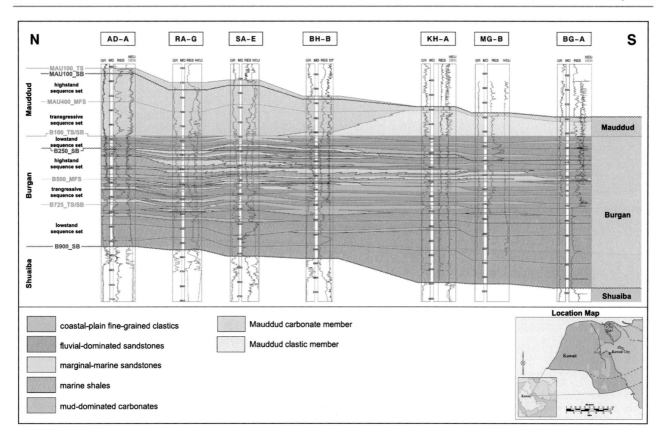

Fig. 6.16 Regional chronostratigraphic correlation from north to south of Kuwait showing facies distribution of Burgan Formation (after Strohmenger et al., 2006)

ranges from 15 to 20%, oil saturation (S_w) from 50 to 90% (Behbehani & Hollis, 2015), and permeability from 1.04 to 22.4 mD (Al-Awadi et al., 2017).

Behbehani and Hollis (2015) investigated the petrophysical properties of Mauddud Formation in Bahrah and Sabriyah fields, north of Kuwait. They found that reservoir quality of Mauddud Formation degrades from Sabriyah to Bahrah oil field. They alluded to this lower reservoir quality in Bahrah due to many factors rather than only facies variability. The reservoir performance (porosity and permeability) is partly affected by depositional facies, its diagenetic processes, structural deformation (tectonism), and timing of oil charge (Behbehani & Hollis, 2015).

Wara Formation

The upper Cretaceous Wara Formation is one of the main reservoirs in Greater Burgan field where oil production represents approximately 10% of the Burgan oil field communitive production (Al-Enzi and Hsie, 1999). It is also a major clastic reservoir in the south of Kuwait at the Wafra oil field. It is separated vertically from the Burgan reservoirs by extensive carbonate, and shale interval of Mauddud Formation and Wara Shales unit, respectively.

In south Kuwait, based on log analysis and well correlation, Wara is divided into three stratigraphic units: Lower, Middle, and Upper (Masarik et al., 2012). The middle Wara represents the best quality reservoir unit which is comprised of interbedded fine to medium-grained sandstone, siltstone, and shale with average porosity of 26%, net permeabilities ranging from 500 to 3000 mD, and average water saturation (S_w) from 18 to 20%, while the Upper and Lower units compose of shales, coals, shaley sandstone, and glauconitic sandstones. Wara reservoir forms a sequence of channel sands (fluvial/tidal) forming stacked channel sandstone, mouth bars, and tidal bars (Fig. 6.18).

The thickness of Wara Formation in Greater Burgan field ranges from 40 to 50 m of which up to 60% of the total thickness comprises sandstones (Alsharhan et al., 2014). In the cleaner sand sections, porosities can reach up to 30% with an average of 24% and permeabilities that can reach up to thousands of millidarcies.

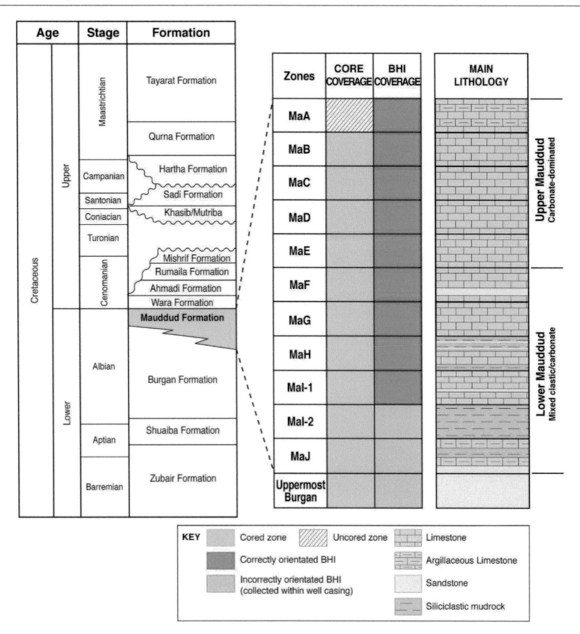

Fig. 6.17 The Late Albian Mauddud Formation reservoir zones of north Kuwait (after Al-Awadi et al., 2017)

6.3.2 Cenozoic Reservoir Rocks

Cenozoic sedimentary succession in Kuwait is divided into upper and lower units, Kuwait and Hasa groups, respectively. Kuwait group comprises a major reservoir, the Miocene Lower Fars sandstone formation, and Hasa Group comprises Paleogene–Eocene Umm Radhuma carbonate Formation.

6.3.2.1 Umm Er Radhuma Formation

The first discovery of heavy oil took place in 1954 in Umm Er Radhuma Formation which is informally referred to as the First, Second, and Third Eocene reservoirs (Fig. 6.19). This

formation range in thickness from 424 to 500 m (Alsharhan and Nairn, 1997). The Wafra oil field, in the Divided Zone between Kuwait and Saudi Arabia, is the only field that produces from this reservoir. The First and Second Eocene reservoirs are the main reservoirs in the Wafra oil field, and both are capped by Anhydrite units (Fig. 6.19), Rus Formation (First Anhydrite) and Second Anhydrite of Paleocene age (Saller et al., 2014).

The First Eocene is composed of dolomite with minor, but locally abundant, gypsum and anhydrite. Diagenesis has played a major role in modifying depositional pore types. The First Eocene reservoir is characterized by high porosity values that range 30–50% which might be related to early

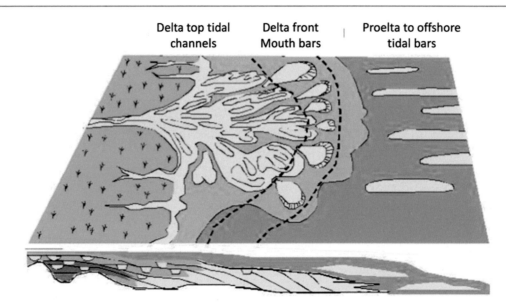

Fig. 6.18 Depositional model for Wara reservoir in Wafra oil field, south of Kuwait (after Masarik et al., 2012)

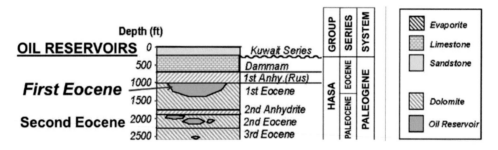

Fig. 6.19 Stratigraphic column for the Wafra Field showing reservoir at the Wafra Field. Contour interval is 20 ft (6 m) location of oil reservoirs (modified after Saller et al., 2014)

dolomitization, shallow burial, and early heavy oil emplacement, while permeability is commonly 100-2000md (Saller et al., 2014). The oil gravity is 19° API, with a 4.43% sulfur content (Nelson, 1968).

The Second Eocene reservoir is composed of dolomites, anhydritic dolostone, and anhydrite and generally shows high porosity values around 30% on density-neutron porosity logs and 20–45% from core plugs measurements. Leaching and dissolution processes have enhanced the predominantly intracrystalline pores forming sometimes millimeter to centimeter size vugs. Vuggy porosity is very common in the lower Second Eocene reservoir and may locally contribute to the high porosity values. Permeability is good to excellent and usually in the 10–100 s of millidarcies range (Wani and Al-Kabli, 2005). The oil gravity is 20° API, with a 4.43% sulfur content (Nelson, 1968).

6.3.2.2 Lower Fars Formation

The Miocene sandstone reservoir, Lower Fars Formation, is a long-established productive reservoir of heavy oil in

northern Kuwait. The formation is under development in the Ratqa oil field, and it shows great potential in other oil fields north of Kuwait such as Raudhatain, Sabiriyah, Bahrah, and Mutriba oil fields (Fig. 6.1). The formation is divided into 2 sandstone layers F1 and F2 separated by a thin layer of shaly sandstone (Fig. 6.20). The two reservoir units F1 and F2 refer to the first and second Lower Fars reservoir sand, respectively (Abdul Razak et al., 2018). The upper reservoir unit F1 is found at shallow depths between 400 and 800ft. The F1 layer is composed dominantly of fine to coarse, poorly sorted, thinly laminated fluvial channel sandstone interbedded with silty shale and very fine to medium sandy bioturbated shale and minor argillaceous and bioturbated sandstone. While the F2 layer consists of fine to medium, subtly fining upward, moderately well-sorted sandstone (Abdul Razak, et al., 2018).

The Lower Fars reservoir has a uniform gross thickness in the Ratqa oil field of about 150ft with a total net thickness that ranges 10-100ft. The average recorded API of F1 units ranges 12–18° API, while the F2 unit has a lower average

Fig. 6.20 Type log of Lower Fars reservoir from north of Kuwait showing the subdivision of Lower Fars into upper unit F1 and lower unit F2 separated by thin shale layer and capped by capping shale (After, Abdul Razak et al., 2018)

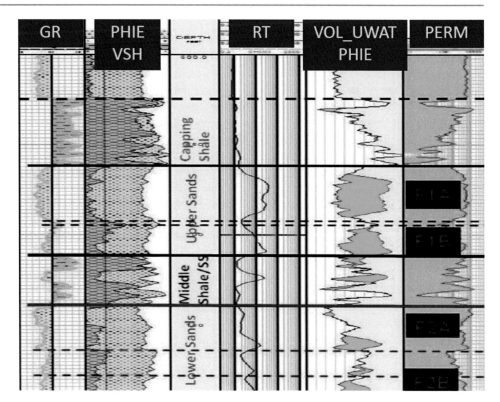

API of 11–15° API. The average computed porosity ranges from 25 to 35% which is in close agreement with core measured porosity. The reservoir is characterized by average permeability that ranges 1–4 Darcy (Al-Ali et al., 2017).

6.4 Seal Rocks

Most of the hydrocarbon in Kuwait oil fields are sealed and capped mainly by shale rocks (e.g., Burgan and Minagish reservoirs) and a few reservoirs by evaporites (e.g., Jurassic Najmah/Sargelu and Eocene Umm Er Radhuma reservoirs). The hydrocarbon potential of the Paleozoic section, in Kuwait, is limited so it will not be discussed. Here, we discuss the major seal rocks of the main hydrocarbon reservoirs of Mesozoic and Cenozoic ages.

6.4.1 Mesozoic Seal Rocks

6.4.1.1 Kra Al-Maru and Jilh Formations

The small quantity of hydrocarbon that is found in the Triassic section in Kuwait (Kra Al-Maru reservoir) is sealed locally by the intraformational halite and anhydrite, while the salt unit in middle Jilh acts as a regional seal for the Triassic system (Husain et al., 2009).

6.4.1.2 Dhruma Formation

This formation seals the oil and gas accumulations of the underlying carbonate reservoirs of the Upper Marrat play. It is composed of argillaceous limestone and fossiliferous shale (Husain et al., 2018a).

6.4.1.3 Gotnia and Hith Formations

Both Gotnia and Hith Formations mark the end of the Jurassic section in Kuwait. The Gotnia Formation (Kimmeridgian age) deposited in a hypersaline, very shallow and semi-confined lagoonal basin that extended from southern Iraq to the Rimthan structure, toward the southwest border between Kuwait and Saudi Arabia. It consists of 4 cycles of halite and anhydrite with occasional occurrences of limestone streaks encapsulated between halite and/or anhydrite (Ali, 1995; Yousif and Nouman 1997). It acts as a regional seal for Upper Jurassic Najmah and Sargelu reservoirs in Kuwait.

In the Arabian Peninsula region, Hith Formation caps the largest conventional carbonate reservoir worldwide, the Arab Formation. Hith Formation (Tithonian age) was deposited in an evaporitic pool and composes of massive anhydrite interbedded and intermixed with argillaceous limestone with minor shales. In the west of Kuwait, Hith Formation reaches a thickness more than 1100 ft and thins to approximately 200 ft toward northeast of Kuwait.

Al-Hajeri (2017) have discussed the integrity of Gotnia and Hith Formations, in Kuwait, based on formation water geochemistry study and suggested that the anomalously abnormal Na-rich formation water in the Cretaceous rocks might be attributed to the vertical migration of saline water from Gotnia and/or pre-Gotnia formations. This might indicate that both Gotnia and Hith Formations are breached by reactivated faults by the Late Cretaceous tectonism.

6.4.1.4 Lower Makhul Shale

Makhul Formation overlies the Jurassic Hith formation. The lower part of Makhul, which consists of heavily bioturbated organic-rich lime mudstone and shale, provides an effective regional top seal for the carbonate reservoirs in the Hith play.

6.4.1.5 Ratawi Formation

Ratawi Formation is divided into 2 members, Lower Ratawi limestone and Upper Ratawi shale. The upper shale member seals the sandstone and limestone streak reservoirs within Ratawi Shale member. It also seals off the major hydrocarbon accumulation in the lower Ratawi limestone and the early Cretaceous Minagish limestone reservoir (Alsharhan and Nairn, 1997).

6.4.1.6 Zubair Formation

This formation is interpreted as a complex section of deltaic origin evolving to estuarine settings with tidal influence. It is dominated by interbedded sand and shale units. The shale act as a good seal for hydrocarbon in the sand reservoirs.

6.4.1.7 Burgan Formation

Despite being the largest oil reservoirs in Kuwait, the interbedded shale within Burgan Formation efficiently plays as a seal for the two major sandstone reservoirs (Third and Forth sand members) (Alsharhan et al., 2014).

6.4.1.8 Wara Formation

The Lower Wara unit, defined as the "Wara Shales", consists of extensive shale interval deposited in marine settings that provide good sealing capacities. This shaley unit provides a regional seal for the underlying Mauddud Formation (Alsharhan et al., 2014).

6.4.1.9 Ahmadi Formation

Ahmadi Formation is the regional seal for the giant hydrocarbon accumulation in Burgan, Mauddud, and Wara reservoirs in Kuwait. It comprises brown shale and limestone (Alsharhan et al., 2014). Its thickness varies across Kuwait, where in the north it reaches 420 feet and about 165 feet in the south. It is divided into Upper member dominated by shale and Lower member which is carbonate-dominated.

6.4.2 Cenozoic Seal Rocks

6.4.2.1 Rus Formation

As discussed earlier in the chapter, there are two reservoirs in the Cenozoic rocks of Kuwait, Umm Er Radhuma (locally known as first and second Eocene reservoirs) and Lower Fars. First Eocene reservoir is capped by Rus Formation (also locally known as first Anhydrite) which consists of anhydrite, subordinate limestone, shale, and marl (Alsharhan and Nairan, 1997). This formation considered to be highly effective in south of Kuwait in the Wafra oil field. The Second Eocene reservoir is capped by this layer of Anhydrite.

6.4.2.2 Lower Fars Formation

The two main reservoir intervals of the Lower Fars Formation contain heavy oil in north Kuwait and are capped by a thick shale unit, and it seems to act as a regional seal for the reservoir. This is indicated as no commercial hydrocarbons are reported above this unit in Kuwait (Abdul Razak et al., 2018).

6.5 Unconventional Petroleum System of Kuwait

Conventional and unconventional resources are fundamentally different in that the traditional components of source, seal, reservoir, and trap must be present for conventional "discontinuous" hydrocarbons to be accumulated. Timing of trap formation relative to the generation of hydrocarbon in the source, migration from source to reservoir, and final entrapment is critical and must be juxtaposed. On the other hand, unconventional resources are continuous, organic-rich mud rocks or carbonates, consisting of self-sourced hydrocarbons trapped within fine-grained, low permeability. As a result of the combined effects of high petroleum viscosity and low matrix permeability, unconventional resources cannot be produced without stimulation (Ma & Holditch, 2016).

Kuwait Oil Company (KOC) displayed interest in commercially exploiting unconventional hydrocarbon reserves and started laying significant emphasis on the exploration and development of unconventional resources. Despite the relatively higher cost of production from unconventional resources and the availability of enormous conventional reserves, the rise in global demand for natural gas and the significant potential of unconventional prospects in Kuwait derived the plan of exploitation toward unconventional oil and gas. However, opportunities for commercial development of unconventional resources, such as fractured carbonates, in Kuwait must be balanced with the technological, environmental, and economic challenges. In

May 2008, Kuwait Oil Company started free gas production from the Jurassic naturally fractured carbonate reservoir which produces sour gas and light oil (API = 48). The main development challenges of the Jurassic gas field are the depth of the reservoir, its high pressure (10,000 psi) and temperature (275 °C), and low porosity and permeability, in addition to the high H_2S (5%) and CO_2 (5%) concentration of the well fluids (Al Qaoud, 2012). The Middle Jurassic Najmah-Sargelu Formations and the Lower Cretaceous Makhul Formation are among the most important unconventional targets in recent exploration efforts in Kuwait (Al Bahar et al., 2019; Husain et al., 2018b). The unconventional shale oil/gas potential within the Najmah Formation was detected West of Kuwait in Dharif, Abduliyah, Minagish, and Umm Gudair Fields (Al Murakhi, 2019).

Recently, the unconventional resources in Kuwait have been classified into 3 categories (Al Baher et al., 2019) based on organic geochemical properties of source and crude and reservoir parameters (Fig. 6.21).

Class-I Unconventional Kerogen: Defines organic-rich/ Kerogen prospects that are further subdivided into two subcategories: Class IA: Self-sourced kerogen/organic-rich intervals including Makhul, Najmah, and Base Gotnia kerogen; and Class IB: Kerogen and Limestone intercalations of Lower Najmah, developed currently as a fractured limestone.

Class-II Unconventional Tight: incorporates tight clastic or carbonate formations with low organic material content,

for example the Marrat tight limestone. Permeability is less than 0.1 millidarcy (mD).

Class-III Unconventional Viscous Oil: it involves both immobile and mobile oil subdivided into Class IIIA: Mobile: it relates to mobile viscous oil of less than or equal to 10,000 centipoise viscosity, current development of Lower Fars under production appraisal; and Class IIIB: Immobile: it relates to immobile viscous oil exceeding 10,000 centipoise viscosity and it includes tar mat and bitumen (e.g., Tayarat Formation).

6.6 Hydrocarbon Thermal Maturation, Generation, Migration, and Entrapment

As discussed earlier, the Jurassic Najmah and Lower Cretaceous Sulaiy (Makhul) Formations are believed to be the main source rocks in Kuwait. Maturity levels of Jurassic source rocks vary throughout Kuwait due to their depth variations. Generally, maturity levels in the south of Kuwait (e.g., Burgan oil field) are lower than in northern oil fields (e.g., Ratqa oil field) because of the difference in burial depths (Abdullah, 2001). The shallowest formation in the north of Kuwait is in the oil-generation phase while in the south, Jurassic source rocks are at the early stage of oil generation.

Abdullah et al. (1997) and Abdullah (2001) evaluated the burial and thermal history of the main Cretaceous and

Fig. 6.21 Classification of unconventional resources (after Al-Bahar, et al., 2019)

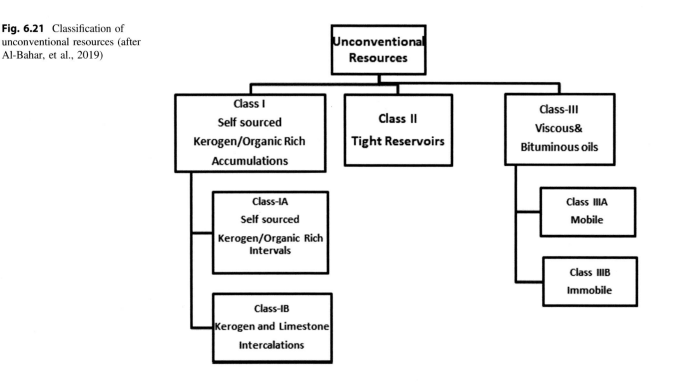

Fig. 6.22 The burial history curve shows that most oil generation of Sulaiy Formation occurred during Late Cretaceous and Early Paleogene and most oil expulsion during Paleogene (After Abdullah et al., 1997)

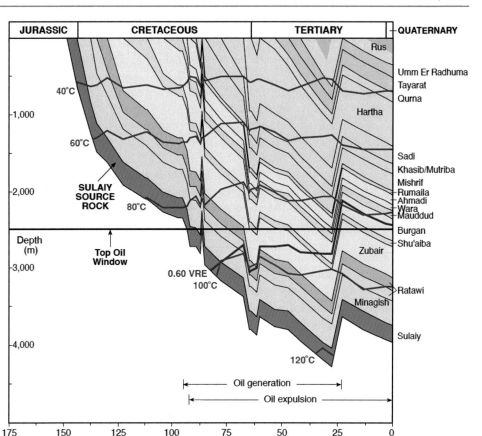

Jurassic source rocks in Kuwait, respectively. The study concluded that most oil generation occurred during Late Cretaceous and Early Paleogene, while oil expulsion took place during Paleogene. Figure 6.22 shows that Sulaiy source rock reached its peak oil generation around 75 Ma (Late Cretaceous) at a temperature of 100° and a depth of around 3,000 m. And it reached its peak expulsion during Paleocene (62 Ma).

The Jurassic source rocks of Kuwait began oil generation in Late Cretaceous to Eocene. A study based on samples from the Burgan oil field showed that deeper Jurassic horizons (Marrat Formation) began to generate oil around 110 Ma (Late Cretaceous), while Najmah Formation (shallowest Jurassic formation) started generating oil around 45 Ma (Eocene) (Fig. 6.23) (Abdullah, 2001). Jurassic source rocks are capped by thick anhydrite and evaporite sections of Gotnia and Hith Formations which might have prevented vertical migration to Cretaceous reservoirs. However, results from oil analysis of Cretaceous and Jurassic reservoirs of Kaufman et al. (1998) showed similar characteristics which might suggest that both were charged by the same source or from a different source of very similar compositions. A study by Al-Hajeri (2017) concluded that Gotnia and Hith Formations might be breached by fracture systems during Late Cretaceous tectonism which might have

facilitated vertical migration of Jurassic hydrocarbon to Cretaceous reservoirs.

Since peak oil expulsion took place during Late Cretaceous and Early Paleogene (Table 6.1) synchronously with Zagros Orogeny, these tectonic stresses along with micro-fracturing of source rock due to intensive oil generation might have enhanced the migration of hydrocarbon out of the source rocks (Abdullah & Connan, 2002). The synchronous events of oil expulsion and trap formation during Late Cretaceous and Early Paleogene could be one of the main reasons for the huge hydrocarbon reserves in Kuwait.

6.7 Conclusions

The first commercial oil discovery in the state of Kuwait was made in 1938 in the Burgan oil field. Since then, several oil fields were discovered which made Kuwait one of the major oil-producing countries worldwide with proven crude oil reserves of about 100 billion barrels (BP Statistical Review of World Energy 2021).

Several geological factors are responsible for the prolific hydrocarbon reserves in Kuwait. The tectonics, subsidence histories, and the diagenetic processes that prevailed, on the

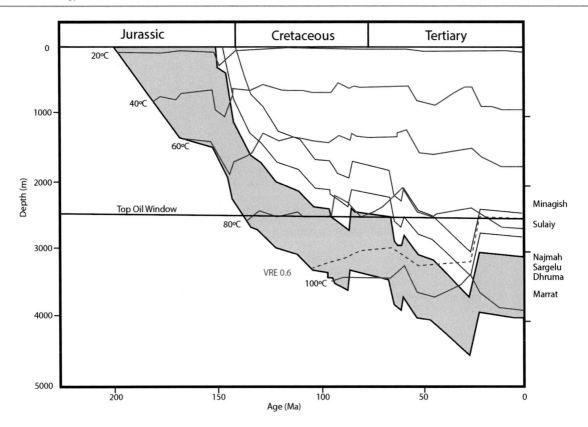

Fig. 6.23 Burial history diagram for one of Burgan oil field wells. The shaded area represents the Jurassic source rocks (modified after Abdullah, 2001)

Table 6.1 hydrocarbon generation and expulsion of Najmah/Sargelu and Sulaiy (Makhul) source rocks in Kuwait (after Alsharhan et al., 2014)

Formation	Hydrocarbon generation	Hydrocarbon expulsion
Najmah/Sargelu	Around 100 Ma (Cenomanian)	Expulsion becomes efficient at 60 Ma to 50 Ma (Selandian–Lutetian)
Sulaiy (Makhul)	Around 60 Ma (Selandian)	Expulsion of oil is expected at 40–30 Ma (Bartonian–Rupelian)

eastern part of the Arabian Peninsula during its geological history, are the main drivers for the exceptional geological conditions. These geological conditions are manifested by the widespread geographic distribution of rich, mature source rocks, high-quality reservoir rocks, and highly efficient seal rocks. Along with the formation of large structural traps during or after oil and gas generation, which helped in the formation of multiple petroleum systems in Kuwait. There are more than 20 oil fields across the country, however, most of the hydrocarbon production comes from the Greater Burgan field. These fields resemble very gentle four-closure anticlines interpreted as drape structures over deep-seated faults or as growth structures related to salt tectonics.

There are many stratigraphic intervals of good-quality reservoirs in Kuwait mainly from the Mesozoic Era. Most of the Cretaceous reservoirs are sealed by interbedded and overlying Cretaceous shales, while the Jurassic reservoirs are sealed regionally by salt and anhydrite of Gotnia and Hith Formations. The most famous reservoir is the siliciclastic reservoir of Burgan Formation which contributes to most of the hydrocarbon production in the state of Kuwait along with other several Cretaceous reservoirs which accounts for about 80% of hydrocarbon production in Kuwait.

Although source rocks are well distributed along the stratigraphic column of Kuwait, from Silurian up to Cretaceous, the richest source rocks are dominant in Jurassic and Lower Cretaceous. Types of kerogens from these source rocks are types II and II–III, a mixture of marine and terrestrial organic matter with TOC values ranging from moderate to excellent. Organic geochemical analysis of the Jurassic and Cretaceous source rocks which are mainly composed of limestones and calcareous shales shows the significant potential for oil generation in Kuwait. Oil generation of the Jurassic and Cretaceous source rocks started in Late Cretaceous–Early Tertiary synchronously with structure trap formation.

Kuwait Oil Company (KOC) displayed interest in commercially exploiting unconventional hydrocarbon reserves. The Middle Jurassic Najmah-Sargelu Formations and the Lower Cretaceous Makhul Formation are among the most important unconventional targets in recent exploration efforts in Kuwait.

References

Abdullah, F. H. A. (2001). A preliminary evaluation of Jurassic source rock potential in Kuwait. *Journal of Petroleum Geology, 24*(3), 361–378.

Abdullah, F. H. A., & Connan, J. (2002). Geochemical study of some Cretaceous rocks from Kuwait: Comparison with oils from Cretaceous and Jurassic reservoirs. *Organic Geochemistry, v. 11*, 125–148.

Abdullah, F. H. A., & Kinghorn, R. R. F. (1996). A preliminary evaluation of Lower and Middle Cretaceous source rocks in Kuwait. *Journal of Petroleum Geology, v. 19*, 461–480.

Abdullah, F. H. A., Peter, J. R., Ormerod, M. P., & Kinghorn, R. F. (1997). Thermal history of the Lower and Middle Cretaceous source rocks in Kuwait. *GeoArabia, 2*(2), 151–164.

Abdullah, F. H., Carpentier, B., Kowalewski, I., Buchem, F., & Huc, A. (2005). Organic matter identification in source and reservoir carbonate in the Lower Cretaceous Mauddud Formation in Kuwait. *GeoArabia, 10*(4), 17–34.

Abdullah, et al. (2017). Petrophysics and hydrocarbon potential of Paleozoic rocks in Kuwait. *Journal of Asian Earth Sciences, v. 148*, 105–120. https://doi.org/10.1016/j.jseaes.2017.07.057

Abdullah, F. H., Shaaban, F., Al-Khamis, A., Khalaf, F., Bahman, F., & Akbar, B. (2021). Evaluation of Paleozoic source rocks in Kuwait. *AAPG Bulletin, 105*(4), 721–748.

Abdul Razak, M. H., Al-Jenaie, J., & Moubarak, H. (2018). Understanding the Reservoir Architecture of Lower Fars Formation in North Kuwait through Seismic Reservoir Characterization, Adapted from oral presentation given at the GEO 2018 13th Middle East Geosciences Conference and Exhibition, Manama, Bahrain. (5–8 Mar 2018).

Abu-Ali, M. A., Rudkiewicz, J. L. L., McGillivray, J. G., & Behar, F. (1999). Paleozoic petroleum system of Central Saudi Arabia. *GeoArabia, 4*, 321–336.

Al-Ajmi, H., Brayshaw, A. C., Barwise, A. G., & Gaur, R. S. (2001). The Minagish field tar mat, Kuwait: Its formation, distribution and impact on water flood. *GeoArabia, 6*(1), 7–24.

Al-Ali, Y., Wang, G., Al-Mula, Y., Hussein, M., Choudhary, P., & Ahmed, F. (2017). Evaluation of the first cycle of steam stimulation in North Kuwait heavy oil thermal pilots, SPE 186182-MS.

Al-Awadi, M., Haines, T., Bertouche, M., Bonin, A., Fuchs, M., Deville De Periere, M. (2017). Predictability of the sedimentological make-up and reservoir quality in the Mauddud formation using FMI Logs-a case study from a North Kuwait field, SPE-187630-MS.

Al-Bahar, M., Vandana Suresh, V., Al-Sahlan, G., & Al-Ajmi, M. A. (2019) Classifications of Unconventional Resources in Kuwait. Search and Discovery Article #80657. Adapted from extended abstract based on oral presentation given at the GEO 2018 13th Middle East Geosciences Conference and Exhibition, Manama, Bahrain. (5–8 Mar 2018).

Al-Eidan, A. J., Wethington, W. B., & Davies, R. B. (2001). Upper Burgan reservoir distribution, northern Kuwait: Impact on reservoir development. *GeoArabia, v. 6*(2), 179–208.

Al-Eidan, A. J., Narahari, S. R., Al-Awadi, M., Al-Ajmi, N. H., Pattnaik, C., Al-Ateeqi, K., Al-Houli, M. M., Kidambi, V. K., De Keyser, T. (2010). Jurassic tight carbonate gas fields of North Kuwait: Exploration to early development. In *Paper presented at the SPE Deep Gas Conference and Exhibition*, Manama, Bahrain. https://doi.org/10.2118/130914-MS.

Al-Enezi, K., & Hsie, J. (1999). Modeling and scale up of a heterogeneous reservoir, Wara Sand, Greater Burgan Field, SPE 53206.

Al-Enezi, S.A., AlAli, S., AlKamiss, A. (2018). The Jurassic petroleum system of Kuwait. AAPG Datapages/Search and Discovery Article #90319 at GEO 2018 13th Middle East Geosciences Conference and Exhibition, Manama, Bahrain. (5–8 Mar 2018).

Al-Enezi, B. (2019). Geological description for unlocking potential of Najmah-Sargelu reservoir in greater Burgan Field. In *European Association of Geoscientists & Engineers, Conference Proceedings, Third EAGE WIPIC Workshop: Reservoir Management in Carbonates, Nov 2019* (Vol. 2019, pp. 1–8).

Al-Failakawi, F., Al-Muraikhi, R., Al-Shamali, A., Al Qattan, A., Belgodere, C., Marti, F., & Quttainah, R. (2018). Fracture characterization of Najmah-Sargelu tight carbonates reservoir using geomechanical attributes in Minagish field, West Kuwait Search and Discovery Article #20447 (2019) Posted January 14, 2019. Adapted from oral presentation given at the GEO 2018 13th Middle East Geosciences Conference and Exhibition, Manama, Bahrain. (5–8 Mar).

Al-Fares, A. A., Bouman, M., & Jeans, P. (1998). A new look at the middle to Lower Cretaceous stratigraphy, offshore Kuwait. *GeoArabia, 3*(4), 543–560.

Al-Hajeri, B. (2017). Application of formation water geochemistry to assess seal integrity of the Gotnia Formation, Kuwait. *Arabian Journal of Geosciences, v. 10*(3), 1–10. https://doi.org/10.1007/s12517-017-2842-2

Ali, M. A. (1995). Gotnia salt and its structural implications in Kuwait. In M. I. Al-Husseini, (Ed.), *Middle east petroleum geosciences, GEO '94: Gulf PetroLink, Bahrain* (Vol. 1, pp. 133–142).

Al-Jallal, I.A. (1995). The Khuff Formation—its reservoir potential in Saudi Arabia and other Gulf countries. In Al-Husseini, M.I. (Ed.), *Middle East Petroleum Geosciences Conference Geo-94, Gulf Petrolink, Bahrain* (Vol. 1, pp. 103–119).

Al-Khamiss, A., AdulMalik, S., & Abdel Hameed, W. (2012). Compositional basin model of Kuwait–leads for yet to find potential. In *Abstract, AAPG Search and Discovery Article AAPG Hedberg Conference Petroleum Systems*. Modeling the Past, Planning the Future, Nice, France. (1–5 Oct).

Al-Murakhi, R. (2019). West Kuwait unconventional hydrocarbon potential evaluation of the Najmah formation - phase 1 project summary. Search and Discovery Article #80658 (2019) Posted January 14, 2019. In *Adapted from oral presentation given at the GEO 2018 13th Middle East Geosciences Conference and Exhibition, Manama, Bahrain* (5–8 Mar 2018).

Al-Qaoud, B. N. (2012). Challenges in Sour Gas handling for Kuwait Jurassic Sour Gas. Paper presented at the SPE Middle East Unconventional Gas Conference and Exhibition, Abu Dhabi, UAE, January 2012. SPE-154452-MS.

Alsharhan, A. S., Strohmenger, C. J., Abdullah, F. H., Al Sahlan, G. (2014). Mesozoic stratigraphic evolution and hydrocarbon habitats of Kuwait. In L. Marlow, C. Kendall, & L. Yose, (Eds.), *Petroleum systems of the Tethyan region: AAPG Memoir* (Vol. 106, pp. 541–611).

Alsharhan, A. S., & Nairn, A. E. M. (1997). Sedimentary basins and petroleum geology of the Middle East (p. 942). Elsevier Publishing Company. ISBN 0-444-82465-0.

Alsharhan, A. (1994). Albian clastics in the Western Arabian gulf region: A sedimentological and petroleum-geological interpretation. *Journal of Petroleum Geology, 17*(3), 279–300. https://doi.org/10.1111/j.1747-5457.1994.tb00135.x

Al-Tendail, S. B., Taal, A., Al-Adwani, T., Abu Ghneej, A., Arasu, R., Barr, D., & Clews, P. (2012). Reviving a classic exploration play in Kuwait: The lower cretaceous Ratawi formation, search and discovery Article #10436.

Al-Wazzan, H. (2021). 3D forward stratigraphic modelling of the Lower Jurassic carbonate systems of Kuwait. *Marine and Petroleum Geology, 123,* 104699–.https://doi.org/10.1016/j.marpetgeo.2020.104699

Andriany, R., & Al-Khamiss, A. (2011). Permo-triassic source rock assessment by inherited of the triassic oils in Kuwait, Third Arabian Plate Geology Workshop Permo-Triassic (Khuff) Petroleum System of the Arabian Plate Kuwait City, Kuwait, 28 Nov-1 Decr 2011

Aqrawi, A. M., & Badics, B. (2015). Geochemical characterisation, volumetric assessment and shale-oil/gas potential of the Middle Jurassic-Lower Cretaceous source rocks of NE Arabian Plate. *GeoArabia, 20*(3), 99–140.

Arasu, R. T., Nath, P. K., Khan, B., Ebrahim, M., Rahaman, M., Bader, S., & Abu-Ghneej, A. F. N. (2012). Stratigraphic Features Within the Ratawi Shale Member of the Lower Cretaceous Ratawi Formation and Their Hydrocarbon Prospectivity in Sudai –Abdali Area, North Kuwait.` Paper presented at the 2012 SEG Annual Meeting, Las Vegas, Nevada. (Nov 2012).

Behbehani, S., & Hollis, C. (2015). Controls on petrophysical properties of the Mauddud formation, Bahra and Sabriyah Fields, Kuwait, IPTC-18511-MS.

Boix, C., Di Simone, S., Deville De Periere, M., et al. (2014). New insights on the Mauddud formation (Upper Albian): Characterisation of sedimentology, diagenesis and reservoir quality. Abstract presented at the AAPG ICE, Istanbul. (14–17 Sept 2014).

Bou-Rabee, F. (1996). Geologic and tectonic history of Kuwait as inferred from seismic data. *Journal of Petroleum Science and Engineering, 16* (pp. 151–167).

Chakrabarti, B., Al-Wadi, M., Abu Hebiel, H., Al-Enezi, A. (2011). Diagenetic controls on carbonate reservoir quality of Jurassic Middle Marrat formation in Burgan Field, Kuwait. Search and Discovery Article #20121 (2011) Posted November 28, 2011. Adapted from extended abstract prepared in conjunction with poster presentation at AAPG International Conference and Exhibition, Milan, Italy. (23–26 Oct 2011).

Grader, A., Al-Jallad, O., Kayali, A. (2019). Micro porosity within the organic matter and its impact on assessment of unconventional potential of a Kerogen rich Najmah formation in Kuwait. Presentation at the Abu Dhabi International Petroleum Exhibition & Conference held in Abu Dhabi, UAE. (11–14 Nov 2019).

Hawie, N., Al-Wazzan, H., Al-Ali, S., Al-Sahlan, G. (2021). De-risking hydrocarbon exploration in lower Jurassic carbonate systems of Kuwait through forward stratigraphic models. *Marine and Petroleum Geology, 123.* ISSN 0264-8172.

Husain, R., Sajer, A., Al-Khamiss, A., Iqbal, M., & Al-Zeabout, N. (2009). Triassic petroleum system of Kuwait, AAPG Search and Discovery Article #90090©2009 AAPG Annual Convention and Exhibition, Denver, Colorado. (7–10 Jun 2009).

Husain, R., Khan, D., Sajer, A., Alammar, N. (2011). Khuff formation in Kuwait: an overview: Conference paper, Third Arabian Plate Geology Workshop Permo-Triassic (Khuff) Petroleum System of the Arabian Plate Kuwait City, Kuwait. (28 Nov–1 Dec 2011).

Husain, R., Al-Wadi, M., Al-Ali, S., Al-Abdullah, M., Al-Fadhli, H., (2018a). Unlocking the potential of unconventional oil and gas resources in Kuwait. Paper presented at the Abu Dhabi International Petroleum Exhibition & Conference, Abu Dhabi, UAE. SPE-192604-MS. (Nov 2018a).

Husain, R., Al-Owihan, H., Mulyono, R., Ahmed, S. M., Al-Fadhli, H., Al-Kandari, A., Al-Ali, S., Al-Wadi, M. (2018b). Hydrocarbon seals in Jurassic carbonate and evaporite sequences in Kuwait: Implications for exploration, second EAGE/AAPG Hydrocarbon Seals of the Middle East Workshop, Abu Dhabi, UAE. (17–19 Apr 2018b).

Inan, S., AbuAli, M. A., Qathami, S., Hakami, A. A., & Shammari, S. H. (2016). Hydrocarbon potential of the Silurian Qusaiba Shales of Saudi Arabia in a regional context of the Paleozoic Shales in the Arabian Platform. In *Conference paper, AAPG GTW The Source Rocks of the Middle east, Abu Dhabi–UAE.* (Jan 2016).

Konert, G., Afifi, A. M., Al-Hajri, S. A., & Droste, H. J. (2001). Paleozoic stratigraphy and hydrocarbon habitat of the Arabian Plate. *GeoArabia, 6,* 407–442.

Kaufman, R. L., Kabir, C. S., Abdul-Rahman, B., Quttainah, R., Dashti, H., Pederson, M. S., & Moon, M. S. (1998). Characterizing the Greater Burgan field using geochemical and other field data. In *SPE 49216 Annual Technical Conference and Exhibition, New Orleans, Louisiana.*

Laboun, A. A. (1987a). Unayzah formation: A new permo-carboniferous unit in Arabia. *The American Association of Petroleum Geologists Bulletin, 71*(1), 29–38.

Laboun, A. A. (1987b). Unayzah formation-a new permian-carboniferous unit in Saudi Arabia. *American Association of Petroleum Geologists Bulletin, 71,* 29–38.

Ma, Y. Z., & Holditch, S. A. (2016). Unconventional oil and gas resources handbook evaluation and development. *Gulf Professional Publishing.* https://doi.org/10.1016/C2014-0-01377-9

Mahmoud, M. D., Vaslet, D., & Al-Husseini, M. I. (1992). The Lower Silurian Qalibah Formation of Saudi Arabia—An important hydrocarbon source rock. *American Association of Petroleum Geologists Bulletin, 76*(10), 1491–1506.

Masarik, C., Gonzalez-Mieres, R., Jamil, F., Buza, J. (2012). Identifying new opportunities in a mature reservoir: Wara formation, Wafra Field, Saudi Arabia/Kuwait Partitioned Zone, SPE 160873.

McGillivray, J. G., & Husseini, M. I. (1992). The Paleozoic petroleum geology of Saudi Arabia. *American Association of Petroleum Geologists Bulletin, 76,* 1473–1490.

Nath, P. K., Sunil Kumar Singh, S. K., Abu Taleb, R., Prasad, R., Badruzzaman Khan, B., & Bader, S. (2012). Characterization and modeling of tight fractured carbonate reservoir of Najmah-Sargelu formation, Kuwait search and discovery Article #41059 (2012) Posted October 29, 2012. Adapted from extended abstract prepared in conjunction with oral presentation at AAPG International Convention and Exhibition, Singapore. (16–19 Sept 2012).

Nath, P. K., Singh, S. K., Ye, L., Al-Ajmi, A. S., Bhukta, S. K., Al-Otaibi, A. H. (2014). Reservoir characterization and stratitructural play of Minagish Formation, SE Kuwait. In *International Petroleum Technology Conference.* OnePetro.

Nelson, P. H. (1968). Wafra field, Kuwait-Saudi Arabia Neutral zone. SPE 2nd Region. In *Technical Symposium Proceedings, Dhahran, Saudi Arabia* (pp. 101–120).

Parmjit, S., Husain, R., Sajer, A., Al-Fares, A. M. (2010). Estimating pre-Khuff thickness and delineating basement configuration in Dibdibba Trough, Kuwait-An integrated model based study, Paper presented at the 2010 SEG Annual Meeting, Denver, Colorado, October 2010.

Qabazard, S. A., Al Tameemi, A., & Abdullah, F. H. (2000). Petrographical, mineralogical and geochemical studies of the Quaternary carbonates, Al Khiran area, southern coast of Kuwait & Lower Cretaceous Middle Minagish Formation, Kuwait southern oil fields. A Thesis in Geology. 2000.

Rabie, A., Husain, R., & Al-Fares, A. M. (2014). Unconventional Petrophysical Workflows for Evaluation of Shale Plays: A Case Study from Kuwait. Search and Discovery Article #80370 (2014) Posted March 31, 2014 *Adapted from extended abstract prepared in conjunction with poster presentation at GEO-2014, 11th Middle East Geosciences Conference and Exhibition. (10–12 Ma 2014).

Saller, A. H., Pollitt, D., & Dickson, J. A. D. (2014). Diagenesis and porosity development in the First Eocene reservoir at the giant Wafra Field, Partitioned Zone, Saudi Arabia and Kuwait. *AAPG Bulletin, 98*(6), 1185–1212.

Sharland, P. R., Archer, R., Casey, D. M., Davies, R. B., Hall, S. H., Heward, A. P., Horbury, A. D., & Simmons, M. D. (2001). Arabian Plate sequence stratigraphy: GeoArabia Special Publication 2, Gulf Petrolink, Manama, Bahrain (p. 371).

Siddiqui, M. A., Al-Omair, F. S., & Al-Salali. Y. Z. (2008). Production from Najmah-Sargelu Reservoirs: A Formation Damage Perspective. AAPG Search and Discovery Article #90077 at GEO 2008 Middle East Conference and Exhibition, Manama, Bahrain.

SinghaRay, D., Mukherjee, P., Matar, S., Alessandroni, M., Joussineau, G., Erwan Perfetti, E., Maux, T. L., Al-Sammak, I., & Al Matar, B. (2015). Characterization and modeling of natural fractures in a tight Najmah-Sargelu formation, Dharif-Abduliyah Field, Kuwait. In *Conference: 77th EAGE Conference and Exhibition, Madrid* (1–4 June 2015).

Snasiri, F., Abdulrazzaq, E., Tirkey, N., Kotecha, R., Gazi, N.H., Al-Othman, M., Al-Sabea, S., & Ali, F. (2015). Excellent stimulation results in deep carbonate reservoir in Marrat formation in Magwa structure in Greater Burgan Field Kuwait. Paper presented at the SPE Kuwait Oil and Gas Show and Conference, Mishref, Kuwait, October 2015. Paper Number: SPE-175342-MS.

Strohmenger, C. J., Al-Anzi, M. S., Pevear, D. R., Ylagan, R. F., Kosanke, T. H., Ferguson, G. S., Cassiani, D. H., & Douban, A. F. (2003). Reservoir quality and K-Ar age dating of the pre-Khuff section of Kuwait: GeoArabia. *Bahrain, 8*(4), 601–620.

Strohmenger, C. J., Patterson, P. E., Al-Sahlan, G., Mitchell, J. C., Feldman, H. R., Demko, T. M., Weller, R. W., Lehmann, P. J., McCrimmon, G. G., Broomhall, R. W., & Al-Ajmi, N. (2006). Sequence stratigraphy and reservoir architecture of the Burgan and Mauddud formations. In P. M. Harris, & L. J. Weber, (Eds.), Giant hydrocarbon reservoirs of the world: From rocks to reservoir characterization and modeling (Vol. 88, pp. 213–245). AAPG Memoir 88/SEPM Special Publication, Boulder.

Tanoli, S. K., Husain, R., Sajer, A. A. (2008). Facies in the Unayzah Formation and the basal Khuff clastics in subsurface, northern Kuwait. *GeoArabia, 13*(4), 15–40.

Wani, M. R., & Al-Kabli, S. K. (2005). Sequence stratigraphy and reservoir characterization of the 2nd Eocene Dolomite Reservoir, Wafra Field, Divided Zone, Kuwait-Saudi Arabia, Paper SPE 92827 presented at the SPE Middle East Oil and Gas Show and Conference, Bahrain. (12–15 Mar 2005).

Wender, L. E., Bryant, J. W., Dickens, M. F., Neville, A. S., & Al-Moqbel, A. M. (1998). Paleozoic (Unayzah) hydrocarbon geology of the Ghawar area, eastern Saudi Arabia. In: Al-Husseini, M.I. (Ed.), *Middle East Petroleum Geosciences Conference Geo-97*, (Vol. 3, pp. 273–301). Gulf Petrolink.

Yousif, S., & Nouman, G. (1997). Jurassic geology of Kuwait. *GeoArabia, 2*(1), 91–110. https://doi.org/10.2113/geoarabia020191

Seismicity of Kuwait

7

Abd el-aziz Khairy Abd el-aal, Farah Al-Jeri, and Abdullah Al-Enezi

Abstract

This chapter deals with all the precious documented recently published and unpublished studies that address the seismic situation and earthquakes in the State of Kuwait. Kuwait is geographically and geologically situated in the northeastern part of the Arabian Peninsula. In addition to being close to the famous Zagros belt of earthquakes, the local seismic sources inside Kuwait make it always vulnerable to earthquakes. We will review the instrumental and historical seismic records and the Kuwait National Seismic Network, including Data acquisition, data analysis, and data analysis. This chapter will also highlight all the recent seismic studies conducted in the Kuwait region. The induced seismicity, the seismic sources affecting Kuwait, as well as determining the types of faults using focal mechanism technique, specifying the seismic crustal models and ground motion attenuation inside Kuwait are being reviewed.

7.1 Introduction

The earthquake is one of the main common phenomena of natural risk. They have got consequences on nature, human existence, infrastructure, and buildings, (Bolt, 1993; Burton et al., 1978; Chung & Bernreuter, 1980; Hall, 2011; Kanamori, 2003). It represents 51% of the total damage of all-natural disasters. An earthquake is an unexpected slipping or motion of a portion of the earth's crust or violent moving of massive rocks known as plates (Fig. 7.1) underneath the earth's surface as a result of an unexpected release

of stresses. The movement of tectonic plates releases stress that builds up along geological faults, causing earthquakes (Bath, 1979; Davison, 2014; Gubbins, 1990). Geologically speaking, a fault is a deep fracture that indicates the boundary between two tectonic plates, or it may be within the same plate. The main earthquake belts are located on the great faults between tectonic plates on the surface of the globe (https://en.wikipedia.org/wiki/Earthquake) as shown in Fig. 7.2. Many earthquake belts run along with coastal areas. Seismically, the point on the Earth's surface is fundamentally called the epicenter where the intensity of the earthquake and the devastation are strongest and greatest. The quantity of energy released by an earthquake is calculated by the seismic magnitude scale known as the Richter magnitude scale (Gutenberg & Richter, 1942; Richter, 1935). A big earthquake happens every few months somewhere in the world, but micro/small tremors occur constantly.

The earthquake (tremor or a quake) creates seismic waves (body waves and surface waves) (Bormann, 2012). The seismic activity or seismic record of an area indicates the frequency, type and magnitude of earthquakes over a certain period of time. The seismometer is used as a tool for measuring seismic recordings (Bath, 1979; Bormann, 2012). Moment magnitude is the most common of the various seismic scales in which earthquakes greater than magnitude of approximately 5 are reported. (Bormann, 2012; Hanks & Kanamori, 1979; Kanamori, 1977). The smaller earthquakes with magnitudes less than 5 are monitored through the national seismic networks and their magnitudes are calculated ordinarily on the local scale, commonly named the Richter scale (Gutenberg & Richter, 1942; Richter, 1935). Smaller earthquakes up to magnitude 3 are mostly imperceptible while the big earthquakes up to 6 and over potentially cause severe damage in/around the epicenter and over great distances, depending on their focal depth. The largest earthquakes throughout the ages are those whose magnitude exceeds 9, despite the fact that there is no restriction to the

A. el-aziz K. Abd el-aal (✉) · F. Al-Jeri · A. Al-Enezi
Kuwait Institute for Scientific Research (KISR), Safat, Kuwait
e-mail: akabdelaal@kisr.edu.kw; dewaky@yahoo.com

A. el-aziz K. Abd el-aal
National Research Institute of Astronomy and Geophysics, Helwan, Egypt

© The Author(s) 2023
A. el-aziz K. Abd el-aal et al. (eds.), *The Geology of Kuwait*, Regional Geology Reviews,
https://doi.org/10.1007/978-3-031-16727-0_7

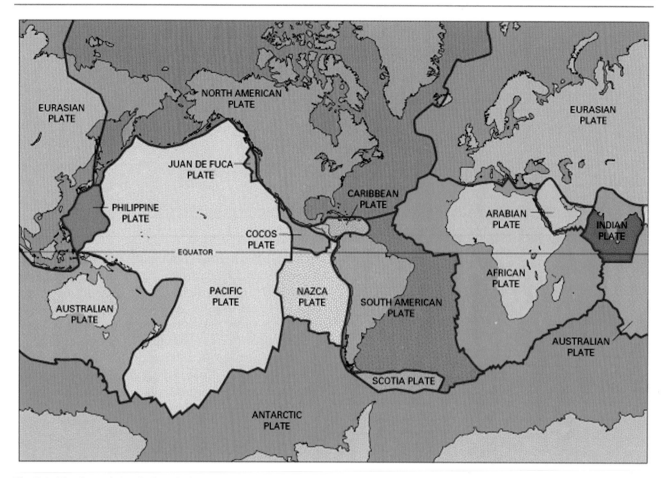

Fig. 7.1 The figure shows the boundaries of tectonic plates and also the continuous slow movement of these tectonic plates, the outer part of the Earth is what causes earthquakes and volcanoes

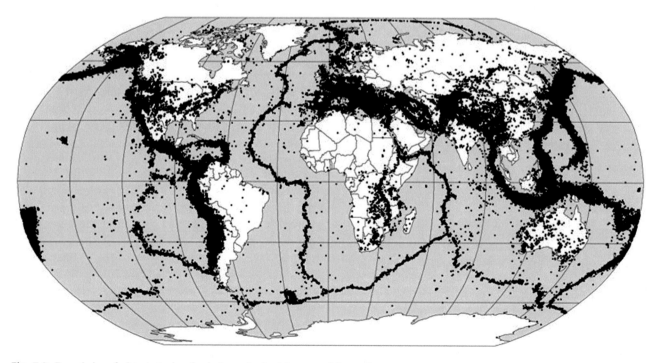

Fig. 7.2 Boundaries of plate tectonics, fire belts, and seismicity map of the earth

possible magnitude. When the hypocenter of a massive earthquake is positioned far from shore, the sea floor may shift sufficiently to cause a tsunami (Dudley & Lee, 1988; Levin & Nosov, 2009).

Earthquakes are classified as either tectonic or man-made (induced) events. Human and daily activities may cause induced earthquakes either directly or indirectly (Abd el-aal et al., 2020, 2021b). Earthquakes can also trigger landslides, and occasionally volcanic activity (Bath, 1979; Bormann, 2012). Earthquakes are generated by natural and man-made sources, including.

- Tectonic movement
  - Plate boundaries termed inter-plate earthquakes
  - Mid-continent, termed intraplate earthquakes
  - Mid-ocean ridge
- Volcanoes
- Reservoir induced
- Nuclear explosions

Scientists are researching ways to predict earthquakes, but their predictions are not always accurate (Allen, 1982; Kanamori, 2003; Lee et al., 2003). From this point, the earthquake occurrences still cannot be predicted reliably, but the characteristics of the earthquake's physical effects can be estimated (i.e., ground shaking, ground failure, surface fault rupture, landslides, regional tectonic deformation, tsunami waves, and aftershocks). Assessment of earthquake occurrence sequence hazard for any certain area plays an important role in proposing measures to minimize earthquake damage and to anticipate the future safe development of the strategic projects.

Particular attention should be paid to the assessment and the mitigation of the earthquake risk in seismically active areas, where monumental sites, human activities, and major investments are concentrated. The economic losses incurred by a devastating earthquake may cause a serious economic disruption with all its detrimental social consequences. Therefore regional as well as detailed urban planning in a seismic active area should rely on the proper knowledge of geological, geophysical, seismological, social and economical implications of the earthquake in the region.

An earthquake is a geodynamic phenomenon. The present seismic activity and the other geodynamic phenomena related to it (e.g., deformation and ruptures of the crust, volcanoes, geothermal manifestation, topographic features, etc.) are results of a relatively recent geologic process, which are usually, called active tectonic (Bath, 1979; Bormann, 2012). Very important information about tectonics can be obtained from the distribution of seismic activity. The main earthquake activity took place along the plate boundaries and recent active faults (Fig. 7.2).

7.2 Historical Earthquakes in Kuwait

Historically, and also by reviewing historical studies and manuscripts, where they never mentioned any strong earthquakes inside Kuwait, the international and regional seismic monitoring centers have recorded a number of small/medium earthquakes in Kuwait as shown in Table 7.1 and Fig. 7.3 (Abdelmeinum et al., 2004). In 1931, the first largest earthquake was recorded inside Kuwait by global monitoring stations and it was large and adequate enough to be monitored with a magnitude of 4.8 on the Richter scale located in the Boubyan island region. Although recent seismic studies after the establishment of the Kuwait network for seismic monitoring (KNSN) did not show seismic activity in the region in which the earthquake was recorded. This indicates that there may be an error in the coordinates and location of the earthquake. In 1973 an earthquake with magnitude 4.6 struck Kuwait and this event was located in the Arabian Gulf (Kuwait side). An earthquake with magnitude 4.1 took place in the same area (Abdelmeinum et al., 2004) in 1976. This was followed by two more earthquakes that were recorded by international agencies with magnitude 3.2 and 3.8 in the same year. The first event was located in northeast Kuwait and the second one was located in southeast Kuwait. After one year in 1977, there was an earthquake with magnitude 4.5 located in southeastern Kuwait (Table 7.1). In 1993 there was one of the most significant earthquakes that hit Kuwait is the Managish earthquake nearby Kuwait Saudi border in Southwestern Kuwait with magnitude 4.8 on the Richter scale (Abdelmeinum et al., 2004). It affected the surrounding areas, although the epicenter of the earthquake is about 150 km away from Kuwait City, but many of its residents have felt it, and some of the residents of the upper floors building have even been terrified, another earthquake in 1997 hit the same area, with a magnitude of 4.3 on the Richter scale and was felt by many citizens and residents. It was recorded by all the stations of the Kuwaiti National Seismic Network.

7.3 Instrumental Earthquake Recording in Kuwait

7.3.1 History of Seismic Recording

Instrumental recording of earthquakes started in Kuwait (1997). The Kuwaiti Government financed the Kuwait Institute for Scientific Research (KISR) to construct and deploy the Kuwait National Seismic Network (KNSN), which covers the whole Kuwaiti territory. KISR upgraded the data communication system from radio telemetry to

Table 7.1 Historical earthquakes recorded in Kuwait

Year	Month	Day	Time			Latitude	Longitude	Depth	Magnitude
			Hour	Minutes	Seconds				
1931	7	5	0	17	57	29.9	48.5	33	4.8
1973	3	14	1	16	44	28.5	49.1	5	4.6
1976	1	2	4	30	32	28.6	48.9	33	4.1
1976	9	26	0	12	47	29.9	47.3	33	3.2
1976	9	27	2	23	57	28.9	48.2	33	3.8
1977	1	16	20	31	26	28.8	48.1	5	4.5
1993	6	2	22	1	48	29	47.6	10	4.8
1997	9	18	20	24	51	28.9	47.5	10	3.9
1997	12	30	18	18	33	28.7	47.5	10	4.3

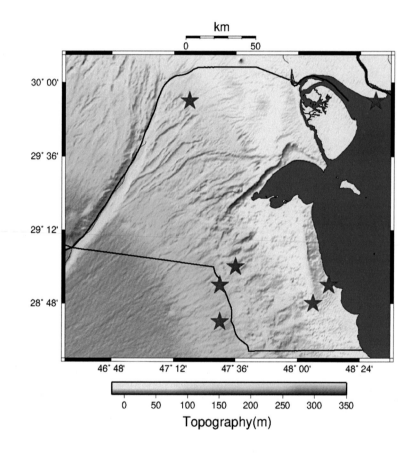

Fig. 7.3 Historical earthquakes recorded in Kuwait by international seismic center

telephone leased lines to increase the efficiency of the KNSN.

In 1997, the KNSN was installed with seven short-period seismic stations and one broadband seismic station. Recently, the seismic stations of the KNSN were completely updated by replacing five short-period sensors with broadband sensors (KISR, 2019).

This network has enhanced the detectability level to record the microearthquakes and define their location precisely. Also, the KNSN contributes to define new seismic sources, which were not known before (KISR, 2019). From

the analysis of KNSN data, it was clear that the recording system was enhanced and developed in the last decade, and this development should be continued to achieve a complete record of earthquakes.

7.3.2 Kuwait National Seismic Network (KNSN)

7.3.2.1 KNSN Configuration

The KNSN network consists of the main center at KISR venue. With the beginning of the network's work, the recorded seismic data was sent from field monitoring stations to the main center through wireless communications and recently, mobile phone lines were used in the data transmission process. The earthquake data received is processed and analyzed in order to derive the earthquake parameters, then saved and archived (Abd el-aal et al., 2020, 2021a, b; Al-Enzi et al., 2007; Gok et al., 2006; Gu et al., 2017, 2018; KISR, 2019).

Figure 7.4 illustrates the distribution of the field seismic monitoring stations which are installed in/around seismic sources as well as to cover Kuwait's country area as much as possible. Table 7.2 contains the sensor type, station code, coordinates and full name of the KNSN filed stations. At present, the KNSN network has equipped with two short-period sensors and six broadband sensors. The broadband sensors are from REFTEK 151-120A seismometer model and the short-period sensors are from SS-1 Range seismometer. Field seismic monitoring stations transfer continuous seismic dataset to the KNSN center at KISR venue over TCP/IP connection using a 5-G router. The main acquisition system at the KNSN receives seismic waveform data from both the Kinemetrics and Reftek acquisition platforms (Abd el-aal et al. 2020, 2021a, b, c; Al-Enzi et al., 2007; KISR, 2019).

7.3.2.2 Data Acquisition

In the last five years, the Kuwait Earthquake Network (KNSN) has been fully and comprehensively developed and updated, where the SeisComP seismological software (https://www.seiscomp.de/) is used as a full-featured real-time seismic acquisition, processing, and analysis platform in the KNSN network (Fig. 7.5). The SeisComP system package has been established by Gempa (www.gempa.de) and GFZ German Research Centre for Geosciences (https://www.gfz-potsdam.de/en/) for earthquake data acquisition, distribution, processing and interactive analysis (KISR, 2019). The seismic data transmission protocol in SeisComP is the SeedLink. Among the SeisComP components are (1) automatic detection of earthquake events, (2) detection and determination of earthquake location and magnitude (3) automatic and manual processing facilities, which are enhanced by graphical user interface (GUIs) for visualization, seismic data quality control as well as rapid event review.

Features of SeisComP

The features of SeisComP (https://www.seiscomp.de/) include seismic data acquisition, waveform dataset recording and distribution, waveform real-time data exchange,

Fig. 7.4 Distribution of Kuwait National Seismic Network's stations

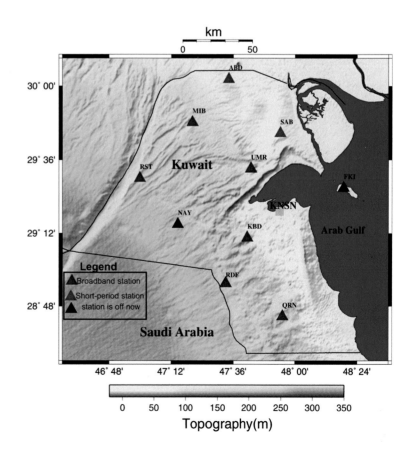

Table 7.2 Stations code and location of Kuwait National Seismic Network's Velocity-meter Units

Station	Code	Latitude	Longitude	Altitude (meter)	Sensor	Digitizer
KBD	9011	29 10.585	47 41.486	113	Reftek 151B	130S
QRN	9012	28 44.600	47 55.119	139	Reftek 151B	130S
RDF	9013	28 55.476	47 33.069	179	Reftek 151B	130S
RST	9015	29 30.024	46 59.87	216	Reftek 151B	130S
MIB	9016	29 48.135	47 20.402	120	Reftek 151B	130S
UMR	9017	29 33.201	47 42.972	85	Reftek 151B	130S
SAB	9018	29 44.616	47 54.240	53	Kinemetrics SS-1	130S
ABD	9019	30 02.193	47 34.178	62	Kinemetrics SS-1	130S

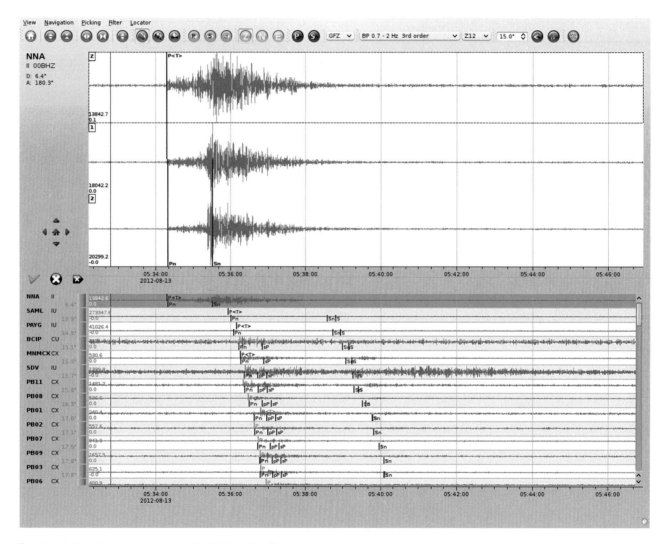

Fig. 7.5 Earthquake data processing using SeisComP software

monitoring and health network status, data processing, issuing of earthquake alerts and messages, complete seismic waveform archiving in miniSEED format, automatic and manual earthquake identification, event parameter archiving, and simple access to relevant details about waveforms, stations, and recent earthquakes (KISR, 2019). Throughout the seismic data analysis steps, the SeisComP initiates an input file that lists phases of each earthquake. The Hypoinverse program and other location software codes utilize in SeisComP with a set of files that give information related to the stations to be used in analysis, which is the crustal model that is suitable for earthquake location, as well as any delays, weights, or any corrections, should be to apply.

7.3.2.3 Data Processing and Analysis

The Kuwait Network KNSN recorded approximately more than 1027 local earthquakes since its inception in 2021, 548 of them before using the SeisComP platform in 2018. Before 2018, the SEISAN program (http://seisan.info/) was used to visualize and analyze earthquake data and save it after deriving seismic parameters, including the location and magnitude of the earthquake, and one of the appropriate programs from the HYPOINVERSE, HYPOCENTER (Lienert & Havskov, 1995) and HYPO71 (Lee, 1990) and was used for hypocenter detection (Al-Enzi et al., 2007). The P-wave and S-wave phase arrival times will be detected and selected using the waveform data from the KNSN in this analysis. Typically, using S-arrivals significantly improves earthquake solution accuracy especially location and focal depth. Because each of the sensors used has 3-components, the S-arrivals phases could be recognized without ambiguity. Consequently, the clear impulses of S-phases were mainly identified. Definitely, all earthquake events were essentially located using at least 5 P-wave picks and 2 S-wave picks. The large-time residuals resulting from phase readings of selected stations are always checked. Subsequently, the best events with (ERHs) and (ERZs) of less than 2 km have been selected for issuing the bulletin. Commonly, The RMS of the located earthquakes does not increase more than 0.3 s. The fundamental earthquake parameters (i.e., origin time, arrival time, travel time, location, depth and magnitude) of each earthquake were determined and archived.

7.3.2.4 Magnitude Scales in Kuwait

One of the most essential problems that seismologists face is detecting and selecting a perfect magnitude scale. Nowadays the SeisComP system used in KNSN has numerous adequate magnitude scales which can be used. Before SeisComP era, in the KNSN network, the duration (M_D) and local (M_L) magnitude scales are mainly chosen and used (Al-Enzi et al., 2007). In KNSN network, there was no standard magnitude scale therefore; duration magnitude scale was mainly used (Lee & Stewart, 1981):

$$M_D = 2\log D - 00.87 + 00.0035\Delta$$

where D refers to the total duration in seconds. The Δ expresses the epicentral distance in km. The duration of an earthquake is considered as the time from the first arrival onset to the time when the earthquake amplitude becomes the same as the pre-event amplitude. For each earthquake event, the announced magnitude is determined using all the magnitudes specified for each station. The local magnitude (M_L) relation that is used is the Hutton and Boore (1987) relation, which is close to the original Richter definition:

$$M_L = Log(amp) + 1.11Log(D) + 0.00189D - 2.09.$$

where amp is the amplitude in nanometers and D is the hypocentral distance in kilometers. The scaling relationship between M_L magnitudes and with the body wave magnitude (M_b) of the National Earthquake Information Center (NEIC) of the United States Geological Survey (USGS) was fitted. M_L is related to M_b by the following relation (Al-Enzi et al., 2007):

$$M_b = 0.463M_L + 2.449$$

7.4 Seismic Sources in and Around Kuwait

7.4.1 Local Seismic Sources

Modern and ancient seismic studies indicate that Kuwait is characterized by weak to moderate seismic activity in general. (Abd el-aal et al. 2020, 2021a, b, c; Al-enzi et al., 2007; Gu et al., 2017; Pasyanos et al., 2007). Before 1900, there are no documented manuscripts or seismic records for the State of Kuwait. During the twentieth century and before the establishment of the Kuwait Network for Seismic Monitoring (KNSN), some earthquakes were recorded inside Kuwait by the international seismological centers (Abdelmeinum et al., 2004). The dates and locations of these earthquakes are mentioned in detail in Sect. 7.2. The largest earthquake that was detected inside Kuwait was that in 1993 with a magnitude of 4.7 (Bou-Rabee & Nur, 2002). Some losses were reported in the infrastructure and public facilities, in addition to the extent of fear and panic that the citizens experienced as a result of their feeling of the earthquake (Abd el-aal et al. 2020; Gu et al., 2017; Bou-Rabee & Nur, 2002; Pasyanos et al., 2007). Figure 7.6 shows the seismicity of Kuwait.

Since the work of the network in 1997, there has been a very great possibility to monitor micro/small earthquakes with high accuracy inside Kuwait, which made it easy to identify and detect the local seismic sources inside Kuwait. The KNSN network has recorded more than 1,027

Fig. 7.6 Local seismic sources
in Kuwait

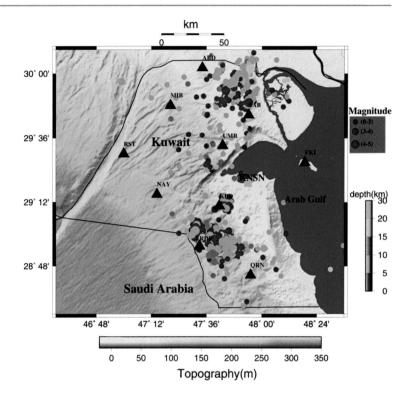

micro/small local earthquakes so far. By plotting the earth-quake data on maps inside Kuwait, it was possible to iden-tify two regions, one in the southwest of Kuwait named Minagish_Umm Gudair area and the other in the northeast of Kuwait, called Raudhatain-Sabriya area where earthquakes occur continuously (Fig. 7.6). Seismic studies conducted in Kuwait are still ongoing to find out the origin of these earthquakes and link their occurrence to the oil extraction process and from these studies (Abd el-aal et al. 2020, 2022; Al-enzi et al., 2007; Bou-Rabee, 2000; Bou-Rabee & Nur, 2002; Bou-Rabee et al., 2004; Gu et al., 2017; Pasyanos et al., 2007).

Although the phenomenon of earthquake swarming (se-quence) is rare in Kuwait and did not occur only once, in the year 2019, where the KNSN recorded about 56 micro/small earthquakes that occurred on two consecutive days starting on the 15th of November (Fig. 7.7). The KNSN monitored 56 micro/small induced earthquakes sequence in the north-ern part of Kuwait having largest magnitude of 4.1 on the Richter scale, which hit Raudhatain-Sabriya area on 15 November 2019, at 23:47:26 Kuwait local time (Fig. 7.6). This earthquake occurred 7.4 km below the earth's surface. On the same day, the KNSN also recorded an earthquake of magnitude 4.0 on the Richter scale, which hit northern part of Kuwait. The earthquake hit at 14:47:37, Kuwait local time; it occurred 1.5 km below the surface of the earth. Many citizens and residents felt it in various parts of Kuwait. These earthquakes did not pose a threat.

The natural, artificial, and induced seismic activity recorded by KNSN from 1997 to 2019 in Kuwait (Fig. 7.6) reflects the incredible increase in the number of smaller earthquakes. This large number of events could be attributed to the 2019 earthquake sequence and road and infrastructure blasting which are defined as artificial earthquakes. Although the KNSN network has only a few seismic stations distributed in many remote sites, still the stations in opera-tion give good azimuth coverage all over Kuwait's territory.

7.4.2 Regional Seismic Sources

Due to its geographical location in the northeastern part of the Arabian Peninsula, being the closest to the Zagros seis-mic belt, Kuwait is more affected by the earthquakes that occur in Iran and the Zagros belt (Fig. 7.8). Therefore, as a result of the collision between the Arab plate and the Eur-asian plate, the rate of earth crust shortening is about 5 mm/year in the northwest of the belt, where it increases at a rate of 10 mm/year in southeast direction along the Zagros belt (Allen et al., 2004; Vernant et al., 2004). Small, med-ium, strong and destructive earthquakes occur along the Zagros belt and at different focal depths, including deep and shallow (Fig. 7.9). The Zagros belt extends from eastern Turkey to Oman in NW–SE direction with a length of 1500 km and a width of 200 to 300 km (Berberian, 1995; Hessami et al., 2003; Jackson & McKenzie, 1984).

Fig. 7.7 Location map of earthquakes sequence took place in 2019 in the northern part of Kuwait (after Abd el-aal et al., 2020)

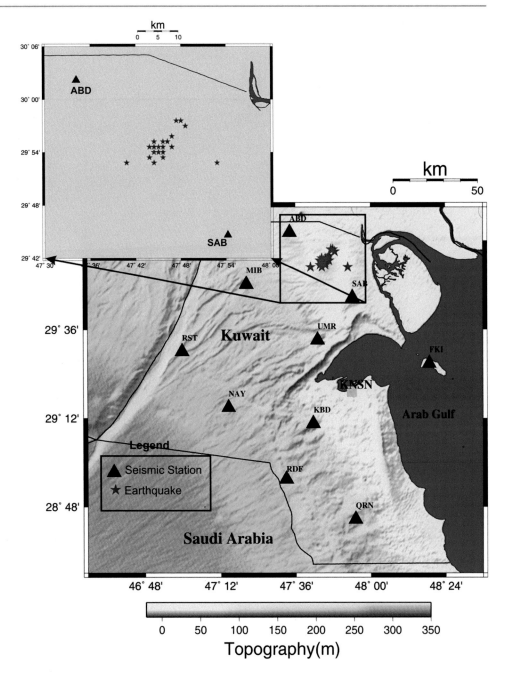

In general, the Zagros Thrust Belt (Zagros FTB) remains the largest seismic belt affecting Kuwait. As a result of the continuous movement on the Zagros FTB belt, earthquakes occur frequently, and their magnitudes range from moderate to strong and sometimes destructive.

The earthquakes recorded in the Zagros (FTB) region are characterized by being of shallow depths up to 20 km, especially in the north, and they increase in depth as we move towards the south (Engdahl et al., 2006; Hatzfeld et al., 2003; Kalaneh & Agh-Atabai, 2016; Talebian & Jackson, 2004; Tatar et al., 2004). Several scientists (e.g.

Berberian, 1995; Jackson, 1980; Jackson & McKenzie, 1984; Kadinsky-Cade & Barazangi, 1982; Ni & Barazangi, 1986) indicated that the mechanical fault plane solutions of the earthquakes that took place along Zagros FTB demonstrate that most seismic events occur on high-angle thrust faults (40°–50°) with strike parallel to the trend of the fold axes at the ground surface. Berberian (1995) pointed out that the major faults in Zagros FTB belt as shown in Fig. 7.10 are Main Zagros Reverse Fault (MZRF), Main Recent Fault (MRF) which are located in the northeastern boundary of the Zagros, High Zagros Fault (HZF), Mountain Front Fault

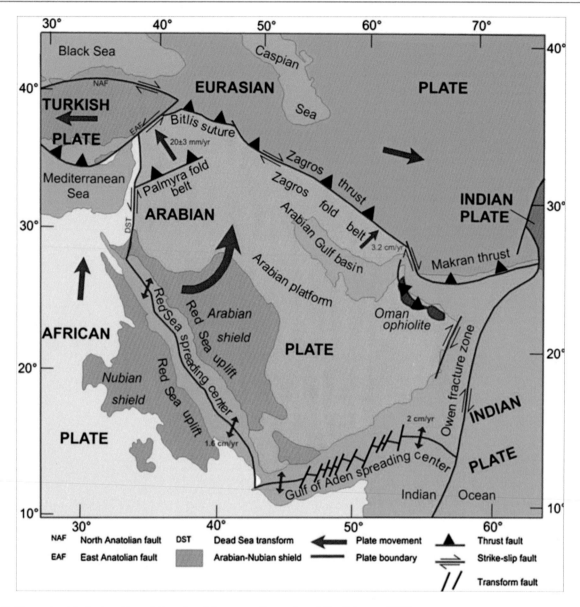

Fig. 7.8 The regional tectonic setting of Kuwait and its surrounding regions (after Johnson, 1998)

(MFF) and Zagros Foredeep Fault (ZFF). The Main Recent Fault (MRF) is a right-lateral strike-slip fault (Berberian, 1995).

Citizens inside Kuwait always feel regional earthquakes, especially the Zagros belt earthquakes, and despite the feeling of the earthquakes, they did not cause any damage, except for people's fear of them as they sometimes shake strongly the buildings (Abd el-aal et al. 2020, 2021c). Definitely, the famous Zagros belt FTB, which always causes moderate to strong earthquakes is far away and not at a small distance from Kuwait, however recently, as a result of the huge urban expansion inside Kuwait, especially the high-rise buildings and towers, people inside these buildings feel ground vibrations more than in the past. This may attribute to several factors, including that the high-rise buildings are

more sensitive to long-period seismic waves, and also the nature and characteristics of the soil and foundation layer beneath buildings play an important role, as it can lead to amplify the seismic waves if they are soft soil.

Earthquakes do not only occur in the Zagros belt zone, but also extend to the western coast of Iran on the Arabian Gulf, as the Kuwait seismic network and also global seismic networks always monitor earthquakes in this region along the coast, and these earthquakes are felt in Kuwait (Abd el-aal et al. 2020, 2021a, b; Engdahl et al., 2006). Earthquakes are also observed and recorded in the Arabian Gulf, but they remain few in number compared to the earthquakes recorded in the Zagros belt. On the other hand, in the western and southern sides of the State of Kuwait, where the Kingdom of Saudi Arabia, earthquakes are rarely detected in

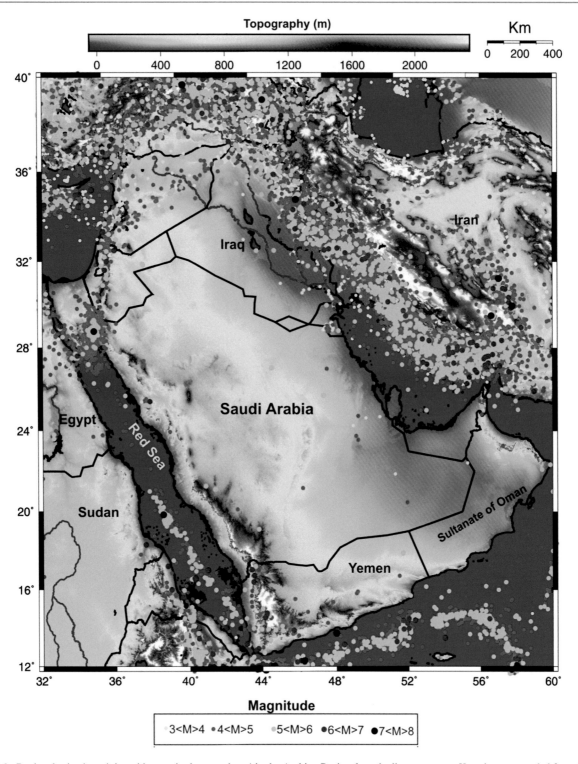

Fig. 7.9 Regional seismic activity with magnitude more than 4 in the Arabian Peninsula and adjacent areas to Kuwait was recorded from 1900 to 2020 (after Abd el-aal et al., 2021b)

this region and they do not pose any seismic hazard to the State of Kuwait.

As for the northern side, where the state of Iraq and the Iranian-Iraqi borders, earthquakes occur, and these

earthquakes are felt in the State of Kuwait, especially if the earthquakes are moderate to strong even with magnitude greater than 5 if they are not close to the borders of Kuwait. Particularly, the Iraqi-Iranian border is considered the

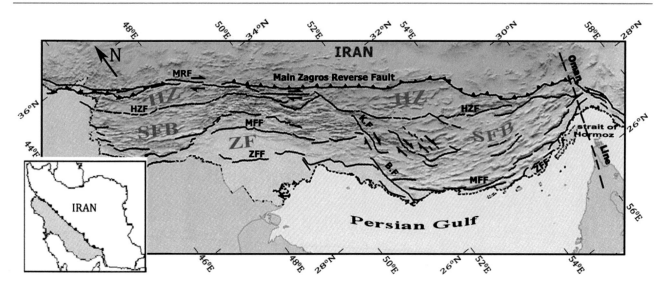

Fig. 7.10 Structure map of the Zagros thrust belt complied from the works of Kalaneh and Agh-Atabai (2016), Talebian and Jackson (2004), Berberian (1995)

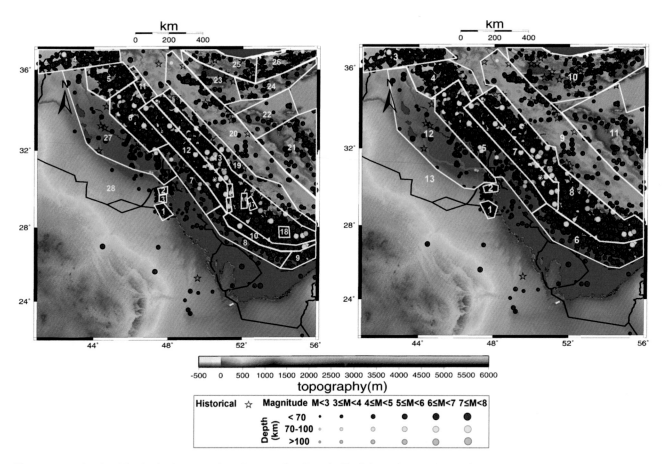

Fig. 7.11 Regional and local seismic sources in and surrounding Kuwait. The left panel represents the first proposed seismotectonic model while the right panel illustrates the second seismotectonic model (after Abd el-aal et al., 2022)

northern extension of the Zagros seismic belt, and therefore many strong earthquakes occur on it (Abd el-aal et al. 2020, 2021a, b, c; Engdahl et al., 2006; Gok et al., 2006; Gu et al., 2017, 2018; Kalaneh & Agh-Atabai, 2016; KISR, 2019).

Abd el-aal et al. (2022) conducted a study to detect and determine the effective local and regional seismic sources in and around Kuwait in two different scenarios of seismotectonic models (Fig. 7.11).

7.4.3 Regional Seismotectonic Setting

One of the most important seismotectonic elements surrounding the State of Kuwait is the Zagros (FTB) Belt, which is the most important tectonic region that greatly affects the State of Kuwait from the eastern side (Abd el-aal et al. 2020, 2021c; Engdahl et al., 2006; Gok et al., 2006; Gu et al., 2017, 2018; Kalaneh & Agh-Atabai, 2016; KISR, 2019). The Zagros FTB is a long zone of about 1,800 km and wide 300 km of the destroyed crusty rocks (Fig. 7.12) that were created at the front of the collision between the Eurasian tectonic plate and the Arabian tectonic plate within the Cretaceous/early Miocene epic (Garzica et al., 2019; Hessami, 2002; Hessami et al., 2003, 2006; Kalaneh & Agh-Atabai, 2016; Vergés et al., 2011). The Zagros FTB zone accommodates one of the largest oil regions in the world which comprise about 49% of the crude oil reserves found in thrust belts (FTBs) (Cooper, 2007).

7.4.4 Crustal Velocity Models in Kuwait

The earthquake velocity model and seismic structure of Kuwait region were studied by the scientists Al-Enzi et al. (2007), Pasyanos et al. (2007). They calculated the seismic velocity and seismic structures using the body-wave (P-and S-waves) waveform dataset of local and teleseismic events. The surface wave dispersion, receiver function and joint inversion were fundamentally used in this study. Indeed, the accuracy of earthquake locations, focal depths, and final network output "the bulletin" relies mainly on the accuracy of crustal models applied to find out the hypocenter

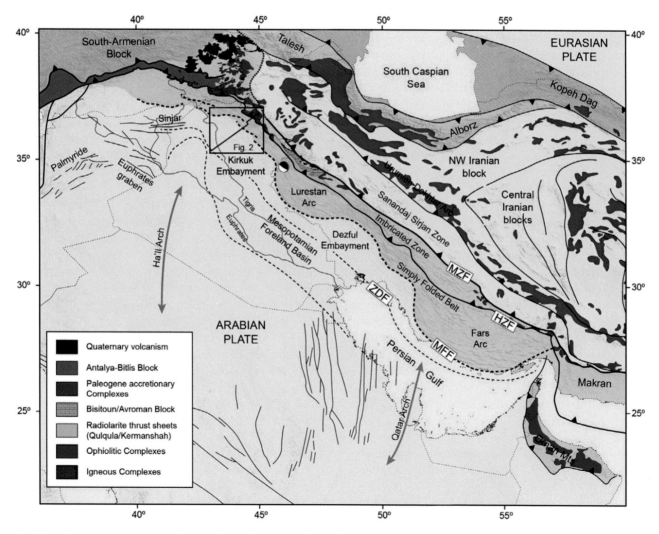

Fig. 7.12 The main tectonic units of Iran deformed between Arabia and Eurasia (after Garzica et al., 2019 which modified from Vergés et al., 2011)

parameters. The variation and difference in the geological structures and its components in the earth's crust and mantle lead to a difference in the velocity values of seismic waves. Al-Enzi et al. (2007) used the surface wave dispersion technique to group surface wave velocity for determining crust and upper mantle structures and seismic velocities in Kuwait region.

The seismic dataset recorded by the KBD broadband seismic station belongs to KNSN network principally used to create surface wave dispersion curves (Al-Enzi et al., 2007; Pasyanos et al., 2007). The use of the recorded seismic data from this station is only due to the absence of any other broadband stations in Kuwait. The authors applied surface wave dispersion using a robust technique called the multiple-filter analysis to calculate Surface Wave Group Velocities (SWGV) (Herrmann, 1973). A huge number of paths from Eurasia, Africa and Middle East were used in the calculations as shown in Fig. 7.13c. Dispersions on paths from the teleseismic and regional earthquakes inside the Kuwait area were measured. However, at the regional distance, most of the measurements were from regional earthquakes recorded from the Zagros seismic belt in Iran (plotted in the red colour paths in Fig. 7.13c). Pasyanos et al. (2001) tomographically used the conjugant gradient technique to invert all of the paths for lateral variation in the group velocity. Inversions were displayed as dispersion maps created separately for both the Rayleigh and Love waves having periods ranging from 7 to 100 s. An illustration map for the group velocity at 15 s Rayleigh waves is plotted in Fig. 7.13d.

It was found that at the KBD station site, there are very slow Love and Rayleigh waves group velocities at short periods, indicating, the presence of dense sediments in this region (Pasyanos et al., 2007) (Fig. 7.14). The authors pointed out there are great matches and comparisons between the 15 s Rayleigh wave group velocity map illustrated in Fig. 7.14d and the sediment thickness illustrated in Fig. 7.13b. Dispersion data were used to invert parameters of the crust and upper mantle in Kuwait. Pasyanos et al., 2007 stated that the obtained results from inversion of surface wave mainly fit with an earth model has a 44 km crustal thickness (Fig. 7.15): It is distributed as 4-km-thick sediments over 40-km-thick crustal layers with a seismic velocity gradient of 0.75 km/s (i.e., 3.0 km/s velocity contrast over 40 km). The upper mantle in the region has a P-wave velocity of 7.8 km/s (i.e., 4.3 km/s S-wave velocity).

The second scheme used by Al-Enzi et al. (2007), Pasyanos et al. (2007) is the Receiver function method (RF). Certainly, the Receiver functions have become famous and very effective in determining the velocity and thickness of the lithosphere and the earth's crust under any seismic recording station (Ammon et al., 1990; Langston, 1979;

Owens et al., 1984). Fundamentally, the technique of Teleseismic Receiver Function (TRF) separates converted P-to-S-waves from the incident P-wave. Thus it is determined by deconvolving the vertical component from the radial component. This approach is sensitive and robust for determining local structures within 100 km of the monitoring station.

Al-Enzi et al. (2007), Pasyanos et al. (2007) calculated the receiver function at station KBD for teleseismic events. The use of epicentral distances for all the earthquakes used in the study ranged between 60 to 85°, gave a steep angle of incidence for the P-waves.

The authors engaged three main groups of earthquake events from in/near the Japanese islands, southwestern Pacific, and from south region of southern Africa (Fig. 7.16).

Al-Enzi et al. (2007) concluded that the results obtained from the inversion of receiver function varied extensively among the stacks from different back azimuths. The study showed that there is a very large difference between inferred velocities in the upper and lower parts of the earth's crust as well as the upper part of the mantle. For this reason the authors (Al-Enzi et al., 2007; Pasyanos et al., 2007) combined two schemes datasets into one inversion approach implementing the joint inversion results (Fig. 7.17).

The KUW1 velocity model was mainly constructed by Al-Enzi et al. (2007) for the Kuwait region. The KUW1 Model is illustrated in green colour in Fig. 7.18 with details in Table 7.3.

Al-Enzi et al. (2007), Pasyanos et al. (2007) concluded that a thick sedimentary column was found, which were classified into two layers. A 17-km-thick upper crust with P-wave seismic velocity equal to 5.89 km/s and S-wave seismic velocity equal to 3.40 km/s exists in the crystalline crust, while the middle crust has a 9-km-thick with P-wave seismic velocity equal to 6.41 and S-wave seismic velocity equal to 3.70, and the lower crust with P-wave velocity = 6.95 km/s and S-wave velocity = 3.90 km/s has 11-km-thick. The total crustal thickness was 45 km. The study shows that beneath the Moho zone, the P-and S-seismic velocities in the upper mantle have 7.84 km/s and 4.4 km/s (Al-Enzi et al., 2007). The results were mainly compared to other seismic velocity profiles for Kuwait region (Fig. 7.18).

The first two models basically investigate the sedimentary thickness. However, three layers in the region have thicknesses of 0.93, 4.00, and 2.35 km, while the P-wave velocities of 2.53, 4.02, and 5 km/s have been characterized in the Laske sediment model (Laske et al., 2001) (magenta lines). Other seismic models take into account the profile of the entire Earth's crust. The CRUST2.0 seismic model shown in black is the first model (Laske et al., 2001). The CRUST2.0 model basically distinguishes this area as

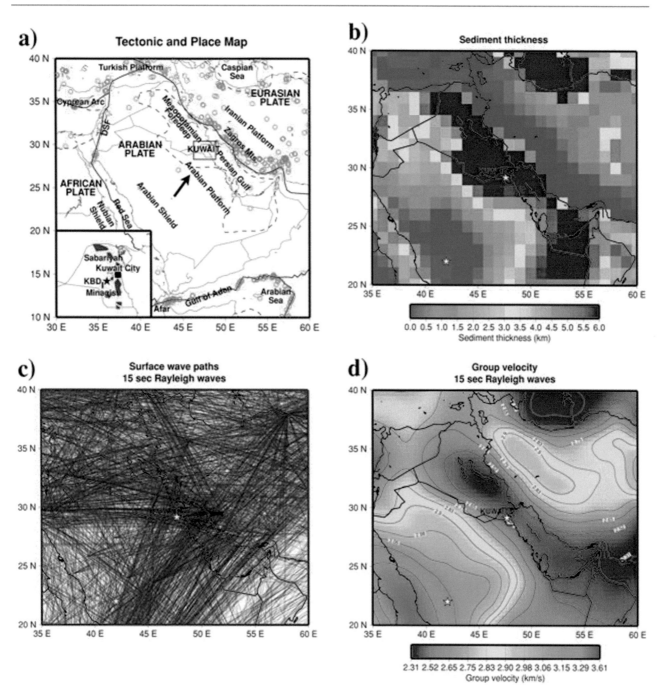

Fig. 7.13 Panel **a** tectonic and structure map of the Arab Peninsula and its surroundings. The star symbol in the map refers to KBD station while the circle symbols represent recent seismicity (events $M \geq 5.0$ from 1990 to present). The dashed lines designated areas have sediment thicknesses of more than 5 km, while the solid lines refer to tectonic plate borders. Panel **b** shoes the thickness map of sediment of the Arab Peninsula and its surroundings from Laske et al. (2001). Panel **c** shows path map of the Rayleigh wave group velocity dispersion measurements at 15 s. The red lines indicate regional paths monitored at station KBD. The Star symbol refers to the place of the profiles illustrated in Fig. 7.14. Panel **d** illustrated the Rayleigh waves group velocities at 15 s obtained from tomographic inversion (after Pasyanos et al., 2007)

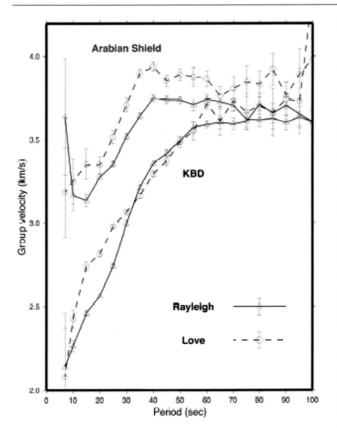

Fig. 7.14 Rayleigh wave group velocities at station KBD (black lines) while the red lines at the Arabian Shield location (after Al-Enzi et al., 2007)

DA, In other words, it is a primary model that has 6 km-thick layers. The crust in the area in this specific model is about 41 km-thick with seismic velocities of 5.88 km/s and 3.28 km/s for P- and S-wave, respectively. Thereupon, the seismic velocities of the mantle below the Moho are of 8.2 km/s and 4.7 km/s.

The crustal model in red colour which is picked from Bou-Rabee (2000) is mainly obtained from CRUST5.1 (Mooney et al., 1998). The sediment thickness in this model is about 7 km, while the thickness of the crust is approximately 40 km; however, the average body-wave seismic velocities of the upper mantle have 8.0 km/s and 4.49 km/s respectively. The crustal model of the Arabian Platform (AP) is an average model for the Arabian Platform and is completely considered from the regional waveform modeling paths from the Zagros seismic belt which was recorded in central Arabia (Rodgers et al., 1999). This model is illustrated on the figure in cyan colour. The model of Pasyanos et al. (2004) is shown in blue is mainly considered from the WENA model.

7.4.5 Focal Mechanism Solutions and Stress Pattern

To understand the nature of seismic sources in Kuwait, focal-mechanism solutions, of the local earthquakes that occurred in Kuwait, are significantly needed. The faults that cause earthquakes are mainly under the Earth's surface, however, they may extend to the ground surface. Outgoing seismic waves are affected by the stress distribution in/near the source, thus imparting asymmetry to distant sites. This of course will lead to the assumption of radiation patterns (Lay & Wallace, 1995). The radiation pattern can be typically defined as a geometric description of the wave amplitude and sensation of the initial motion distribution on the P-and S-wave-fronts near the source (Fig. 7.19). However, the shear dislocation with low order symmetry presents inferring a predictable relationship between radiation pattern of seismic wave motion and orientation of fault-plane, allowing the identification of faulting processes remotely.

The focal-mechanism of earthquake provides details fundamental to understanding the nature and origin of the seismic event. The focal-mechanism solution of earthquake permits stress mode to change in the ground to be identified from far distances at the exact time and location of the earthquake. Accumulation of focal-mechanism solutions (i.e., details about fault slip and fault orientation) have revealed the kinematic behavior of seismic zones, and led to research on the seismological implications of global tectonics (Fig. 7.20).

Al-Enzi et al. (2007) conducted a study to prepare focal-mechanism maps for earthquakes in Kuwait, the available phase readings, or seismograms, of the KNSN were used. Thirty-three earthquakes were selected according to the accuracy of their location, clearness of first pick and number of minimum recording seismic stations (Al-Enzi et al., 2007). Those events were recorded in at least 6 stations. Generally, the accuracy of focal-mechanism determination increases as the number of stations increases. The selected events were grouped into two groups according to focal depth the first group included events that were located at depth of less than 6 km, and the second group included events that were located at depths of more than 6 km.

To construct the focal mechanism of an event's source, the PMAN program was used. In constructing the fault-plane solution for an earthquake, 3 parameters were used the type of polarities of the first onset P-wave from waveform data at numerous stations, source-to-station azimuths for those stations, and takeoff angles (AIN) for the P-wave rays traveling from the earthquake focus towards the station. AIN determination requires knowledge of the P-wave velocity with

Fig. 7.15 Velocity dispersion results of surface wave using inversion technique. The upper lift panel shows the receiver-function data fit while the dispersion data are shown in lower lift panel. The velocity model is illustrated at the right panel (after Al-Enzi et al., 2007)

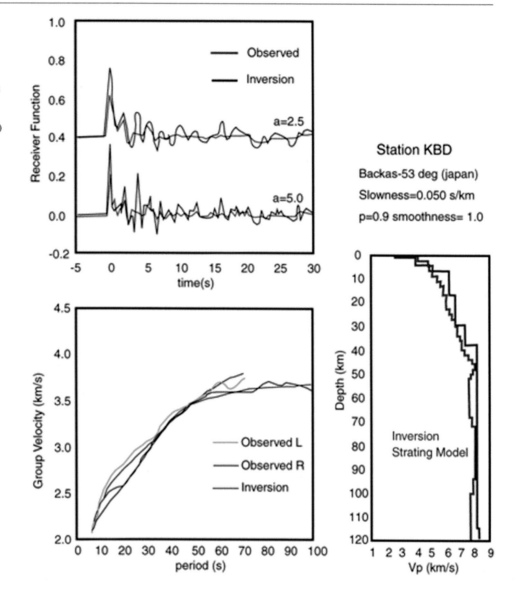

depth. The used seismic velocity model in the source area significantly affects AIN calculation for rays incoming at a certain seismic station. The use of inaccurate seismic velocities model may lead to errors in the separation of fault planes and also not setting the polarity readings of the stations in the correct quadrants. The AIN angles were calculated by using the newest crustal model developed for Kuwait.

Due to the lack of stations that cover the area and the low magnitudes of the recorded events, the fault-plane solutions of individual events were uncontrolled. So, composite fault-plane solutions were made to overcome the uncontrolled fault-plane solutions of individual events. Six composite diagrams were prepared in this study, 2 for the first group (S1 and S2) and 4 (D1 to D4) for the second group (Al-Enzi et al., 2007). Composites indicate to reverse and normal focal mechanisms. Such structures are not observed

in the exposed surface. At the southern region, the composites of events which have depths of more than 6 km (D1, D2 and D3) are distinguished by a normal faulting with a strike-slip element (Fig. 7.21). In turn, the composite (S1) of seismic events located at depths of less than 6 km is distinguished by a reverse faulting with minor component of strike-slip. The two composites (S2 and D4) for the northeastern region are characterized by reverse mechanisms, although the events were located at different depth levels. These mechanisms reflect the domination of compression forces in the northern part of Kuwait, consistent with the general trend of forces in the area toward the subduction zone along the Zagros Mountains (Al-Enzi et al., 2007).

Abd el-aal et al. (2021a) prepared a study regarding the phenomenon of earthquakes sequence that occurred in the northern part of the State of Kuwait, and this phenomenon is rare and occurred for the first time in Kuwait since the

Fig. 7.16 Upper panel; shows
event distribution. Lower panel
illustrates path plot from
teleseismic earthquakes to station
KBD (after Al-Enzi et al., 2007)

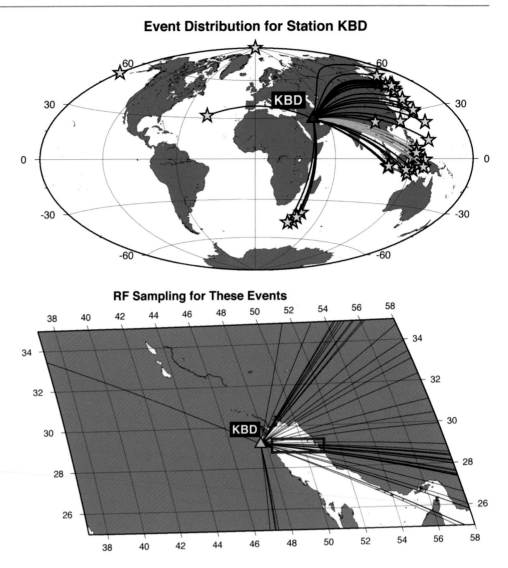

Fig. 7.16 Upper panel; shows event distribution. Lower panel illustrates path plot from teleseismic earthquakes to station KBD (after Al-Enzi et al., 2007)

establishment of the KNSN network. On November 15, 2019, for two consecutive days, the KNSN recorded a number of micro/small earthquakes, numbering more than 56. Only 29 earthquakes were selected from the sequence in terms of the number of recorded stations, their magnitudes, and the accuracy of the location of the selected earthquakes. The study was carried out by Abd el-aal et al. (2021a), including the determination of earthquake properties in terms of source parameters, fault type and fault orientation, as well as the determination of moment magnitudes using modern geophysical methods such as moment Tensor and inversion techniques, in addition to determining the pattern of the stresses in the region.

Abd el-aal et al. (2021a) identified the two largest earthquakes in the group, with moment magnitudes 4.4 and 4.5, respectively. The study indicated that one from the two earthquakes is deeper in focal depth than the other. Fourteen earthquakes were purified and selected for focal mechanism process based on the quality of the waveform and minimum

required station number. The study indicated that the fault causing earthquakes is of the normal fault type with a minor strike-slip element. The study also showed that the fault takes NE-SW direction, and this direction fully corresponds to the distribution, clustering and shape of the earthquakes sequence illustrated in Fig. 7.22 (Abd el-aal et al. 2021a). The researchers stated that the obtained results indicated that this seismic sequence has a human component equivalent to the tectonic component as assigned from the CLVD and the ratio of DC components in addition to the obtained values of the stress drop (Abd el-aal et al. 2021a). All the results obtained from the study indicate that these earthquakes are of the type "triggered seismicity" and also fully agree with the geological and structural situation in the region (Abd el-aal et al. 2021a). The authors also pointed out Kuwait contains many oil production fields, but earthquakes do not occur except in certain places with pre-existing faults in the northern and southern regions and not from others (Abd el-aal et al. 2021a).

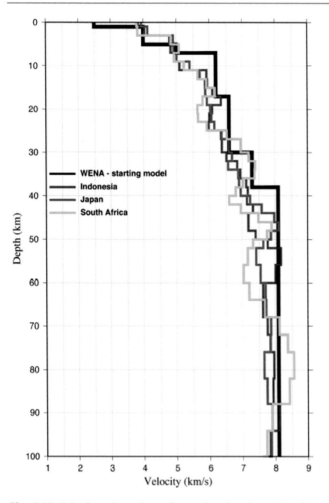

Fig. 7.17 Joint inversion scheme for receiver function and surface wave datasets (after Al-Enzi et al., 2007)

The geological structure and tectonic situation in the region, where the earthquakes sequence took place, may be approximated by considering the stress pattern throughout the release of energy from earthquake events (Abd el-aal et al. 2021a). The authors mainly estimated the stress style from the focal mechanism solutions of the earthquakes sequence presented in Fig. 7.22 by means of the software code-named STRESSINVERSE of Vavrycuk (2014). The software code utilizes Michael's method (Michael, 1984) which is robust and accurate to determine the stress axes direction. This powerful procedure can provide reasonable accuracy of the stress axes estimation, even if the focal mechanism solutions are randomly picked (Fojtikova & Vavrycuk, 2018; Fojtikova et al., 2010). The fault-plane solutions of nearly 14 earthquakes from the earthquake sequence are fundamentally applied to recognize the stress style in the area (Fig. 7.23).

The stress distribution in the region where the earthquake sequence occurred was mainly determined from the focal mechanism solutions of 14 earthquakes by performing the stress tensor procedure. The main stress axes directions are shown in Fig. 7.23, while Table 7.4 contains their obtained values. Therefore, the estimated results in Fig. 7.23 display that the axis σ_1 has a nearly vertical plunge taking direction SSE, while the axis σ_3 shows a vertical plunge in main direction SWW (Fig. 7.23). A schematic diagram of Mohr circle presented in Fig. 7.23 panel (b), demonstrated that the estimated fault planes are principally clustered on the failure criterion area called Mohr–Coulomb, proofing that the analyzed dataset are consistent with the fault instability model (Vavrycuk, 2011, 2014; Vavrycuk et al., 2013). The results obtained from stress tensor pointed to the predictable shape ratio is about 0.53 and the friction is about 0.4 on the faults for the earthquake sequence as revealed in Fig. 7.23 panel (d).

7.5 Conclusion

In this chapter, we review the most important works, outputs and results that specifically pertain to the Kuwait seismicity. There are really a few works on this subject. Kuwait is located on the Arabian Gulf in the northeast corner of the Arab peninsula. Actually, studying and monitoring seismic activity in Kuwait started recently in 1997. There is no record of historical earthquakes in Kuwait. Many scientists believe that the local seismic activity in Kuwait is closely related to the process of extracting oil from Kuwait. Especially the local earthquakes inside Kuwait are characterized by small/minor magnitudes and shallow focal depths.

Seismic studies in Kuwait indicate that there are two regions where local earthquakes occur. The Minagish_Umm Gudair region is located in the southwest of Kuwait, and the Raudhatain-Sabriya region is located in the northeast of Kuwait. Kuwait is constantly exposed to regional earthquakes from the famous Zagros belt.

In this chapter, we summarized most of the works that dealt with the seismological situation in Kuwait, as well as the history of seismic monitoring in Kuwait.

Acknowledgements All thanks and appreciation to the Kuwaiti National Seismic Network for providing us with all seismic reports, published researches, earthquake catalogs and data of the local earthquakes.

Fig. 7.18 The figure shows comparison between the seismic velocity model named KUW1 and different seismic velocity models from regional regions. The seismic velocity model (KUW1) is plotted in green colour. The regional seismic models include the CRUST2.0, A P, WENA, sediment, and Kuwait model which are plotted in black, cyan, blue, magenta, and red respectively. The black arrows in the right panel demonstrate the sedimentary layers from Bou-Rabee (2000), Al-Enzi et al. (2007)

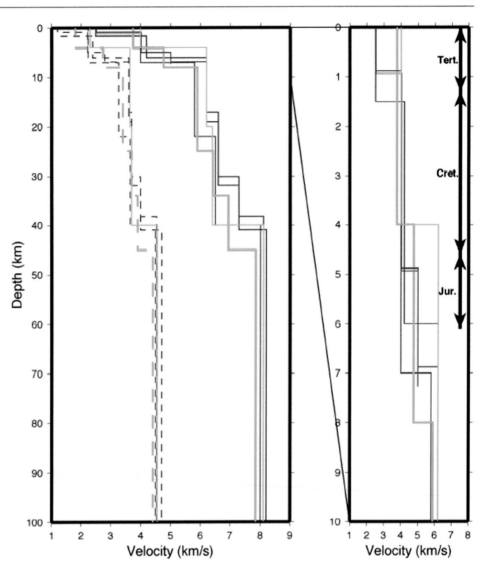

Table 7.3 Outline of the seismic models for Kuwait region and its vicinity. Label refers to colour or symbol used in Fig. 7.18 (after Al-Enzi et al., 2007)

Model	Label	Sediment Thickness (km)	Average Crustal Velocity (km/s) (P-/S-wave)	Crustal Thickness (km)	Upper Mantle Velocity (km/s) (P-/S-wave)
Laske sed	Magenta	7.3	*	*	*
Sed. Profile	Arrow	6.1	*	*	*
CRUST 2.0	Black	6	5.88/3.28	41	8.20/4.7
Kuwait	Red	7	6.18/3.47	40	8/4.49
WENA	Blue	7	6.18/3.31	38	8.1/4.58
AP	Cyan	4	6.07/3.5	40	8.1/4.55
KUW1	Green	8	5.96/3.38	45	7.84/4.44

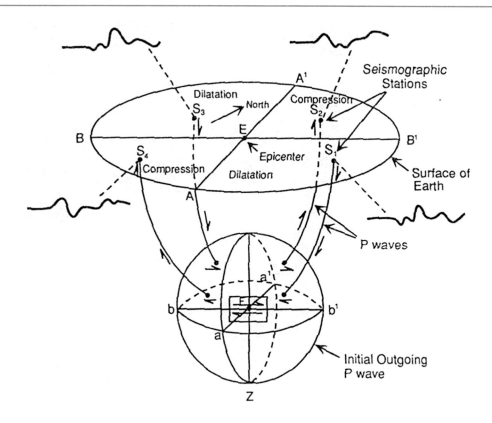

Fig. 7.19 Three-dimensional variation of P-and S-wave amplitude and polarity on a spreading wavefront source

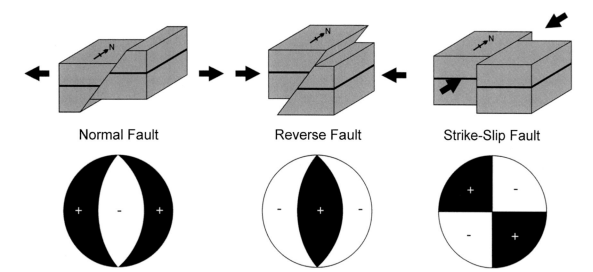

Fig. 7.20 Diagram illustrates the type of faults and their focal mechanism solutions

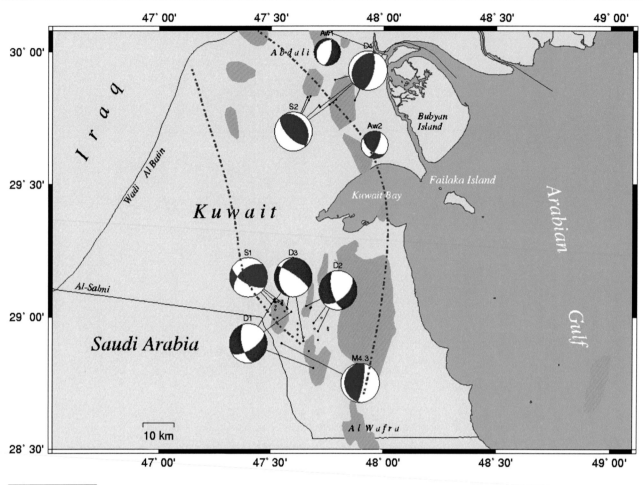

Fig. 7.21 Focal mechanism solutions of the earthquakes in Kuwait (after Al-Enzi et al., 2007)

Fig. 7.22 Focal mechanism
solutions of the earthquake
sequence took place at the
northern part of Kuwait in 2019
(after Abd el-aal et al., 2021a)

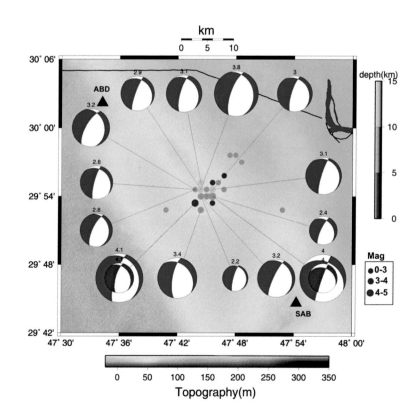

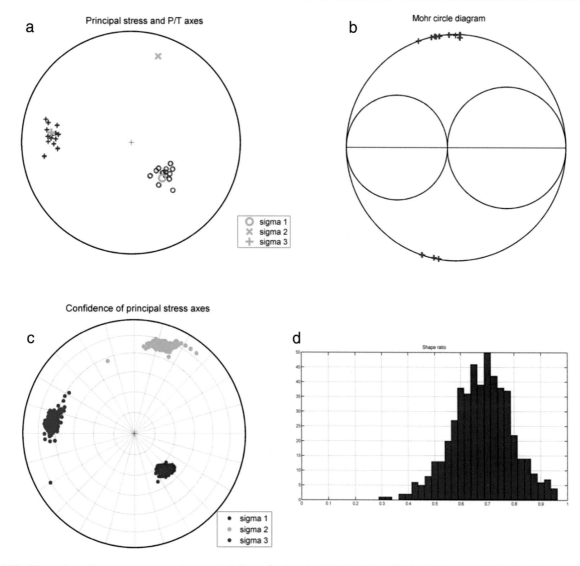

Fig. 7.23 The estimated stress tensor results acquired from focal mechanism solutions in the northern region of Kuwait. The panel (**a**) illustrates the σ_1, σ_2 and σ_3 axes and P- and T axes. Panel (**b**) shows the Mohr's circles with the faults recognized by the blue crosses. Panel (**c**) shows the confidence limits of the main stress axes. Finally, the shape ratio values are mainly plotted in Panel (**d**)

Table 7.4 The azimuth and plunge values of main stress axes of the earthquakes sequence occurred at the northern area of Kuwait

Principle stress axes	Earthquake sequence	
	Azimuth	Plunge
σ_1	138.5°	54.9°
σ_2	17.7°	19.8°
σ_3	276.8°	27.6°

References

Abd el-aal, A. K., Al-Jeri, F., Al-Enezi, A., & Parol, J. A. (2020). Seismological aspects of the 15 November 2019 earthquakes sequence, Kuwait. *Arabian Journal Geoscience, 13*, 941. https://doi.org/10.1007/s12517-020-05919-1.

Abd el-aal, A. K., Parol, J. A., Al-Jeri, F., & Al-Enezi, A. (2021a). Modeling and simulating ground motion from potential seismic sources to Kuwait: Local and regional scenarios. *Geotectonics, 55*. https://doi.org/10.1134/S001685212103002X.

Abd el-aal, A. K., Al-Enezi, A., Sadalla, H., & Al-Jeri, F. (2021b). Tectonic and anthropogenic characteristics of the 15 November 2019 micro earthquakes sequence, Kuwait. *Geotectonics, 55*, 137–152. https://doi.org/10.1134/S0016852121010039.

Abd el-aal, A. K. A., Al-Enezi, A., Al-Jeri, F., & Mostafa, S. I. (2021c). Earthquake deterministic hazard and sensitivity analysis for residential Cities in Kuwait. *Journal of Volcanology and Seismology, 15*(6), 445–462. https://doi.org/10.1134/S0742046321060051.

Abd el-aal, A. K., Al-Enezi, A., Al-Jeri, F., Naser, O., & Alenezi, Mostafa, S. I. (2022). Probabilistic and deaggregation seismic hazard maps for Kuwait. *Submitted to Arabian Journal of Geosciences.*

Abdelmeinum, M. M., Abdel-Fattah, R., & Agami, D. N. (2004). "Earthquakes" Kuwait Research for Scientific Research publishing in Arabic language.

Al-enzi, A., Sadeq, A., & Abdel-Fattah, R. (2007). Assessment of the Seismic Hazard for the state of Kuwait. Internal report EC011C, submitted to KFAS 2007.

Allen, C. R. (1982). Earthquake Prediction—1982 overview. *Bulletin of the Seismological Society of America, 72*(6B), S331–S335.

Allen, M., Jackson, J. A., & Walker, R. (2004). Late Cenozoic reorganization of the Arabia-Eurasia collision and the comparison of short-term and long-term deformation rates. Tectonics 23, TC2008.

Ammon, C. J., Randall, G. E., & Zandt, G. (1990). On the non-uniqueness of receiver function inversions. *Journal of Geophysical Research, 95*, 15303–15318.

Bath, M. (1979). Introduction to Seismology (Second, Revised ed.). Basel: Birkhäuser Basel. ISBN: 9783034852838.

Berberian, M. (1995). Master 'blind' thrust faults hidden under the Zagros folds, active basement tectonics and surface morphotectonics. *Tectonophys, 241*, 193–224.

Burton, I., Kates, R. W., & White, G. F. (1978). *The Environment as Hazard.* New York: Oxford University Press.

Bolt, B. A. (1993). Earthquakes and geological discovery, *Scientific American Library,* ISBN: 978-0-7167-5040-6.

Bormann, P. (Ed.) (2012). New manual of seismological observatory practice (NMSOP-2), IASPEI, GFZ German Research Centre for Geosciences, Potsdam; nmsop.gfz-potsdam.de. https://doi.org/10.2312/GFZ.NMSOP-2.

Bou-Rabee, F. (2000). Seismotectonics and earthquake activity of Kuwait. *Journal Seismology, 4*, 133–141.

Bou-Rabee, F., & Nur, A. (2002). The 1993 M4. 7 Kuwait earthquake: Induced by the burning of the oil fields, Kuwait. *Journal Science Engineering, 29*(2), 155–163.

Bou-Rabee, F., & Abdel-Fattah, R. (2004). Seismological observation in the state of Kuwait. *Kuwait Journal Science Engineering, 31*(1), 175–192.

Cooper, M. (2007). Structural style and hydrocarbon prospectivity in fold and thrust belts: A global review. In A. C. Ries, R. W. Butler & R. H. Graham (Eds.), *deformation of the continental crust: The legacy of mike coward.* Special Publications. 272. London: Geological Society. pp. 447–472. ISBN: 978-1-86239-215-1. Retrieved 2 July 2011.

Chung, D. H., & Bernreuter, D. L. (1980). Regional Relationships Among Earthquake Magnitude Scales., NUREG/CR-1457.

Davison, C. (2014). The founders of seismology. ISBN: 9781107691490.

Dudley, W. C., & Lee, M. (1988: 1st edition) Tsunami! ISBN: 0-8248-1125-9.

Engdahl, E. R., Jackson, J. A., Myers, S. C., Bergman, E. A., & Priestley, K. (2006). Relocation and assessment of seismicity in the Iran region. *Geophysical Journal International, 167*, 761–777.

Fojtikova, L., & Vavrycuk, V. (2018). Tectonic stress regime in the 2003–2004 and 2012–2015 earthquake sequences in the Ubaye Valley, French Alps. Pure Appl Geophys. https://doi.org/10.1007/s00024-018-1792-2.

Fojtíková, L., Vavryčuk, V., Cipciar, A., & Madarás, J. (2010). Focal mechanisms of micro-earthquakes in the Dobrá Voda seismoactive area in the Malé Karpaty Mts. (Little Carpathians), Slovakia. Tectonophysics, 492(1–4), pp.213–229.

Garzica, E., Vergés, J., Sapin, F., Saura, E., Meresse, F., & Ringenbachc, J. C. (2019). Evolution of the NW Zagros Fold-and-Thrust Belt in Kurdistan Region of Iraq from balanced and restored crustal-scale sections and forward modeling. *Journal of Structural Geology, 124*(2019), 51–69.

Gok, R. M., Rodgers, A. J., & Al-Enezi, A. (2006). Seismicity and improved velocity structure in Kuwait. Internal report submitted to KISR, report no: UCRL-TR-218465.

Gu, C., Al-Jeri, F., Al-Enezi, A., Büyüköztürk, O., & Toksöz, M. N. (2017). Source mechanism study of local earthquakes in Kuwait. *Seismological Research Letters, 88*(6), 1465–1471. https://doi.org/10.1785/0220170031.

Gu, C., Prieto, G. A., Al-Enezi, A., Al-Jeri, F., Al-Qazweeni, J., Kamal, K., Kuleli, S., Mordret, A., Büyüköztürk, O., & Toksöz, M. N. (2018). Ground motion in Kuwait from regional and local earthquakes: Potential effects on tall buildings. *Pure Application Geophysics, 175*, 4183.

Gubbins, D. (1990). *Seismology and plate tectonics.* Cambridge University Press. ISBN: 978-0-521-37141-4.

Gutenberg, B., & Richter, C. F. (1942). Magnitude and energy of earthquakes. *Buletin of the Seismology Society of Amercia, 32*, 163–191.

Hall, S. S. (2011). Scientists on trial: At fault? *Nature, 477*(7364), 264–269. Bibcode:2011Natur.477.264H. https://doi.org/10.1038/477264a.

Hanks, T. C., & Kanamori, H. (1979). A moment magnitude scale. *Journal of Geophysical Research, 84*, 2348–2350.

Hatzfeld, D., Tatar, M., Priestley, K., & Ghafory-Ashtiany, M. (2003). Seismological constraints on the crustal structure beneath the Zagros Mountain belt (Iran). *Geophysical Journal International, 155*, 403–410.

Herrmann, R. B. (1973). Some aspects of bandpass filtering of surface waves. *Bulletin of the Seismology Society America, 63*, 663–671.

Hessami, K. (2002). Tectonic history and present-day deformation in the Zagros fold-thrust belt. Acta Universitatis Upsaliensis. Comprehensive Summaries of Uppsala Dissertations from the Faculty of Science and Technology 700. 13 pp. Uppsala. ISBN: 91-554-5285-5.

Hessami, K., Nilforoushan, F., & Talbot, C. J. (2006). Active deformation within the Zagros Mountains deduced from GPS measurements. *Journal of the Geological Society of London, 163*, 143–148.

Hessami, K., Jamali, F., & Tabassi, H. (2003). Major active faults in Iran. Ministry of science, research and technology. *International Institute of Earthquake Engineering and Seismol.* (IIEES), 1:250000 scale map.

Hutton, L. K., & Boore, D. (1987). The M_L scale in Southern California. *Bulletin of the Seismology Society of America, 77*, 2074–2094.

Jackson, J. A. (1980). Error in focal depth determination and the depth of seismicity in Iran and Turkey. *Geophysical Journal of the Royal Astronomical Society, 61,* 285–301.

Jackson, J. A., & McKenzie, D. (1984). Active tectonics of the Alpine-Himalayan Belt between western Turkey and Pakistan. *Geophysical Journal of the Royal Astronomical Society, 77,* 185–264.

Johnson, P. R. (1998). Tectonic map of Saudi Arabia and adjacent areas. Deputy Ministry for Mineral Resources, USGS TR-98-3, Saudi Arabia.

Kadinsky-Cade, K., & Barazangi, M. (1982). Seismotectonics of southern Iran: The Oman Line. *Tectonics, 1,* 389–412.

Kalaneh, S., & Agh-Atabai, M. (2016). Spatial variation of earthquake hazard parameters in the Zagros fold and thrust belt. *SW Iran. Nat Hazards, 2016*(82), 933–946. https://doi.org/10.1007/s11069-016-2227-y.

Kanamori, H. (1977). Energy-release in great earthquakes. *Journal of Geophysical Research, 82,* 2981–2987. https://doi.org/10.1029/Jb082i020p02981.

Kanamori, H. (2003). Earthquake prediction: An overview. International Handbook of Earthquake and Engineering Seismology. 81 Part B. *International association of seismology & physics of the earth's interior,* pp. 1205–1216.

KISR. (2019). *Annual earthquakes bulletin.* Kuwait institute of scientific Research, Kuwait.

Langston, C. A. (1979). Structure under Mount Rainer, Washington, inferred from teleseismic body waves. *Journal of Geophysical Research, 84,* 4749–4762.

Laske, G., Masters, G., & Reif, C. (2001). Crust 2.0: a new global crustal model at 2 × 2 degrees http://mahi.ucsd.edu/Gabi/rem.html.

Lay, T and Wallace, T. (1995). Modern global seismology. *International Geophysics Series,* (Vol. 58). New York: AcademicPress.

Lee, W. (1990). Hypo71pc program. IASPI software library, Vol 1.

Lee, W., Kanamori, H., Jennings, P., & Kisslinger, C. (2003). International Handbook of Earthquake and Engineering Seismology, Part B. Volume 81, Part B, pp. 937–1948.

Lee, W., & Stewart, S. (1981). *Principles and applications of microearthquake networks.* Academic Press.

Levin, B., & Nosov, M. (2009). *Physics of tsunamis.* Springer, Dordrecht 2009. ISBN: 978-1-4020-8855-1.

Lienert, B. R., & Havskov, J. (1995). A computer program for locating earthquakes both locally and globally. *Seismological Research Letters, 66,* 26–36.

Michael, A. J. (1984). Determination of stress from slip data: Faults and folds. *J Geophys Res Solid Earth, 89*(B13), 11517–11526.

Ni, J., & Barazangi, M. (1986). Seismotectonics of Zagros continental collision zone and a comparison with the Himalayas. *Journal of Geophysical Research, 91*(B8), 8205–8218.

Owens, T. J., Zandt, G., & Taylor, S. R. (1984). Seismic evidence for an ancient rift beneath the Cumberland Plateau, Tennessee: A detailed analysis of broadband teleseismic P waveforms. *Journal of Geophysical Research, 89,* 7783–7795.

Pasyanos, M. E., Walter, W. R., & Hazler, S. E. (2001). A surface wave dispersion study of the Middle East and North Africa for monitoring the Comprehensive Nuclear-Test-Ban Treaty, Pure App. *Geophysics, 158,* 1445–1474.

Pasyanos, M. E., Walter, W. R., & Flanagan, M. P., Goldstein, P., & Bhattacharyya, J. (2004). Building and testingan a priori geophysical model for Western Eurasia and North Africa. *Pure and Applied Geophysics, 161,* 235–281.

Pasyanos, M. E., Tkalcic, H., Gok, R., Al-Enezi, A., & Rodgers, A. J. (2007). Seismic structure of Kuwait. *Geophysical Journal International, 2007*(170), 299–312. https://doi.org/10.1111/j.1365-246X.2007.03398.x.

Richter, C. F. (1935). An instrumental earthquake magnitude scale. *Bulletin of the Seismology Society America, 25,* 1–32.

Rodgers, A., Walter, W., Mellors, R., Al-Amri, A. M. S., & Zhangm, Y. S. (1999). Lithospheric structure of the Arabian shield and platform from complete regional waveform modeling and surface wave group velocities, *Geophysical Journal International, 138:* 871–878.

Tatar, M., Hatzfeld, D., & Ghafory-Ashtiany, M. (2004). Tectonics of the central Zagros (Iran) deduced from microearthquake seismicity. *Geophysical Journal International, 156,* 255–266. https://doi.org/10.1111/j.1365-246X.2003.02145.x.

Talebian, M., & Jackson, J. (2004). A reappraisal of earthquake focal mechanisms and active shorting in the Zagros Mountains of Iran. *Geophysical Journal International, 156,* 506–529.

Vavrycuk, V. (2011). Principal earthquakes: Theory and observations from the 2008 West Bohemia sequence. *Earth and Planetary Science Letter, 305,* 290–296. https://doi.org/10.1016/j.epsl.2011.03.002.

Vavrycuk, V. (2014). Iterative joint inversion for stress and fault orientations from focal mechanisms. *GJI, 199*(1), 69–77.

Vavrycuk, V., Bouchaala, F., & Fischer, T. (2013). High-resolution fault image from accurate locations and focal mechanisms of the 2008 sequence earthquakes in West Bohemia, Czech Republic. *Tectonophys, 590,* 189–195. https://doi.org/10.1016/j.tecto.2013.01.025.

Vergés, J., Saura, E., Casciello, E., Fernàndez, A., Jiménez-Munt, I., et al. (2011). Crustal-scale cross-sections across the NW Zagros belt: Implications for the Arabian margin reconstruction. *Geological Magazine, 148*(5–6), 739–761.

Vernant, Ph., Nilforoushan, F., Hatzfeld, D., Abassi, M. R., Vigny, C., Masson, F., Nankali, H., Martinod, J., Ashtiani, A., Bayer, R., Tavakoli, F., & Chery, J. (2004). Present-day crustal deformation and plate kinematics in Middle East constrained by GPS measurements in Iran and northern Oman. *Geophysics Journal Interncational, 157,* 381–398.

Geo- and Environmental Hazard Studies in Kuwait

8

Jasem Mohammed Al-Awadhi, Abd el-aziz Khairy Abd el-aal, Raafat Misak, and Ahmed Abdulhadi

Abstract

Low magnitude Earthquakes are the most natural hazard facing Kuwait, while other environmental challenges such as flooding, dust fallout, land degradation, and aeolian sand movement often arise from human impact as well as natural factors. Because of the rapid socio-economic development in the last five decades in Kuwait, these issues cause environmental and social problems as well as economic disturbance; they are also considered natural disasters for country. The scale and intensity of the geological environment hazards are considerably increasing especially land degradation, and impacting on the harsh structure of desert ecosystem. Due to fragility of the desert environment, human activities exceeding the carrying capability of the geo-environment system can easily lead to geological and environmental hazards; such as runoff, sand and dunes movements and dust fallout causing serial environmental and health impacts. Geographic Information System (GIS) has been used to evaluate the degrees of geological hazard and risk by producing maps for each hazard; seismic, sand potentiality, hydrologic risk, land degradation, and sand drift severity maps are produced.

J. M. Al-Awadhi (✉)
Department of Earth and Environmental Sciences, Faculty of Science, Kuwait University, Kuwait City, Kuwait
e-mail: jasem.alawadhi@ku.edu.kw

A. el-aziz K. Abd el-aal · A. Abdulhadi
Kuwait Institute for Scientific Research (KISR), Safat, Kuwait

A. el-aziz K. Abd el-aal
National Research Institute of Astronomy and Geophysics, Helwan, Egypt

R. Misak
Desert Research Center, Al Matariyyah, Egypt

8.1 Earthquake Hazard in Kuwait

The earthquakes potentially affect the surrounding environment including man-made structures and human life. In general, earthquakes are either classified as tectonic or natural earthquakes or man-made earthquakes (Abd el aal et al., 2020; Bath, 1979; Bormann, 2012). Seismic activity around and inside Kuwait has been studied by some researchers, for example (Abd el-aal et al., 2020; Abd el-aal et al., 2021a, 2021b, 2021c; Al-Enzi et al., 2007; Gu et al., 2017, 2018). The nature of the occurrence of earthquakes near places of oil extraction and production in Kuwait has been studied by these researchers. There are some researchers linking the oil production process and the occurrence of earthquakes in those places, especially since the earthquakes that are recorded near the places of oil production have a shallow depth and also small/medium magnitudes (Abd el-aal et al., 2020, 2021a; Gu et al., 2017).

Kuwait as a result of its proximity to one of the most active places and seismic belts in the world, which is the Zagros seismic belt, moreover, the occurrence of local earthquakes near the places of oil extraction, makes Kuwait more vulnerable to seismic hazards and must be studied in details to reduce its effects on infrastructure and local community. The seismic hazard assessment in and around Kuwait is studied through the well-known and common methods of assessing the seismic hazard, which are the probabilistic (PSHA) and the deterministic (DSHA) seismic hazard approaches.

8.1.1 Probabilistic Seismic Hazard Analysis (PSHA)

The PSHA method employs the principal achievable amount of dataset, including seismic, geophysical, and geological data to construct models of the earthquake producing processes. The probability that the value Z will be exceeded within t years will be obtained through the following formula:

$$P_t(z) = \sum_{j=1}^{k} \int_{m=M_{min}}^{m=M_{max}} P_t(m) \int_{r=R_0}^{R_{max}} P(r)(P(A \geq Z)m,r)dmdr$$

(8.1)

where j refers to the seismogenic zone,
m symbol refers to the moment magnitude,
r represents the distance to the earthquake source.

$P_t(m)$ refers to the probability of occurrence of a magnitude m earthquake in zone j within t years which can be obtained through the following equation:

$$P_t(m) = 1 - \exp\left[\frac{-t}{n(m)}\right]$$

(8.2)

To carry out Probabilistic seismic hazard computation at a particular place, the subsequent input coefficients are required (Cornell, 1968; Reiter, 1990).

1. Updating historical and instrumental earthquake catalogues are fundamental.
2. Seismic zones which describes areas of equal seismic activity and also the faults (Fig. 8.1).
3. The recurrence parameters for the detected earthquake zones, where every seismic zone is represented by a seismic recurrence relation. This relation mentions the chance of seismic event of a given magnitude which occurs anywhere within the seismic source during a certain interval of time. The maximum magnitude is selected from every seismic source, representing the biggest earthquake to be estimated.
4. A ground motion-attenuation models express the attenuation of the seismic waves as a assignment of magnitude and distance. For more details on the inputs of the seismic hazard assessment process, please read (Abd el aal et al., 2015, 2022; Mostafa et al., 2018).

The main products of the process of evaluating and calculating the seismic hazard in the State of Kuwait are maps and curves of the spectral acceleration values with period (Abd el aal et al., 2021c). Spectral Seismic acceleration maps have been drawn to illustrate variation of the level of the

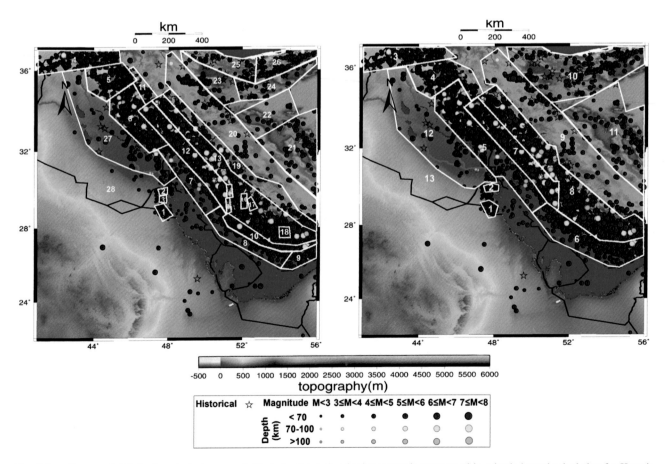

Fig. 8.1 **a** Shows a detailed seismotectonic zones. **b** illustrates the regional seismotectonic zones used in seismic hazard calculation for Kuwait region

calculated seismic parameters considered with either probability of exceedance within specified period or with return period (Abd el aal et al., 2022). The computation was performed at 4 different periods including PGA, 0.1, 1.0, and 4 s (Fig. 8.2).

Abd el aal et al. (2021c) calculated the uniform hazard spectrum (UHS) in the following six main cities in Kuwait, basically in Sabah Al Ahmad, Mubarak Al-Kabeer, Hawally, Al Farwaniyah, and Al Jahra (Fig. 8.3). UHS is mainly used to establish a seismic hazard maps or charts at spectral periods range from 0 to 4 s at return periods of 75, 475 975, and 2475 years on bedrock conditions.

Abd el aal et al. (2021c) calculated the deaggregation charts of seismic hazards to identify the involvement of both source and earthquake magnitude to site location to the hazard calculation at return period of 475 years for six main cities in Kuwait at spectral periods of 0.1 s. The deaggregation calculations are very helpful in detecting prevailing hazardous seismic event at a certain location for use by the civil engineer. The deaggregation charts of ground motion hazards at six sites in Kuwait at the period of 0.1 s for return period of 475 years for the detailed seismotectonic source model are shown in Fig. 8.4. The deaggregation results show that most of the seismic hazardous occur from moderate seismic events with Mw ranging from 4 to 5 situated at very short location from these cities (Abd el aal et al., 2021c).

8.1.2 Deterministic Seismic Hazard Analysis (DSHA)

The hazard was also calculated by deterministic approach method (Abd el aal et al., 2021c). The calculations were made in 6 important places in Kuwait, which are the governorates of Kuwait. Figure 8.5 shows the results obtained by Abd el aal et al. (2021c).

Comparing the DSHA results to the PSHA results for six governorates of Kuwait, it is clear that the results of DSHA are relatively higher as a result of this method being the worst case.

8.2 Dust Fallout in Kuwait

Drought and recurring strong winds increase the frequency of dust fallout in Kuwait. Frequency and intensity of dust storms vary with climatic conditions. Dust and sand storms originating in the flood plain in the south part of Iraq, upwind of Kuwait, are major contributors to the sediments in Kuwait (Khalaf et al., 1980). The drainage of Mesopotamian Central marshes and Lake Hammar was accelerated after the First Gulf War (1991) and by 2003 they were drained to

nearly 10% of their original size. This created a vast region of dried-up areas, creating a new potential dust source. Though the Second Gulf War (March–April 2003) inflicted a massive destruction to the surrounding environment yet a positive aspect was the restoration of the drained marshlands and lakes thus eliminating a potential dust source in proximity to Kuwait's northern border. The Iraqi-Kuwaiti war in 1990 followed by the liberation of Kuwait in 1991 was a localized environmental disaster never witnessed previously in this area. The extensive destruction to the Kuwait desert environment augmented pre-existing sources and created new local dust sources.

Dust fallout is observed in Kuwait throughout the year (13% of daytimes), while during the summer months (April to August) it rises to 25% of daytime. Meteorological data from the Kuwait Airport (July 2000 to March 2010) analyzed by Al-Dousari and Al-Awadhi (2012) for the monthly average dust storm days is shown in Fig. 8.6. In general, dust fall increased during the drought years and decreased during the years with higher rainfall. Except for the Sahara Desert, Kuwait experiences higher dust fall in comparison with the surrounding regional and global areas (Al-Dousari & Al-Awadhi, 2012).

8.2.1 Dust Regional Sources

Five major regional dust sources surrounding Kuwait were identified by Al-Dousari and Al-Awadhi (2012) (Fig. 8.7). These are as follows:

1. Iraqi desert surrounding Kuwait border.
2. The Mesopotamian Flood Plain in Iraq.
3. Saudi Arabia's desert surrounding Kuwait border.
4. Drainage of the southern Iraqi marshes.
5. Iranian dry marshes (*Sabkhas)* at northern coastal area of Arabian Gulf.

8.2.2 Amount of Dust Fallout

Significant quantities of dust are deposited along the trajectory of moderate to strong dust storms blowing across Kuwait that originate from the west, northwest, and north of Kuwait from sources identified in the previous section. Modified dust traps designed by Al-Awadhi (2005) were installed at 42 sites covering eight zones in Kuwait, at a height of 2.4 m above the sediment surface level. The dust fallout data was collected for 14 months from November 2006 till December 2007 (Al-Dousari & Al-Awadhi, 2012) (Table 8.1). Total annual dust fall at the eight zones was 395 tons/km^2 and the monthly average was 33 tons/km^2.

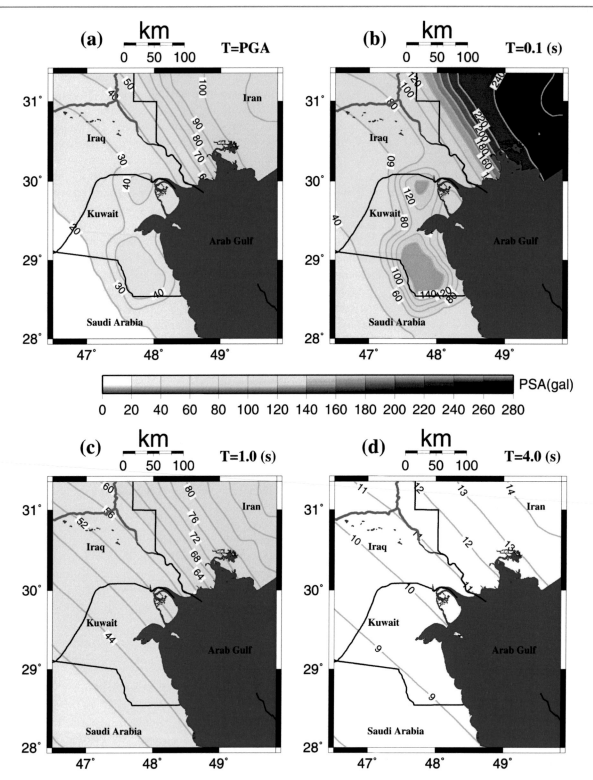

Fig. 8.2 Seismic hazard calculation maps for Kuwait area at periods PGA, 0.1, 1, 4 s for 475 years return period

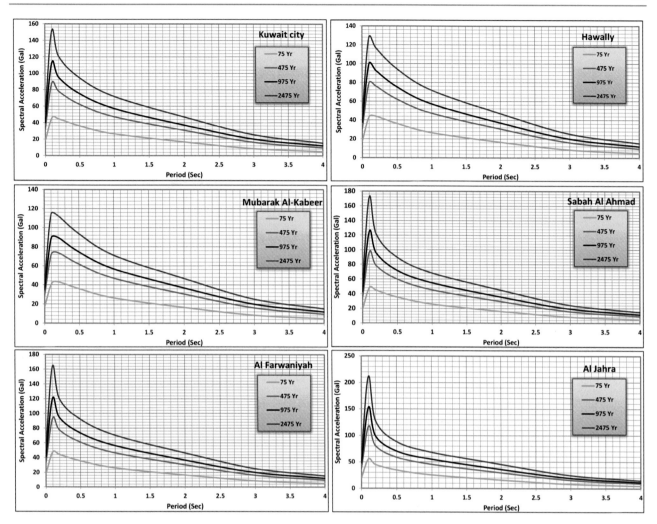

Fig. 8.3 Uniform hazard spectrum (UHS) for six governorates in Kuwait at return period 75, 475, 975 and 2475 years

There is strong quantitative and temporal correlation between the trajectories of major dust storms and the dust deposition recorded within the path at the eight zones. Local topography, existing vegetation, and meteorological conditions at any local region significantly affect the dust fall within and at the vicinity of these zones. Other probable explanation could be related to.

1. Proximity to regional sources (mainly for southwestern desert of Iraq).
2. Massive quantities of mud within depressions, *sabkhas,* and intertidal zones.

Accordingly, grain size and composition vary over dust samples collected from the eight zones. Such variations are primarily due to the origin and accumulation of the sand from a single source or multiple sources, local and regional. For example, analysis of Bubiyan Island dust reveals that it is trimodal with three dominant fractions of coarse silt, very fine silt, and clay; the trimodal distribution suggests several sources. Dust from the open desert in western Kuwait is unimodal with a dominant very coarse silt size fraction. It is coarse with heavy fractions, large size (100 μm), and smooth with sub-angular quartz grain with a few adhering carbonate particles (Fig. 8.3b), while Bubiyan Island dust is fine carbonate (30 μm) mixed with gypsum and bassanite (Fig. 8.8c). Generally, sand particles with size of fine and very fine originate from local sources since they usually transport either by creeping or saltation over short distances only.

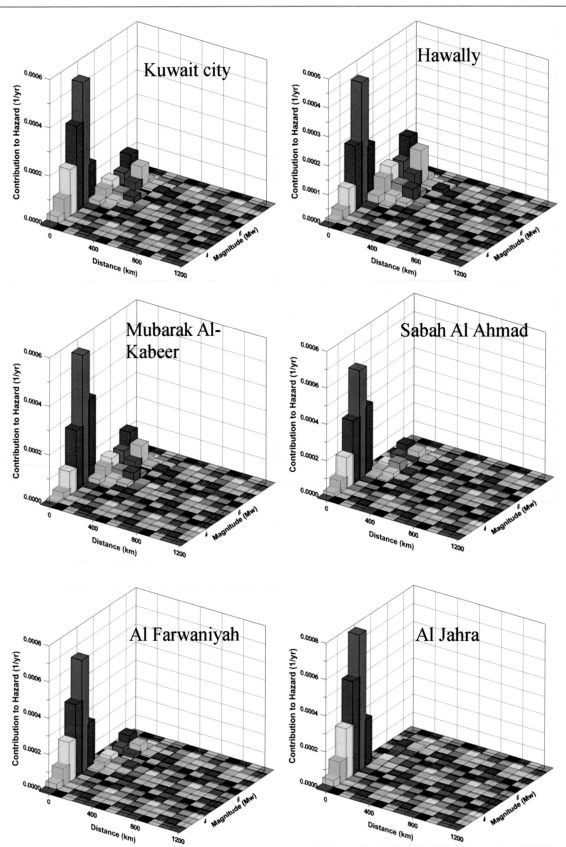

Fig. 8.4 The deaggregation charts at the six governorates of Kuwait for the short spectral period of 0.1 s for return period of 475 years for the detailed seismotectonic source model

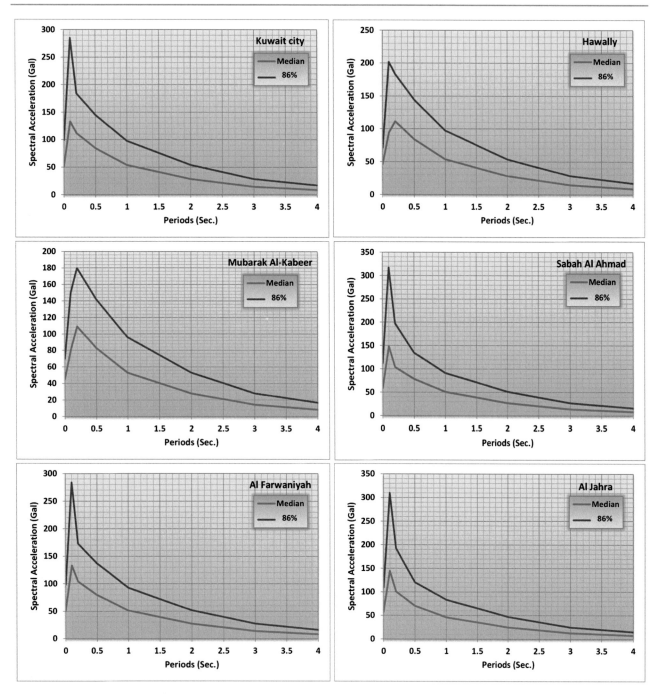

Fig. 8.5 The seismic hazard estimation by DSHA approach at the six governorates of Kuwait (after Abd el aal et al., 2021c)

8.2.3 Mineralogical Characteristics of the Dust

Kuwait dust samples, collected by Al-Dousari and Al-Awadhi (2012), from 5 sectors in Kuwait, investigated by XRD for mineralogical analysis, the results reviled that the dust fallout contains Quartz (35.2%), Calcite (28.5%), Dolomite (10.7%), Carbonates (39.5%), Feldspars (12%), and Clay (4.5%) as the major constituents. Other minerals in

the dust were found in trace or small quantities such as anhydrite, basanite, gypsum, and heavy minerals (Table 8.2).

8.2.4 Elemental Concentration in Dust Fallout

Al-Awadhi and AlShuaibi (2013) collected samples of dust falling over Kuwait city for 12 months from March 2011 to

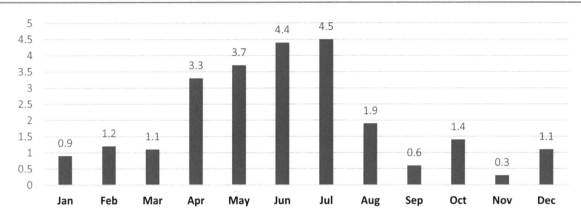

Fig. 8.6 The monthly average number for dust storm days in Kuwait (2000–2010) (after Al-Dousari & Al-Awadhi, 2012)

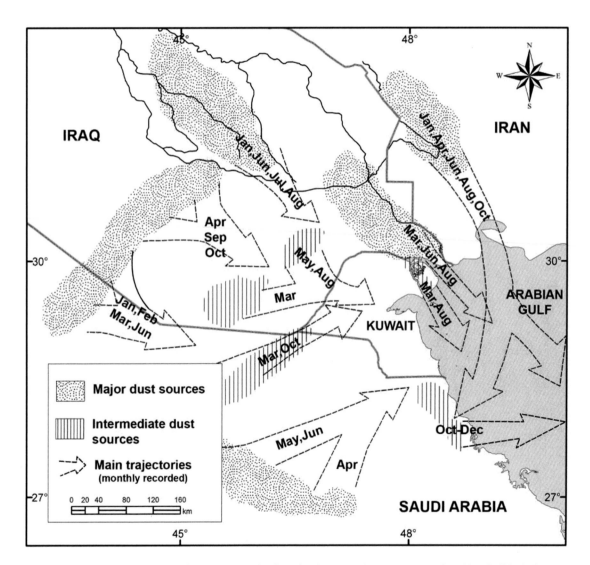

Fig. 8.7 Major and intermediate source areas of dust storms and trajectories for the north-western areas of Arabian Gulf including the study area in reference to dust storm days in Kuwait and satellite images from 2000 to 2010

Table 8.1 Variations of monthly and annual dust fall in eight zones (tons/km^2)

Months↓ Zones→	National Park	Al-Mutla	Al-Liyah	Al-Jahra	Shuwaikh	Warba	Sabiya	Bubiyan	Monthly Total
Jan	0	1	3	2	2	7	6	13	**34**
Feb	0	3	5	3	3	1	6	7	**28**
Mar	0	2	3	3	3	4	6	6	**27**
Apr	1	4	5	4	3	9	6	15	**46**
May	0	2	2	2	6	3	3	7	**25**
Jun	0	2	2	4	9	1	5	4	**27**
Jul	1	7	3	10	8	2	8	8	**47**
Aug	0	2	2	3	5	3	7	9	**32**
Sep	0	2	2	2	4	4	6	9	**29**
Oct	0	1	1	1	2	3	3	6	**17**
Nov	0	1	3	1	2	16	10	10	**43**
Dec	0	1	2	1	2	5	21	8	**40**
Total	**2**	**28**	**33**	**36**	**49**	**58**	**87**	**102**	**395**

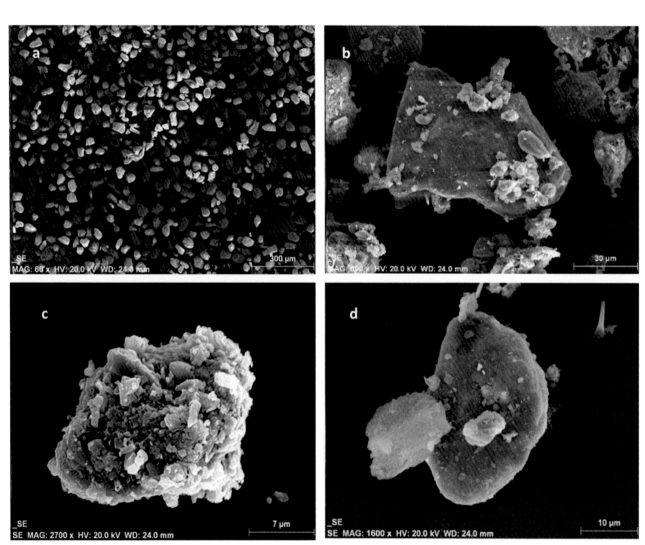

Fig. 8.8 Smooth dust particles within Wester zone. **a** very coarse silt, **b** adhering carbonates particles, (**c, d**) adhering gypsum and bassanite particles

Table 8.2 Kuwait Dust samples analysis by XRD for minerals

Sector	Area	Quartz	Calcite	Dolomite	Carbonates	Feldspars	Clay	Other
Bubiyan	East Kuwait	28	20	14	34	18	5	14
National Park	North Kuwait	38	38	7	45	10	2	5
Warba	NE Kuwait	36	30	11	42	8	5	9
Sabiya	East Kuwait	39	26	11	37	12	6	5
	Average %	**35.25**	**28.5**	**10.75**	**39.5**	**12**	**4.5**	**8.25**

Table 8.3 Elemental contents in dust fallout in Capital city of Kuwait and its surface sediments (mg/kg)

mg/kg	Soil (N = 184)	Dust (N = 120)
V	19.2	56.3
Cr	25.6	131.0
Co	3.7	24.0
Ni	25.6	193.6
Cu	22	97
Zn	18.4	213
Zr	5.1	14.5
Mo	0.2	2.2
Ba	51.6	150
Pb	4.6	32.1
Sr	191	232
Fe	4162	16,136
Mn	121	315
Ti	178	524
Cd	0.4	0.4

February 2012 and analyzed the samples for mean annual elemental concentrations (Table 8.3). They used these data for enrichment factor (EF) analysis, which is indicative of the potential impact of dust falls to the elemental composition of surface sediments where dust from local or regional sources is deposited. The following equation of Sutherland (2000) is used for the computation of EF:

$$EF = (M/N)_{sample}/(M/N)_{baseline} \quad (8.3)$$

EF is the concentration ratio of a metal (M) to its normalizer (N) for both investigated samples and baseline reference samples (i.e., surface sediments).

Magnesium (Mn) was selected as the reference element (normalizer), and $M_{baseline}$ was determined using mean elemental concentrations of the Kuwait soil. Accordingly, Table 8.4 shows the EFs of 16 elements.

Based on contamination degrees using the EF values (Table 8.4), which were suggested by Sutherland (2000), Loska and Wiechuya (2003), the dust is significantly contaminated by Mo and Zn (mean EF > 5), while it is moderately contaminated by Ni, Pb, Co, Fe, Cr, Cu, Ti, Zr, Ba, and V (2 < mean EFs < 5). In general, the dust is less contaminated by Cd and Sr.

8.3 Land Degradation in Kuwait

Land degradation in the country can be attributed to uncontrolled overgrazing, off-road driving, camping, and quarrying. Gulf war in 1990 and its subsequent military movements added more environmental damages including soil compaction and soil pollution in the form of oil lakes. Several reports on land degradation are available (e.g., Khalaf, 1989; Omar, 1991; Zaman, 1997; Brown & Porembski, 1997, 1998; Howle, 1998; Misak et al., 1999; Shahid et al., 1998, 1999, 2003; Al-Dousari et al., 2000; Al-Awadhi et al., 2001; Omar et al., 2005). Aeolian (wind) and fluvial (water) processes, causing soil erosion, soil compaction, sealing and crusting and depletion of desert vegetation cover, are the two principal causes of land degradation in Kuwait. Misak et al. (1999) reported 44% moderate and 32% severe land degradation of the Kuwait desert ecosystem. Al-Dousari et al. (2000) studied the soil compaction in Al Salmi area adjacent to the western border with Saudi Arabia. A map for soil compaction was prepared, which shows that 89.5% was non-compacted whereas 8.8% was highly compacted and only 1.7% was slightly

Table 8.4 Determined Enrichment factors in the dust fallouts in Kuwait city

Enrichment factor

Elements	Mean	Max	Min
V	2.0	8.4	0.2
Cr	3.1	10.3	0.5
Co	3.9	12.9	0.4
Ni	4.4	32.5	0.0
Cu	2.4	19.1	0.4
Zn	6.8	39.9	0.7
Zr	2.0	8.1	0.0
Mo	5.3	29.3	0.4
Ba	2.0	8.0	0.2
Pb	4.2	39.1	0.0
Sr	0.5	1.4	0.1
Fe	3.6	16.4	0.4
Mn	1.0	1.0	1.0
Ti	2.0	9.6	0.4
Cd	0.3	8.3	0.0

compacted. Generally, loss of land productivity is a major factor while assessing land degradation. However, in Kuwait protection of desert ecosystem is of prime importance with a view to preserve environment quality and biodiversity. Cattle grazing in the desert area is historically a cultural activity with hardly any financial benefits. Harshness of Kuwait climate has resulted in minimal pedogenesis with hardly any changes to the sand and gravel parent material.

8.3.1 Manifestation of Land Degradation

8.3.1.1 Vegetation Cover Deterioration

Perennial desert shrubs (*Haloxylon salicornicum*), in the western and southern parts of Kuwait desert and open farmland areas, were recorded in abundance prior to the occupation of Kuwait on August 2, 1990. Coalescence of *Nabkhas* developed rugged vegetative sand sheets, which resulted in dense to moderate vegetation cover. Another survey during February 2010 showed severe degradation with vegetation cover only in the protected areas such as oil fields, military areas, and nature reserves.

8.3.1.2 Soil Erosion

Soil erosion in Kuwait is attributed to Aeolian processes mainly and to Fluvial processes to some extent. The fragile desert ecosystem is prone to degradation by the uncertain, unpredictable, and varying climatic conditions as well as to uncontrolled anthropogenic activities. Soil erosion in the desert areas is accelerated due to scarcity of vegetation cover that provides soil stabilization by deep-rooted shrubs and

trees, and the higher susceptibility of deposits to aeolian and fluvial processes (Khalaf & Al Ajmi, 1993; Al-Awadhi & Misak, 2000).

During the dry season in 2008, for example, with rainfall less than 30 mm, wind erosion caused loss of 10 to 15 cm of topsoil at an estimated rate of around 1,000 m^3 ha^{-1}. Previous studies show that during the 1980s the deposition and erosional areas in the desert was in equilibrium (Khalaf & Al Ajmi, 1993). Previously sand sheets, with dense coverage of perennial shrub *Cyperus conglomeratus,* enhanced the depositional aeolian processes in the southern part of Kuwait, acting as a stabilizing agent to form well-developed sandy soil. Nowadays, these areas have been subjected to severe erosion processes and replaced the surface of the areas by thin smooth sand sheets covered with desert pavements. Also, during the early 1980s several small 3 m high *barchan* dunes were observed in the north areas. In 2010 survey revealed the formation of higher 10 m *barchan* dunes.

Soil erosion due to Fluvial processes is dependent on the rainfall, which averages around 125 mm $year^{-1}$, with extremes of 20 mm to 325 mm $year^{-1}$. Rare intensive precipitations (20 mm day^{-1}) over degraded natural drainages may lead to erosive surface runoff. Fluvial erosion by rain, in Kuwait, formed several gullies with depth ranging from 50 to 150 cm and rills inter-spacings ranging from 20 to 50 m (Misak et al., 2001).

8.3.1.3 Soil Degradation Vulnerability

Three classes (high, moderate, and low) of vulnerability to degradation of eight soil types have been characterized in

Kuwait (KISR, 199): (1) *Torriorthents* (1%) and *Torripsamments* (27%) are classified as highly vulnerable, (2) *Aquisalids* (7%) is moderately vulnerable to degradation, and (3) *Haplocalcids* (8%), *Petrocalcids* (11%), *Haplogypsids* (0.5%), *Calcigypsids* (6%), *Petrogypsids* (33%), and others (7%) are classified as low vulnerability.

8.3.1.4 Enhancement of Sand Encroachment

Droughts, sediment loses and strong winds are the main causes of aeolian sand movements in Kuwait. The prevailing wind direction in Kuwait is NW–SE. Sandstorms, lasting for days and weeks, result in massive sand encroachment and widespread soil loss due to wind erosion. A field survey in 2010 revealed that shifting sands were encroaching farms, oil flow pipelines, roads, and a water storage facility existing across the prevailing wind direction. The thickness of the drifting sand accumulation ranges from 50 to 120 cm.

8.3.1.5 Disruption of Surface Runoff

Building bund walls along desert roads and highways form dams at the upstream side resulting in obstructing the surface runoff flow and disturbing the existing hydrological system, and consequently in recharging of shallow groundwater reservoirs. In addition, when the accumulated mud dries up, it forms a potential source of dust storms during windy days. For example, when the catchment and drainage systems are disturbed, the loss of about 40% of surface runoff water to reach discharging areas is expected.

8.3.2 Causes of Land Degradation

8.3.2.1 Anthropogenic Activities

Significant increase in annual population growth rate, currently at around 3.8%, calls for a delicate balance between population welfare, socio-economic growth and land use to prevent or minimize land degradation and overexploitation of desert resources. Anthropogenic activities include overgrazing (Omar et al., 1999), sandy roads (Al-Awadhi, 2001; Al-Dousari et al., 2000), traditional camping (AlSudairawi & Misak, 1999) and using sand and gravel quarries (Al-Awadhi, 2001). The short and long terms of the impacts related to anthropogenic activities are presented in Table 8.5.

8.3.2.2 The Gulf War

The occupation (August 2, 1990) and liberation (February 26, 1991) war activities destroyed most of the Kuwait desert ecosystem (Holden, 1991; Karrar et al., 1991; Al-Ajmi et al., 1994). The heavy bombing, trench digging, war maneuvres, off-road heavy vehicles including tanks movements during the war resulted in the destruction of the native shrubs and plants, soil compaction and high degree damage to the desert topsoil, making the desert vulnerable to frequent sand movements and dust fallouts. During the war, 727 oil wells were burned, which released massive quantities of SO_2 causing acid rain and form oil pits (Al-Besharah, 1992). In addition, extensive destruction of topsoil and biomass in the Kuwaiti desert resulted due to the removal of mines and unexploded ordinance after the liberation.

Table 8.5 Summary of the anthropogenic activities on land degradation

Activity	Short-term impact	Long-term impact	Nature of the impact
Over-grazing	Reduction in the biomass, soil and sediment disturbance. Potentiality to originate dust fallout and sand movement	Increasing evaporation rate leading to loss of the vegetation species	Over-grazing is mainly associated with camel and sheep in the desert of Kuwait
Off-road vehicles (sand roads) and camping	Soil compaction, decrease in permeability, and lowering vegetation cover. Increase the potentiality erosion by wind and water	Potentiality of sand movement and creating active sand sheets. Reduction fertility of soil	Impacting at least 45% of top soils in the open areas by various degrees of compaction
Gravel quarrying	Disturbing of the armor layer of pebbles and gravel Disruption and rupture of surface and near surface sediments Exposing finer grains to be easily carried by wind Vanishing the vegetation cover	Creating new sources for dust and sand. Disturbance of natural drainage system and loss of running water in quarries	Extensive gravel quarrying operations commenced, without any planning, regulation or governmental oversight

8.3.2.3 Climatological Factors

Land degradation processes are influenced by the following climatological factors:

1. The scarcity and low rainfall (approximately 115 mm/year).
2. High frequent of drought periods (years with lower rainfall than the average). During the drought periods the soil temperature increase and the vegetation cover depletes.
3. The strength of the prevailing NW winds (reaching 30 m s^{-1}) during the summer season. This exceeds the rates of sand transport and topsoil erosion (Al-Awadhi & Misak, 2000).

8.3.3 Assessing Land Degradation Indicators

8.3.3.1 Field Assessment

Al-Awadhi et al. (2005a, 2005b) assessed the extent and magnitude of land degradation in four open areas and recognized seven indicators for land degradation in Kuwait: (1) soil erosion by wind, (2) soil erosion by water, (3) deterioration of vegetation cover, (4) soil crusting and sealing, (5) soil compaction, (6) soil contamination by oil, and (7) soil salinization (Fig. 8.9). The four assessed areas (Al Mutlaa, Al-Sabiya, Sulaibiyah, and Ahmadi-Al-Dahr) showed noticeable variations in vegetation cover, soil erosion by wind/water and soil compaction. Comparing with the extend and magnitude of land degradation in the nearby

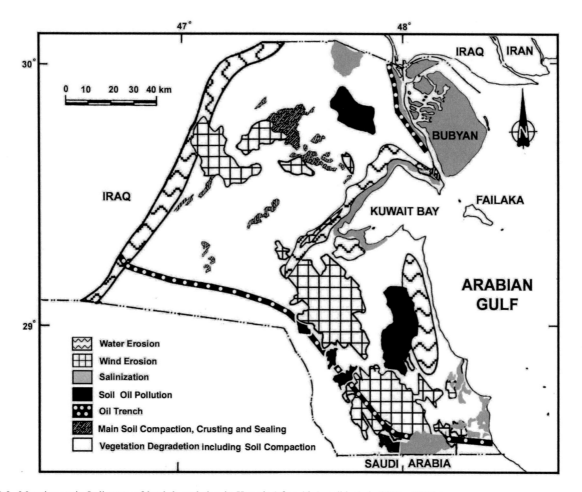

Fig. 8.9 Mapping main Indicators of land degradation in Kuwait (after Al-Awadhi et al., 2005a, 2005b)

protected areas, the bulk density of soils and compaction rate increased with average values of 17 and 61.8%, respectively. Heavy vehicle movement caused soil compaction with remarkably high values with impermeable layers at about 22 cm depth. As a subsequence result, the average infiltration rate in the open areas decreased with an average value of 52.7%.

8.3.3.2 Mapping Land Degradation Hazard Using GIS

Al-Awadhi (2008) integrated a March 2001 Landsat image and nine other Kuwait maps of physical attributes to compose a Geographic Information System (GIS) composite land degradation hazard map. The nine maps are (1) sand drift potential, wind energy, (2) surface sediment type, (3) vegetation density cover, (4) land use type, (5) drainage type, (6) topography change, (7) vegetation type, (8) iso-salinity contour, and (9) salinized areas. The relative weightages of the criteria to land degradation, assessed by Delphi and Analytical Hierarchy Process (AHP), with input from the professional knowledge of local experts, resulted in the map (Fig. 8.10) showing four degrees of land degradation hazard such as very high (15%), high (36.6%), moderate (35%), and low (13.4%).

Table 8.6 shows a comparison between the overall model assessment of land degradation and the field assessment.

8.4 Aeolian Sand Movement

Kuwait land surface was formed during the early pluvial period. Fluvial processes formed water courses and land depressions, which were covered by alluvial sediments in subsequent epochs. Kuwait weather is characterized by hot and dry summer months (May to September), with prevalent north-westerly strong winds (30 m.s^{-1}) and several sandstorms, resulting in aeolian sand depositions, from the upwind high deflation area of the Mesopotamian flood plain, and instability of the fragile ecosystem. The scarcity of deep-rooted shrubs and trees across the open central desert areas results in significant aeolian processes that accelerate soil erosion, extensive and intensive sand encroachment, formations of sand dunes, sand sheets, and sand drifts over areas reserved for defence facilities, oil exploration, groundwater fields, electrical transmission stations, cattle farms, road network, and residential projects. This is a major environmental and socio-economic issue. The annual total sand drift measured in Kuwait is 7.8×10^4 kg.m^{-1}-width

Fig. 8.10 Land degradation hazard map of Kuwait (after Al-Awadhi, 2008)

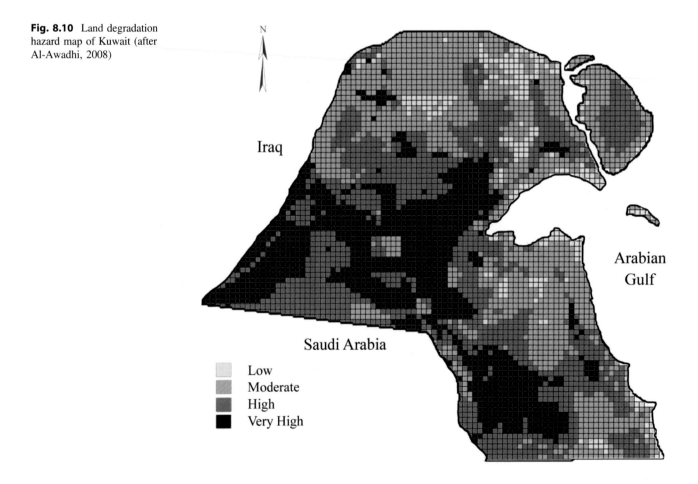

Table 8.6 A comparison between the overall model assessment of land degradation and field assessment

Land degradation classes	Model Assessment (Al-Awadhi, 2008)	Field assessment (source: Al-Awadhi et al., 2005a, 2005b)	
	Class (%)	Class (%)	Plant coverage (%)
High	36.6	25–40	<5.5
Moderate	35	35–45	5.5–25
Low	13.4	10–25	>25

(Al-Awadhi & Misak, 2000). Al-Awadhi and Al-Awadhi (2009) developed a sand transport model, which computed the summer monthly sand transport rate 5,900 kg.m^{-1} towards southeast. The highest sand transport rate occurs within the Al Huwaimiliya sand dune corridor, which crosses the country from NW to SE (Fig. 8.11).

8.4.1 Vulnerability of Soil Classes to Drifting

Kuwait Institute for Scientific Research (1999) used USDA Soil Taxonomy to map eight soil types in Kuwait, namely, in order of abundance, Petrogypsids (33%), Torripsamments (27%), Petrocalcids (11%), Haplocalcids (8%), Aquisalids (7%), Calcigypsids (6%), Torriorthents (1%), Haplogypsids (0.5%), and other miscellaneous types (7%). Vulnerability to degradation and drift during the drought periods varies among the soil types, which can be classified as (a) High:

Torriorthents and Torripsamments; (b) Moderate: Aquisalids (7%); and (c) Low: other types.

8.4.2 Enhancement of Sand Encroachment

Al-Awadhi and Misak (2000) identified 13 mobile sand bodies in Kuwait (Fig. 8.12) and estimated Kuwaiti Dinars (KD) 1.68 million as the annual cost of removal of 4.39×10^6 m^3, 50 cm to 120 cm thick, accumulated sand from facilities within the main passage of the wind corridor, classified as high to very high encroachment zones. Kuwait Oil Company (KOC) expenses for sand removal over a five-year period (2003–2008) were KD 6.2 million. Most facilities are located at the downward side of the Al Huwaimiliya-Wafra wind corridor with the upwind sandy plain as the source of shifting sand.

The socio-economic development of Kuwait associated with the oil boom has come at the cost of damage to the

Fig. 8.11 Spatial variation of sand transport rate (kg.m^{-1}) in Kuwait. Letters indicate zones of sand transport rate; **a** severe, **b** moderate, and **c** slight (after Al-Awadhi & Al-Awadhi, 2009)

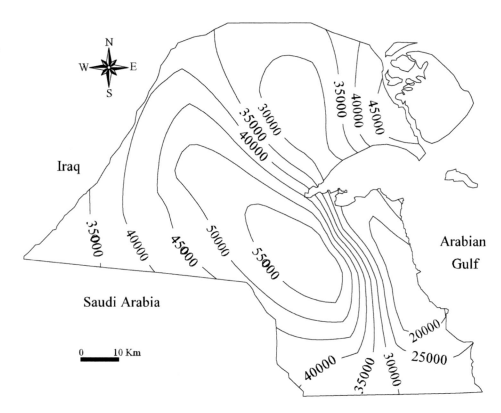

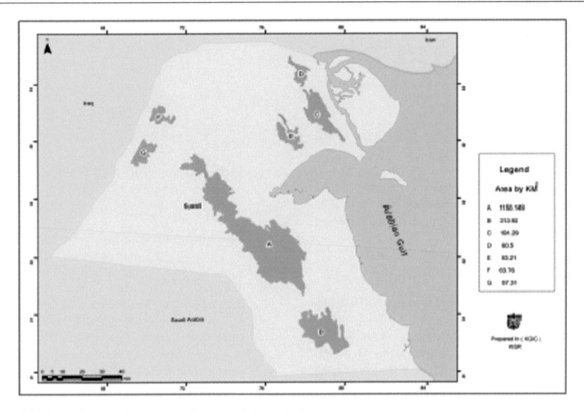

Fig. 8.12 Mobile sand bodies in Kuwait (after Al-Awadhi & Misak, 2000)

fragile ecosystem of the Kuwaiti desert land. The extensive use of gravel from the quarries for infrastructure development damages the vegetative cover, depletes fauna, and disperses fine sand during transportation from the quarries to the construction sites. Existing gravel quarries are estimated to cover an area of 383 km^2 or 2.14% of the total land area of Kuwait. Traditional overgrazing in unprotected areas has resulted in almost total loss of vegetation cover whereas the protected areas are flush with vegetation. Major plant species native to Kuwaiti desert are cyperus conglomeratus, Rhanterium epapposum, Zygophyllum qatarense, Halaxylon salicornica, Panicum turgidum, and Stipagrostis plumose (Fig. 8.13) Omar et al. (2001).

8.4.3 Natural Factors Controlling Aeolian Processes

Surface sediments. Khalaf et al. (1984) estimated that the Kuwait desert, covering 80% of the surface area, is covered by several types of surface sediments. Significant deposits are aeolian (50%), residual gravel, and playa (35.1%). Tidal flats, Coastal *sabkhas*, sand dunes, plain deposits, sandstone, clay, and calcareous rocks cover the other areas.

Wind. Surface wind velocity is critical to aeolian processes and downwind movement of deflated sand over short distances (saltation). During summer months prevalent wind direction (60%) is north-westerly with wind speed reaching 29 m.s^{-1}. Average wind speed in Kuwait is 4.3 m.s^{-1} with highest average 5.1 m.s^{-1} in June and lowest average of 3.2 m.s^{-1} in January (Al-Awadhi & Misak, 2000). Al-Awadhi et al. (2005a, 2005b) analyzed wind data recorded by 8 meteorological stations in Kuwait and computed a value of 354 vector unit (VU) for the sand drift potential (DP). Fryberger's (1979) classification places Kuwait in the intermediate wind energy deserts (DP between 200 and 400 VU).

Surface roughness. Flat residual gravel deposits in the northern areas of Kuwait contribute to the surface roughness, which has a bearing on the sand movement due to wind flow aerodynamics. Surface topography (Fig. 8.14) is a flat to gently rolling desert plain interspersed with low hills, ridges, scarps, and *wadis*. When wind speeds exceed the sheer threshold (>5 m.s^{-1}), finer particles are airlifted from the windward side and residual gravel grains accumulate on the leeward side (Figs. 8.15 and 8.16).

Drainage System. Drainage basins in Kuwait follow regional topography (Fig. 8.17). Northern drainage networks

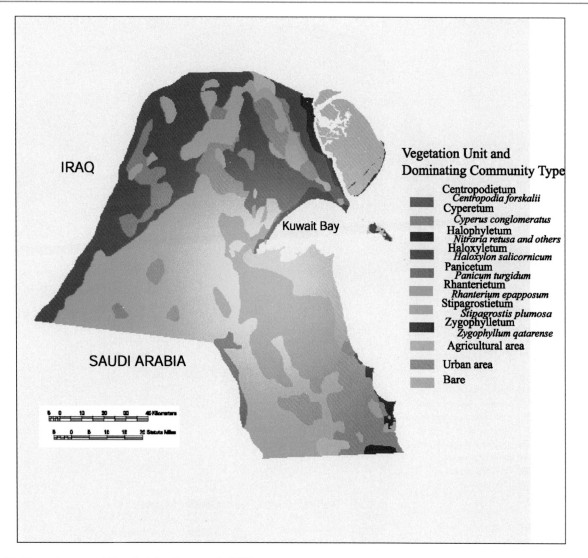

Fig. 8.13 Vegetation map of Kuwait (after Omar et al., 2001)

are well-defined large areas, developed on the gently sloping northeast plain aligned in NW–SE direction. The fine texture, low relief, high-density northern drainage network basins flow into a shallow depression. Southern Kuwait is devoid of any significant drainage network.

While annual rainfall in Kuwait is only 115 mm, wet season (October to March) can experience occasional cloud bursts with rainfall exceeding 20 mm/day with surface runoff transporting sediments within the drainage network. During the hot summer months, the sediments dry up and are used as local sand supply sources.

8.4.4 Mapping Sand Encroachment Hazard Using GIS

Sand encroachment in Kuwait is a perennial problem that requires professional planning by experts to identify all related causes to devise ways to combat it. Considering the spatial extent of sand encroachment issue over all of Kuwait, excluding the urban areas and the islands, Geographical Information System (GIS) technology was used to develop a map showing sand encroachment. Delphi and Analytical Hierarchy Process (AHP) were used by local experts, with

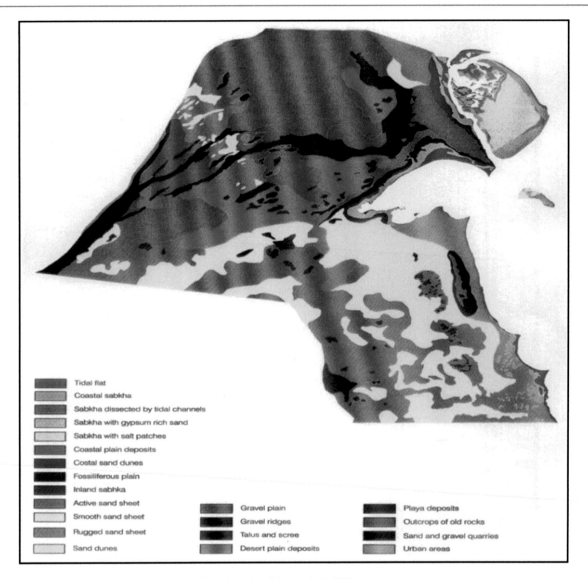

Legend entries:
- Tidal flat
- Coastal sabkha
- Sabkha dissected by tidal channels
- Sabkha with gypsum rich sand
- Sabkha with salt patches
- Coastal plain deposits
- Costal sand dunes
- Fossiliferous plain
- Inland sabhka
- Active sand sheet
- Smooth sand sheet
- Rugged sand sheet
- Sand dunes
- Gravel plain
- Gravel ridges
- Talus and scree
- Desert plain deposits
- Playa deposits
- Outcrops of old rocks
- Sand and gravel quarries
- Urban areas

Fig. 8.14 Surface sediment map of Kuwait (modified after Khalaf & Al-Ajmi, 1993)

extensive knowledge of the area, to quantify and rank the control factors (Al-Hellal & Al-Awadhi, 2006). An AHP-based sand encroachment susceptibility workflow chart to develop an AHP model to determine the sand encroachment susceptibility index is shown in Fig. 8.18.

Al-Awadhi (2008) integrated a March 2001 Landsat image and seven other Kuwait maps of physical attributes to compose a Geographic Information System (GIS) composite sand encroachment susceptibility index map. The nine maps are (X_1) sand drift potential, wind energy, (X_2) surface sediment type, (X_3) vegetation density cover, (X_4) land use type, (X_5) drainage type, (X_6) topography change, and (X_7) vegetation type. The relative weight factor values were entered in Eq. 8.4 to compute the Sand Encroachment Index (SEI).

$$SEI = 0.37X_1 + 0.26X_2 + 0.12X_3 + 0.11X_4 + 0.07X_5 + 0.05X_6 + 0.02X_7$$

$$(8.4)$$

where
  X_1: Sand drift potential (Wind energy);
  X_2: Surface sediment type;
  X_3: Vegetation density cover;
  X_4: Land use type;
  X_5: Drainage density;
  X_6: Topography change;
  X_7: Vegetation type.

Sand Encroachment Susceptibility (SES) zones were computed by solving Eq. 8.4 using Raster Calculator spatial

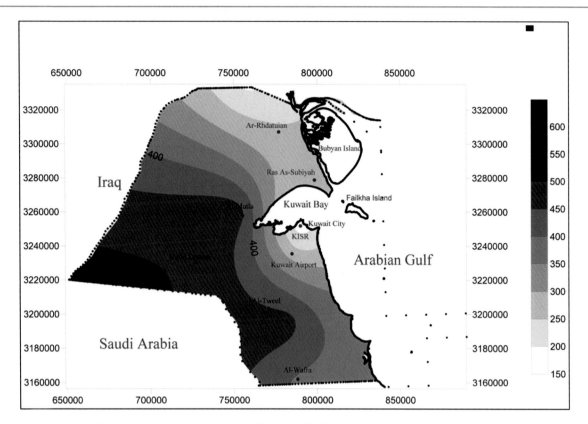

Fig. 8.15 Annual sand drifts potential by wind (after Al-Awadhi et al., 2005b)

analyst function tool in ArcView8 software, which is capable of weighting and combining reclassified raster layers.

Table 8.7 shows the percentages of sand encroachment classes indicated in the SES map of Kuwait (Fig. 8.19).

8.5 Destructive Flash Floods

In Kuwait, flash floods occur in several areas including urban and desert areas. These floods have negative socio-economic, economic, and environmental consequences. Generally, floods usually occur when the rainfall amount reaches 30–40 mm in one event, e.g., February 1993 flash flood (40 mm) with duration 6–8 h, November 1997 flash flood (105 mm) with duration 3–4 h), and November 2018 flash flood (60 mm). Losses of the destructive floods of November 2018 reached KD 180,000,000. These include losses of the oil refineries, stock market, and suspension of work of governmental organizations for three days. Historical flood events in Kuwait took place on December 27, 1934 and November 30, 1954. The following hilly terrain areas (80–220 m above sea level) are considered the main watershed areas in Kuwait: Jal Az Zour hills, Al Rukham hills, Jal Al Liyah Ridge, Ahmadi Ridge, Hills of Wadi Al-Batin and Ritqa-Abdaly. Ahmadi Ridge is the sole watershed in the urban area of Kuwait. It is

about 40 km length and 5 km width (area 200 km²). This ridge is affected by a water divide. Two different drainage systems are developed on both sides of the water divide. Hydrographic basins in Kuwait are distinguished into exterior and interior units. The flash floods cause intensive damage to several residential areas and infrastructures including roads, bridges and tunnels. Death cases were reported on November 11, 1997 and November 9, 2018.

The geomorphologic, hydromorphologic features, and ground elevation are among the factors controlling the conditions of flash floods in Kuwait. Hill shade image (Fig. 8.20a) and the Digital Elevation Model (Fig. 8.20b) show at least 7 hydromorphologic features. These are Wadi Al Batin, Dibdiba ridges and wadis, Raudtain Depression, Jal Az Zour Hilly Terrain, Ahmadi Ridge, Rugged terrain and Shaqiah Playas. Digital Elevation Model offers important information about the ground elevation of the wadis and watersheds. For example, the ground elevation of Wadi Al Batin main channel is close to 280 m above sea level, while that of the main channel of Wadi Al Ahmadi is close to 40 m above sea level.

Hill shade value ranges between 0 and 247, Jal Az Zour Hilly Terrain has the lowest value while Wadi Al Batin has the highest value (Fig. 8.21a) shows hill shade map, while (Fig. 8.21b) shows Digital Elevation Model for Kuwait.

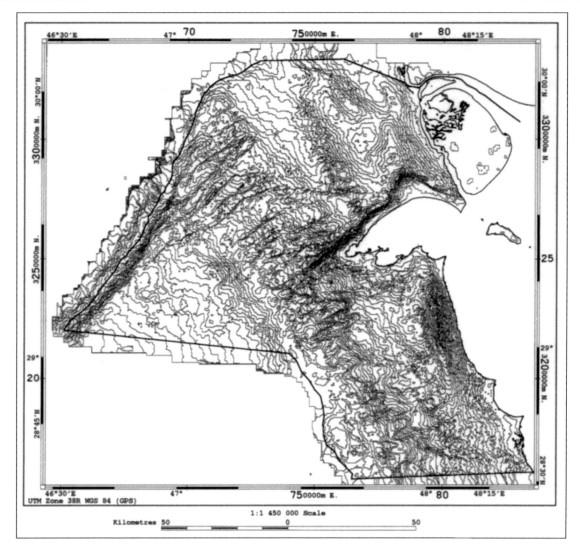

Fig. 8.16 Topographic map of Kuwait; contour interval 5 m (after Misak et al., 2000)

8.5.1 Impact of Flash Floods

Flash floods in Kuwait have several types of damages. These are

- **Socio-economic**
  - Great threats to human life.
  - Physical damage to buildings and infrastructures (roads, power lines, drainage systems, and other utilities)
  - Traffic problems, ground collapse, and death cases (floods of 11 November 1997 & November 2018
- **Economic**
  - Delay and disruption of different human activities.
  - Inundation of large areas resulting in transportation delay.
  - High cost for maintaining the damage of flash floods.

- Loss of millions of cubic meters of fresh water (rainwater).
- **Environmental**
  - Severe water erosion (removal of soil materials by runoff water).
  - Uprooting trees and tear out natural vegetation (grasses and shrubs).
  - Disruption and wearing away of desert surface.
  - Damaging of wildlife habitats.

8.5.2 Lessons Learned from the Floods of Nov. 2018

Floods of November 2018 were the worst during the last fifty years. Floods were caused by heavy rain storms close to

Fig. 8.17 Paleodrainage map of Kuwait (after Misak et al., 2000)

300 mm. It caused severe damage to the existing infrastructures and development plans. Highways, bridges, tunnels, fences, houses, dams, dykes, water wells, storm water drainage systems, and oil facilities were damaged by the runoff water. Figure 8.22 shows damages of November 2018 flash floods in several areas.

Economic losses of the destructive floods reached KD 180,000,000 (Al Hemoud, 2018). These include losses of the oil refineries, stock market, and suspension of work for the governmental organization for three days. In addition, huge number of water ponds were resulted from the rainstorms of November 2018. These ponds were distributed in several areas in south part of Kuwait. The total area of these ponds reached 6.79 km^2, while the size and the depth varied between 0.0176 and 2.2 Km^2 and 0.5 m to 5 m (about 2 m in average), respectively. The total volume of water in these water ponds attained about 13 million m^3.

The following are the most significant lessons which are learned from the 2018 November floods.

- Watershed management should start at the upper reaches of drainage basins (secondary channels).
- Big risk (unsafe) to construct check dams to the east of King Fahd Road.
- Rainwater storage pits are safe and cost-effective measures for flash floods management.
- Combination between well-established check dams and large-size rainwater storage pits (2000 m^3) is appreciated under specific conditions.
- Establishing an Early Warning System (EWS) for flash floods is the most effective tool for risk reduction.
- Maintaining soil crusts and soil compaction enhances infiltration of rainwater into the soil (measure of runoff management).

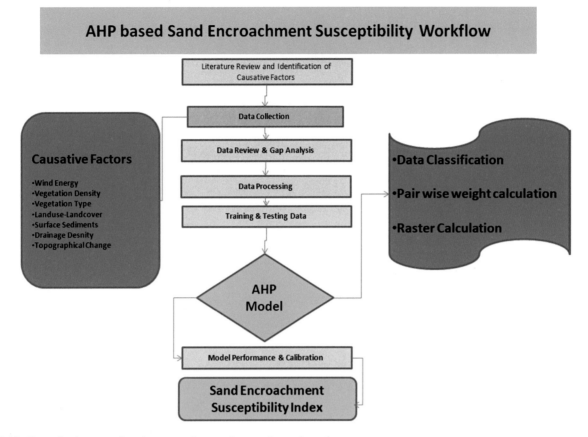

Fig. 8.18 Generalized process flowchart to produce sand encroachment hazard map

Table 8.7 Sand Encroachment Classes (SEC) and Sand Encroachment Index (SEI)

Sand Encroachment Classes (SEC)	Sand Encroachment Index (SEI)	%	Area (km²)
Very low	0–48	5	782
Low	48–59	14	2,343
Moderate	59–69	24	4,190
High	69–79	22	3,841
Very high	79–100	35	6,060

– The surface runoff created on the slopes and embankments of highways is a significant source of flooding.

8.5.3 Surface Hydrologic Maps

A surface hydrologic map was produced by Misak et al. (2013). Accordingly, hydrographic basins (wadis) in Kuwait are divided into two main categories as follows:

- **Exterior** set of basins where the discharge happens around water bodies, i.e., such type of basins are formed in Khor As Sabiyah, Kuwait Bay, and Arabian Gulf.

- **Interior** set of basins where the discharge happens in inland hollows (playas) and plains. Such types of basins are differentiated into several units, exhibiting wide ranges in size and landforms, e.g., Ritqa, Abdaly, Raudtain-Umm El Eish collectors (locally called khabari), Al Liyah-Umm Al Rimmam, Umm Ruwaysat, Wadi Al Batin, Dibidibah, and Kabd-wafra areas.

Figure 8.23 shows three maps of drainage basins of Kuwait.

In Kuwait, runoff water has different directions including east, north, and west (Fig. 8.24).

Based on the analyses of data and information of flood events, the geography and morphology of drainage basins,

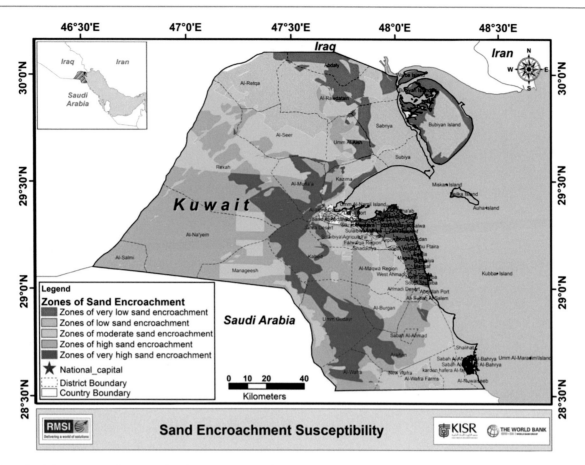

Fig. 8.19 Sand Encroachment Susceptibility (SES) map of Kuwait

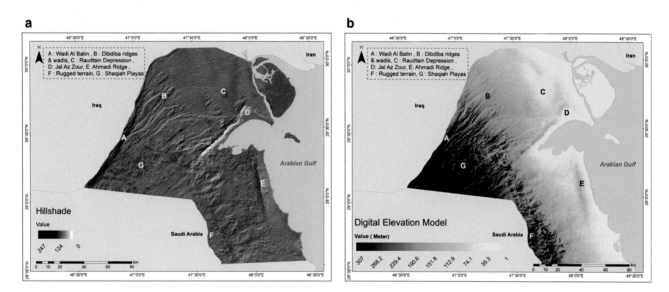

Fig. 8.20 Hillshade map showing significant hydromorphologic features (**a**) and Digital Elevation Model (**b**) showing significant features

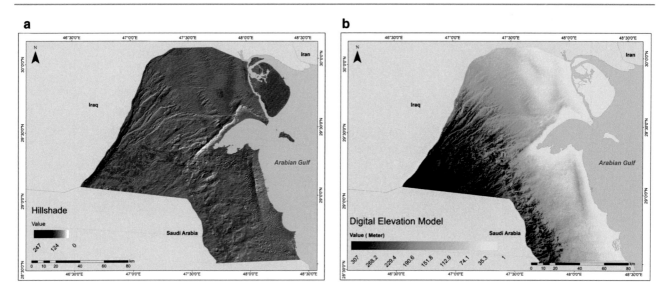

Fig. 8.21 Hill shade map (**a**) and Digital Elevation Model (**b**) of Kuwait

Fig. 8.22 Soil erosion in Fahaheel area (**a**), Destruction of dykes and dams of Wadi Al Ahmadi (**b**), Destruction of storm water drainage system of Wadi Al Ahmadi (**c**)

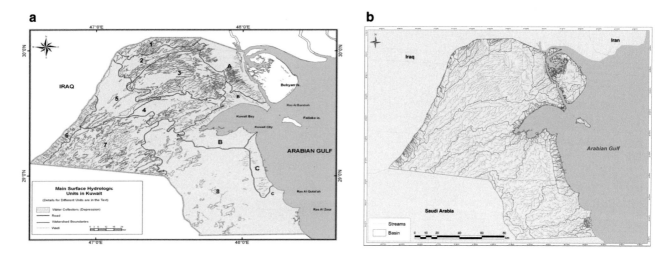

Fig. 8.23 Drainage basins in Kuwait (**a**) (after Misak et al., 2013) and Drainage basins in Kuwait as 2018 (**b**) (modified of Misak et al., 2013)

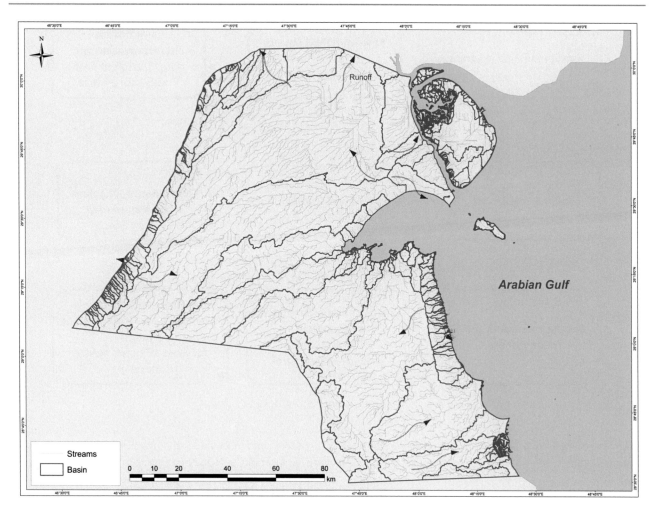

Fig. 8.24 Maps showing the drainage basins and the runoff directions (2018), and arrows indicate the runoff direction

the drainage wadis are classified, on hazard bases into three categories. These are as follows:

1. Extremely dangerous
   Extremely dangerous basins include the following:
   – Eastern basins of Ahmadi Ridge (8 basins including Wadi Al Ahmadi which covers about 17 km^2).
   – Western basins of Ahmadi Ridge (Burgan playa).
   – Wadi Arifjan (affects Sabah Al Ahmad Residential City).
   – Wadi Al Batin.
2. Dangerous
   Dangerous basins include the following:
   – Kabd-Wafra basins.
   – Ritqa-Abdali basins.
   – Several basins cutting Jal Az Zour Hilly Terrain, e.g., Wadi Al Auga.
3. Moderately dangerous
   Moderately dangerous basins include the following:

   – Liyah basins.
   – Al Rukham basins.
   – Dibdiba plain basins.
   – Raudtain basin.

8.5.4 Proposed Risk Mitigation Plan (RMP) for Flash Floods in Kuwait

In Kuwait, several land use types increase the risks of flash floods. Examples of these land use types are off-road vehicles and rangelands grazing. Both land uses result in soil compaction and degradation of vegetation. As mentioned in relevant literature, the amount of infiltrating rate of rainwater in compacted soils may reduce by 40–100% of its natural rate if the soil is not compared. Consequently, the runoff water and the associated soil erosion increase. So, management of rangeland and off road traffic is one of the key approaches for risk reduction of flash floods.

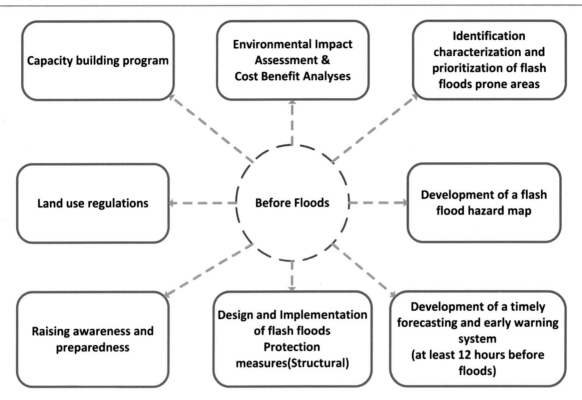

Fig. 8.25 Main approaches of the Proposed Risk Mitigation Plan (PRMP) for flash floods in Kuwait

The proposed Risk Mitigation Plan consists of number of strategic approaches. These approaches are shown in Fig. 8.25.

Before floods, a set of 8 strategic approaches and measures are proposed. These include 1—Identification, characterization, and prioritization of flash floods prone areas, 2—Development of a flash flood hazard map, 3—Development of a timely forecasting and early warning system (at least 12 h), 4—Design and implementation of flash floods structural measures (dams, dykes, and reservoirs), 5—Raising awareness and preparedness, **6**—Land use regulations, 7—Capacity building program, and 8—Environmental Impact assessment and cost–benefit analyses.

During floods phase, approaches include 1—Follow up media for update weather conditions and emergency activities, 2—Evacuation activities, 3—Monitoring the conditions of floods (magnitude, rainfall, depth, and velocity of water), and 3-Follow up water harvesting activities. After floods phase, approaches include Damage Assessment and losses estimation as well as restoration program.

8.6   Conclusion

This chapter deals with meticulously and carefully all the natural hazards facing the State of Kuwait, as well as the technical techniques of mitigating their effects. The chapter reviewed and accounted the earthquake hazard and the modern geophysical methods used to calculate the seismic hazard values for the State of Kuwait. The risks of dust falling on Kuwait were studied with a high efficiency, and the results were reviewed. Land degradation in Kuwait were also studied and referred to in this chapter, and a review of the most important results published in this regard. The Aeolian sand movement and destructive flash Floods were also studied in this chapter and their most important topics and results are presented.

References

Abd el-aal, A. K., Al-Jeri, F., Al-Enezi, A., & Parol, J. A. (2020). Seismological aspects of the 15 November 2019 earthquakes sequence, Kuwait. *Arabian Journal Geosciencem 13*, 941. https://doi.org/10.1007/s12517-020-05919-1.

Abd el-aal, A. K., Parol, J. A., Al-Jeri, F., & Al-Enezi, A. (2021a). Modeling and simulating ground motion from potential seismic sources to Kuwait: Local and regional scenarios. *Geotectonics, 55*. https://doi.org/10.1134/S001685212103002X.

Abd el-aal, A. K., Al-Enezi, A., Sadalla, H., & Al-Jeri, F. (2021b). Tectonic and anthropogenic characteristics of the 15 November 2019 micro earthquakes sequence, Kuwait. *Geotectonics, 55*, 137–152. https://doi.org/10.1134/S0016852121010039.

Abd el-aal, A. K. A., Al-Enezi, A., Al-Jeri, F., & Mostafa, S. I. (2021c). Earthquake deterministic hazard and sensitivity analysis for residential Cities in Kuwait. *Journal of Volcanology and Seismology, 15*(6), 445–462. https://doi.org/10.1134/S0742046321060051.

Abd el-aal, A. K., Al-Enezi, A., Al-Jeri, F., Naser, O., Alenezi, & Mostafa, S. I. (2022). Probabilistic and deaggregation seismic hazard scenarios for Kuwait. *Submitted to Arabian Journal Geoscience.*

Abd el-aal, A. K., Kamal, H., Abdelhay, M., & Elzahaby, K. (2015). Probabilistic and stochastic seismic hazard assessment for wind turbine tower sites in Zafarana Wind Farm, Gulf of Suez, Egypt. *Bulletin of Engineering Geology Environment, 74*(4), 1225–1241.

Al-Ajmi, D., Misak, R., Al-Dousari, A., & Al-Enezi, A. (1994). Impact of the Iraqi war machinery and ground fortifications on the surface sediments and aeolian processes in Kuwait. In *Proceedings International Conference on the Effect of the Iraqi Aggression on the State of Kuwait*, Kuwait, 2–6 April, 3, 229–248.

Al-Awadhi, J. (2005). Dust fallout and characteristics in Kuwait: A case study. *Kuwait Journal of Science and Engineering, 32*, 135–152.

Al-Awadhi, J. M., & AlShuaibi, A. A. (2013). Dust fallout in Kuwait city: Deposition and characterization. *Science of the Total Environment, 461–462*, 139–148.

Al-Awadhi, J., & Al-Awadhi, A. (2009). Modeling the aeolian sand transport for the desert of Kuwait-Constraints by field observations. *Journal of Arid Environments, 73*, 987–995.

Al-Awadhi, J. (2008). Mapping land degradation hazard in Kuwait: Using Delphi and AHP methods. *Kuwait Journal of Science and Engineering, 35*(1A), 71–91.

Al-Awadhi, J. M. (2001). Impact of gravel quarrying on the desert environment of Kuwait. *Environmental Geology, 41*(3–4), 365–371.

Al-Awadhi, J. M., & Misak, R. (2000). Field assessment of aeolian sand processes and sand control measures in Kuwait. *Kuwait Journal of Science and Engineering, 27*(1), 159–176.

Al-Awadhi, J. M., Al-Dousari, A., & Al-Enezi, A. (2001). Barchan dunes in northern Kuwait. *Arab Gulf Journal of Scientific Research, 18*(1), 32–40.

Al-Awadhi, J. M., Omar, S. A., & Misak, R. F. (2005a). Land degradation indicators in Kuwait. *Land Degradation and Development, 16*(2), 163–176.

Al-Awadhi, J. M., Al-Hellal, A., & Al-Enezi, A. (2005b). Sand drift potential in the desert of Kuwait. *Journal of Arid Environments, 63*, 425–438.

Al-Besharah, J. (1992). The Kuwait oil fires and oil lakes—facts and numbers. *Proc* (pp. 12–15). University of Birmingham.

Al-Dousari, A. M., & Al-Awadhi, J. M. (2012). Dust fallout in northern Kuwait, major sources and characteristics. *Kuwait Journal of Science and Engineering, 39*(2A), 171–187.

Al-Dousari, A. M., Misak, R., & Shahid, S. A. (2000). Soil compaction and sealing in AL-Salmi area, Western Kuwait. *Land Degradation and Development, 11*, 401–418.

Al-Enzi, A., Sadeq, A., & Abdel-Fattah, R. (2007). Assessment of the Seismic Hazard for the state of Kuwait. Int rep EC011C, submitted to KFAS 2007.

Al-Hellal, A., & Al-Awadhi, J. M. (2006). Assessment of sand encroachment in Kuwait Using GIS. *Environmental Geology, 49*, 960–967.

Al-Sudairawi, M., & Misak, R. (1999). Challenges and problems confronting the sustainable development in the desert of Kuwait. In *Proceedings. 9th International Conference Environmental Protection Is a Must*, Alexandria, Egypt.

Bath, M. (1979). Introduction to Seismology (Second, Revised ed.). Basel: Birkhäuser Basel. ISBN: 9783034852838.

Bormann, P. (Ed.) (2012). New manual of seismological observatory practice (NMSOP-2), IASPEI, GFZ German Research Centre for Geosciences, Potsdam; nmsop.gfz-potsdam.de. https://doi.org/10.2312/GFZ.NMSOP-2.

Brown, G., & Porembski, S. (1997). The maintenance of species diversity by miniature dunes in a sand-depleted Haloxylon salicornicum community in Kuwait. *Journal of Arid Environments, 37*, 461–473.

Cornell, C. A. (1968). Engineering seismic risk analysis. *Bulletin of the Seismological Society of America, 18*, 1583–1606.

Gu, C., Al-Jeri, F., Al-Enezi, A., Büyüköztürk, O., & Toksöz, M. N. (2017). Source mechanism study of local earthquakes in Kuwait. *Seismological Research Letters, 88*(6), 1465–1471.

Gu, C., Prieto, G. A., Al-Enezi, A., Al-Jeri, F., Al-Qazweeni, J., Kamal, K., Kuleli, S., Mordret, A., Büyüköztürk, O., & Toksöz, M. N. (2018). Ground motion in Kuwait from regional and local earthquakes: Potential effects on tall buildings. *Pure Application Geophysics, 175*, 4183.

Hamoud, N. (2018). Al-Alati Int. *Journal of Engineering Research and Application, 8*(4) www.ijera.com ISSN: 2248-9622, (Part—III) April 2018, pp. 06–17.

Holden, C. (1991). Kuwait's unjust deserts: Damage to its desert. *Science, 251*, 1175–1181.

Howle, S. (1998). *The long term environmental consequences of the Gulf War in northeastern Kuwait.* M.Sc. Thesis, University of Massachusetts, Boston, Massachusetts, USA.

Karrar, G., Batanouny, K. H., & Mian, M. A. (1991). A rapid assessment of the impact of the Iraqi-Kuwait conflict on terrestrial ecosystem, part II. The State of Kuwait. Report prepared for International Fund for Agricultural Development, Rome, Italy.

Khalaf, F. I. (1989). Desertification and aeolian processes in the Kuwait desert. *Journal of Arid Environments, 16*, 125–145.

Khalaf, F. I., Al-Kadi, A., & Al-Saleh, S. (1980). Dust fallout in Kuwait. Kuwait Institute for Scientific Research, Final report No. KISR/PPI 108/EES-RF-8016, Kuwait.

Khalaf, F. I., & Al-Ajmi, D. (1993). Aeolian processes and sand encroachment problems in Kuwait. *Geomorphology, 6*, 111–134.

Khalaf, F. I., Gharib, I. M., & Al-Hashash, M. (1984). Types and characteristics of the recent surface deposits of Kuwait. *Arabian Gulf Journal of Arid Environments, 7*, 9–33.

Kuwait Institute for Scientific Research. (1999). Soil survey for the state of Kuwait, vol 1. Executive summary. AACM International, Adelaide, Australia, CD-ROM.

Loska, K., & Wiechula, D. (2003). Application of principle component analysis for the estimation of source of heavy metal contamination in surface sediments from the Rybnik Reservoir. *Chemosphere, 51*, 723–733.

Misak, R., Khalaf, F., & Omar, S. (2013). Managing the hazards of drought and shifting sands in drylands (the case of Kuwait), Developments in Soil Classification, land use planning and Policy Implications, pp. 703–729 editors Shabbir A. Shahid, Faisal Taha, Mahmoud Adbulftatah.

Misak, R., Al-Awadhi, J. M., & Al-Sudairawi, M. (1999). Assessment and controlling land degradation in Kuwaiti desert ecosystem. In *Proceeding Conference on the Impact of Environmental Pollution on the Development in the Gulf Region*, Kuwait, 15–17 March.

Misak, R., Kwarteng, A., Al-Sudairawi, M., Omar, S., Al-Awadhi, J., Al-Obaid, E., Shahid, S., & Kerdi, Z. (2000). *Controlling land degradation in severalareas of Kuwait, Phase 1: mapping and assessment.* Kuwait Institute For Scientific Research, Report No. KISR 6005, Kuwait.

Mostafa, S. I., Abd el-aal, A. K., & El-Eraki, M. A. (2018). Multi scenario seismic hazard assessment for Egypt. *Journal Seismology.* https://doi.org/10.1007/s10950-018-9728-y.

Omar, S. A., Shahid, S. A., & Misak, R. (1999). *Assessing damage magnitude and recovery of the terrestrial eco-system.* Follow up of natural and induced desert recovery. Kuwait Institute for Scientific Research, Kuwait.

Omar, S. A. (1991). Dynamics of range plants following 10 years of protection in arid rangelands of Kuwait. *Journal of Arid Environments, 21*, 99–111.

Omar, S. A. S., Bhat, N. R., Shahid, S. A., & Assem, A. (2005). Land and vegetation degradation in war-affected areas in the Sabah Al-Ahmad Nature Reserve of Kuwait: A case study of Umm. *Ar. Rimam. Journal of Arid Environments, 62*, 475–490.

Omar, S. A. S., Misak, R., King, P., Shahid, S. A., Abo-Rizq, H., Grealish, G., & Roy, W. (2001). Mapping the vegetation of Kuwait through reconnaissance soil survey. *Journal of Arid Environments, 48*, 341–355.

Reiter, L. (1990). *Earthquake hazard analysis*. Columbia University Press, 254 pp.

Shahid, S. A., Omar, S. A. S., & Al-Ghawas, S. (1999). Indicators of desertification in Kuwait and their possible management. *Desertification Control Bulletin, 34*, 61–66.

Shahid, S. A., Omar, S. A. S., Grealish, G., King, P., El-Gawad, M., & Al-Mesabahi, A. (1998). Salinization as an early warning of land degradation in Kuwait. *Problems of Desert Development, 5*, 8–12.

Shahid, S. A., Omar, S. A. S., Misak, R., & Abo-Rizq, H. (2003). Land resource stresses and degradation in the arid environment of Kuwait: An overview. In A. S. Alsharhan, W. W. Wood, A. S. Goudie, A. Fowler, & E. M. Abdellatif (Eds.), *Desertification in the third Millennium*. Swets and Zeitlinger Publisher, Lisse, The Netherlands, pp. 351–360.

Sutherland, R. A. (2000). Bed sediment associated trace metals in an urban st ream, Oahu. *Hawaii. Environmental Geology, 39*(6), 611–627.

Zaman, S. (1997). Effects of rainfall and grazing on vegetation yield and cover of two arid rangelands in Kuwait. *Environmental Conservation, 24*, 344–350.

Groundwater in Kuwait

Raafat Misak and Wafaa Hussain

Abstract

The fresh and brackish groundwater resources in the State of Kuwait are restricted to two main water-bearing formations (aquifers). These are the Dammam Formation and Kuwait Group. The Kuwait Group aquifer is generally unconfined, i.e., water table condition, whereas the Dammam fractured limestone Formation is a confined-semi confined aquifer. The quality of groundwater in Kuwait varies from brackish in the southwest to brine in the northeast of Kuwait. Fresh groundwater bodies of TDS less than 1000 mg/l occur on saline groundwater of TDS 100,000 mg/l in the north and the northeast, e.g. Raudhatain and Umm Al-Aish water fields. Generally, the water table varies from zero at the Arabian Gulf Coast to about 90 m below the surface in the southwest. Significant ongoing and future groundwater projects include monitoring groundwater level and water quality, establishing hydrological, geological and hydrochemical databases, reducing groundwater levels, long-term monitoring for groundwater quality e.g. Raudhtain and Um-Al Aish freshwater reservoirs and environment treatment of groundwater reservoirs. *The current study discusses the following parts: groundwater quality, groundwater geology, aquifer systems, Al-Raudhatain freshwater field, groundwater misuse and consequences (case of Wafra Agricultural Area), and* Monitoring *water ponds and saline soils, Al Wafra Agricultural Area (2008–2011).*

R. Misak (✉)
Desert Research Center Egypt, Al Matariyyah, Egypt
e-mail: raafat@vision-kuwait.com

W. Hussain
Ministry of Education, Kuwait City, Kuwait

9.1 Introduction

The rainfall in Kuwait is irregular and scanty, about 130 mm/yr in average. One of the natural hazards in Kuwait is Drought. In the last fifty years, several dry seasons that Kuwait experienced. For instance, during 2007–2008 and 2008–2009, the total rainfall was 35 and 65 mm, respectively. Some groundwater fields are seasonally recharged by rainfall and runoff water. These include Al-Raudhatain and Umm Al-Aish water fields.

In Kuwait, the water supplies include desalinized seawater (70% of water supplies), groundwater, and treated sewage water. Table 9.1 shows the installed capacity (MIG) and daily average of gross consumption (MIG) for desalinated water for 1980, 1990, 2000, 2020, and 2020. The installed capacity ranges from 100 Million Imperial Gallon in (1980) to 683.3 Million Imperial Gallon (2020).

Figures 9.1, 9.2 and 9.3 present information on the capacity distillation plants' and consumption of freshwater and brackish water production.

In Kuwait, there are five wastewater treatment plants operating, with a total capacity of 239 million m^3/y. The generation of wastewater in Kuwait is 154.6 m^3/capita/y, of which approximately 75% is treated (Aleisa & Al Shayji, 2019).

The usable groundwater in Kuwait is generally brackish to saline, except for some isolated freshwater lenses in Al-Raudhatain and Umm Al-Aish, north of Kuwait. Fresh and brackish groundwater resources are limited to the Kuwait Group and the Dammam Formation aquifers. The first aquifer is generally unconfined, whereas the second one is a confined-semi confined. The groundwater table range from zero at the Gulf coast to about 90 m below the surface in the southwest (Hussain, 2004). Detailed descriptions of the groundwater type and aquifers are given by many authors (Parsons Corporation, 1964; Al-Hamad, 1964; Bergstrom & Aten, 1964; Hantush, 1970; Senay, 1973; Omar et al., 1981; Al-Ruwaih, 1984, 1985, 2000;

Table 9.1 Statistical indicators of distilled and freshwater (Ministry of Electricity and Water, 2021)

Year	Installed capacity (MIG)	Daily average of gross consumption (MIG)
1980	100	64.1
1990	252	130.3
2000	283.2	241.7
2010	423.1	367.5
2020	683.3	457.6

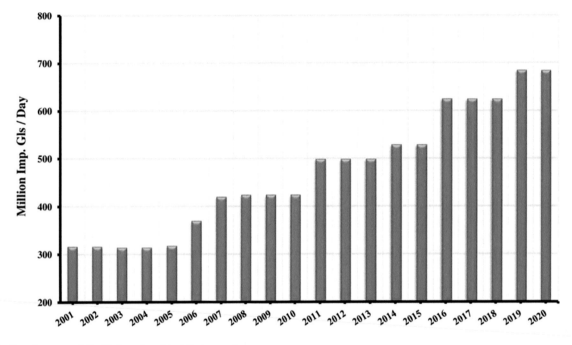

Fig. 9.1 Development of distillation plants' installed capacity

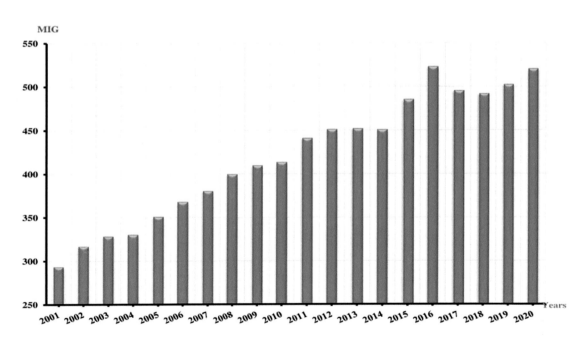

Fig. 9.2 Maximum daily gross consumption of freshwater

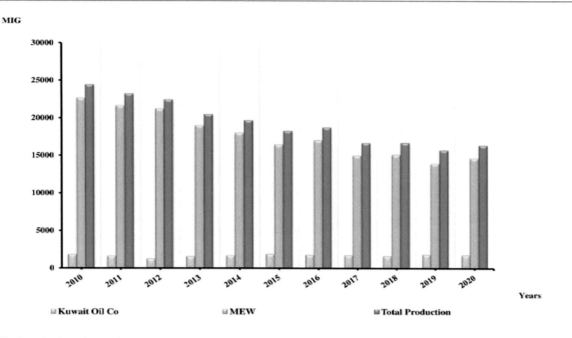

Fig. 9.3 Total production of groundwater

Al-Ruwaih, 1987; Abusada, 1988; Al-Murad, 1994; Mukhopadhyay et al., 1994; Al-Ruwaih et al., 1998; Al-Awadi et al., 1998; Al-Sulaimi et al., 2000; Al-Fahad & Al-Senafy, 2000; Viswanathan, et al., 2002; Al-Sulaimi & Al-Ruwaih, 2004; Khaled Hadi et al., 2018; Bhandary et al., 2018).

In Kuwait, the desalinated water meets almost 100% of the freshwater needs of Kuwait. It is provided to the local public consumers at the rate of almost $0.67 per m^3 while for industrial consumers at the rate of $0.21 per m^3 (Al-Rashed et al., 1998).

Significant ongoing and future groundwater projects include monitoring groundwater level and water quality, establishing hydrological, geological and hydrochemical databases, reducing groundwater levels, long-term monitoring for groundwater quality e.g. Raudhtain and Um-Al Aish freshwater reservoirs and environment treatment of groundwater reservoirs (UNCC, United Nations Compensation Committee) claim No. 5000256a.

This Chapter consists of the following parts: 1. Groundwater quality, 2. Groundwater geology, 3. Aquifer systems, 4. Al-Raudhatain freshwater field, 5. Groundwater misuse and consequences (case of Al Wafra Agricultural Area) and 6. Monitoring *water ponds and saline soils, Al Wafra Agricultural Area (2008–2011).*

9.2 Groundwater Quality

9.2.1 Al-Raudhtain Field (Freshwater)

Geographically, Al-Raudhtain freshwater field exists at the northeastern parts of Kuwait. The water of this field is considered as a mixture of predominantly fossil water recharged during Recent-Pleistocene, and replenished from infiltration of seasonal rainfall. The mixture of these waters gives an average salinity of 625 mg/l. The total potential of this field is 1057 m^3/d and the total volume of usable water in storage is approximately 739,762 m^3/d, half of it with salinity less than 1000 mg/l. The maximum saturated thickness of water containing less than 1000 mg/l of TDS is about 33 m, while water containing less than 2000 mg/l of TDS is about 10 to 20 m (Omer et al., 1981; Al-Ruwaih, 1987).

9.2.2 Dammam Formation Aquifer

The Dammam Formation aquifer constitutes a significant water source in Kuwait. Its quality reduces from southwestern to the central part of Kuwait. The TDS change from less than 3000 mg/l to 10,000 mg/l, respectively. The recharge source of the aquifer is from infrequent precipitation on the outcrop in Saudi Arabia and it flows from the north and the east. In the same direction, the mineralization of the water increases, and the chemical contents change from bicarbonate to sulfate to chloride.

9.3 Groundwater Geology

As mentioned in literature, the geologic sequence of Kuwait was mainly influenced by the stable shelf conditions of the Arabian Plate. The oldest exposed sedimentary rocks in Kuwait is Dammam Formation (Fig. 9.4).

The Tertiary–Quaternary sediments are divided into the Kuwait and the Hasa Groups. The first includes three formations, i.e., Dibdibba, Fars and Char. While the second group includes Dammam, Rus, and Radhuma Formations. The Mesozoic (Late Cretaceous) rocks are carbonate (Al-Ruwaih, 2001; Al-Sharhan & Naim, 1997). Holoce-Eocene lithostratigraphic column and groundwater conditions are shown in Fig. 9.5.

9.3.1 Kuwait Group

The Kuwait Group contains clastic sediments of the Miocene–Pleistocene age. Based on lithological bases, Kuwait Group can be divided into three units (Owen & Naser, 1958). These are from base to top are Ghar, Lower Fars, and Dibdibba Formations. The thickness of the Dibdibba Formation varies from a few meters to about 183 m in the northern part of Kuwait (Omar et al., 1981). The sediment

content of Lower Fars Formation is shale, sandstone, conglomerate, and thin fossiliferous limestone. Its thickness varies from 60 m in the west of Kuwait to 180 m north of Kuwait. The Ghar Formation contains marine to terrestrial sand, silt, and gravel, with thickness reaches 183 m (Al-Ruwaih, 1999). Figure 9.6 shows sheets and sections of Kuwait Group.

9.3.2 Hasa Group

Hasa Group belongs to Middle Eocene–Paleocene and made of three formations (Al-Sulaimi et al., 1992). These are Dammam, Rus, and Radhuma. The Dammam Formation (Middle–upper Eocene) is the most significant aquifer in Kuwait.

The thickness of the Dammam Formation varies from 120 m in the southwest of Kuwait to 280 m in Sabriya in the north.

9.4 Aquifer Systems

The Tertiary–Quaternary sequence is the most significant aquifer system in Kuwait. Two separate water-bearing formations were identified, i.e., the upper Kuwait Group (clastic sediments) and the lower Dammam Formation (fractured limestone). Both aquifers are separated by a layer of cherts and/or clay (Al-Ruwaih & Hadi, 2005). Kuwait stretches over the discharge section of a hydrological system in which groundwater is recharged by infiltrating precipitation mostly through Hasa Group outcrops in the north-northeastern part of Saudi Arabia. Under the natural conditions of the

Fig. 9.4 Dammam Formation, Ahmadi Quarry, 2018, **a** surface section, **b** Cover of recent sediments, and **c** vertical fractures

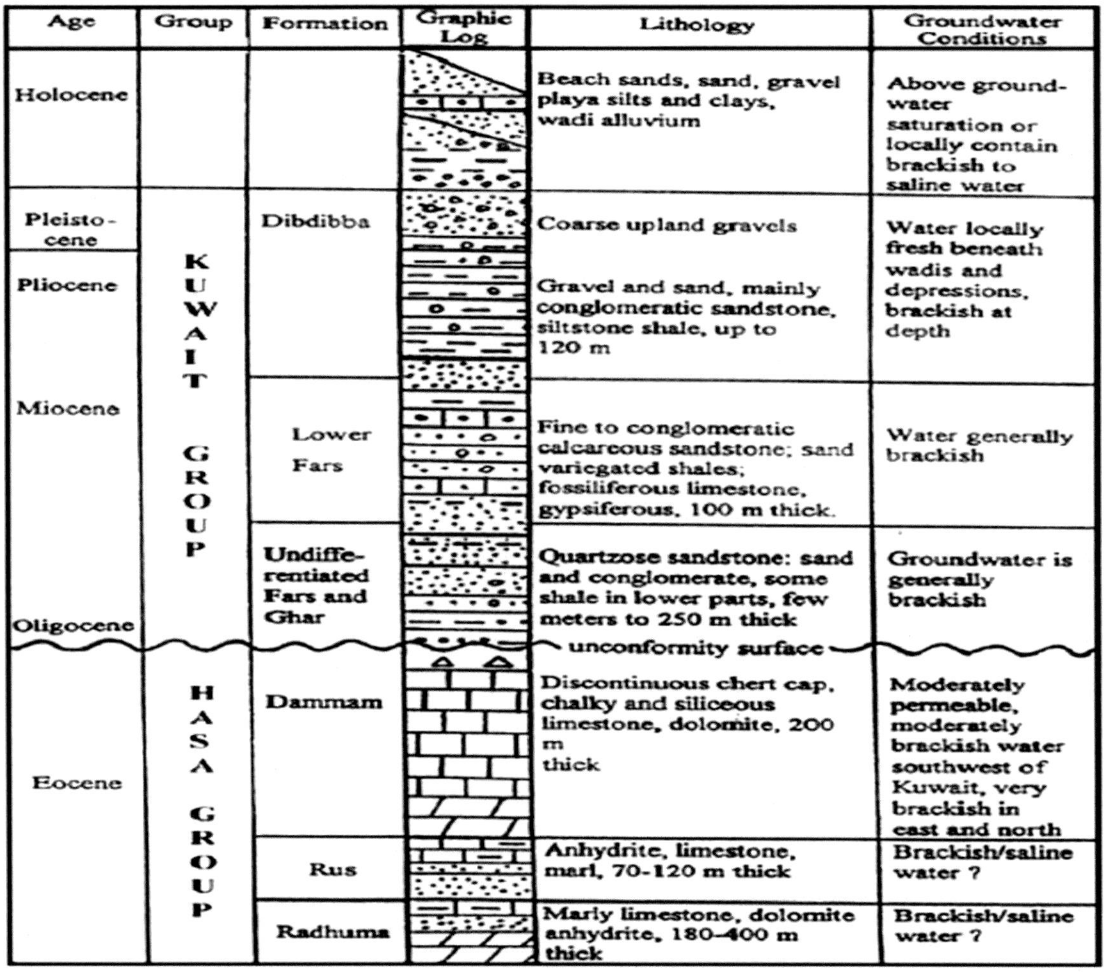

Age	Group	Formation	Graphic Log	Lithology	Groundwater Conditions
Holocene				Beach sands, sand, gravel playa silts and clays, wadi alluvium	Above ground-water saturation or locally contain brackish to saline water
Pleisto-cene	K U W A I T G R O U P	Dibdibba		Coarse upland gravels	Water locally fresh beneath wadis and depressions, brackish at depth
Pliocene				Gravel and sand, mainly conglomeratic sandstone, siltstone shale, up to 120 m	
Miocene		Lower Fars		Fine to conglomeratic calcareous sandstone; sand variegated shales; fossiliferous limestone, gypsiferous. 100 m thick.	Water generally brackish
Oligocene		Undiffe-rentiated Fars and Ghar		Quartzose sandstone; sand and conglomerate, some shale in lower parts, few meters to 250 m thick	Groundwater is generally brackish
				~ unconformity surface ~	
Eocene	H A S A G R O U P	Dammam		Discontinuous chert cap, chalky and siliceous limestone, dolomite, 200 m thick	Moderately permeable, moderately brackish water southwest of Kuwait, very brackish in east and north
		Rus		Anhydrite, limestone, marl, 70-120 m thick	Brackish/saline water ?
		Radhuma		Marly limestone, dolomite anhydrite, 180-400 m thick	Brackish/saline water ?

Fig. 9.5 Lithostratigraphic representation of the Tertiary–Quaternary sediments of Kuwait

hydrological events, the groundwater quality has composition that differs from southwest to northeast. It ranges from brackish to very saline, with some freshwater lenses within a saline groundwater in the north of Kuwait. In the southwest, the groundwater contains total dissolved salts equal to 3000 mg/l. This is increased toward northeast to about 100,000 mg/l.

Groundwater exploration and exploitation in Kuwait have been limited to the Neogene-Quarternary and the Eocene systems. Drilling data, geophysical logs, and groundwater analyses have made it possible to define the conditions, dimensions, and major characteristics of the aquifers and aquitards in most of Kuwait's territory. Figure 9.7 shows the surface geology of Kuwait.

The saturated part of the Kuwait Group and the underlying Dammam Formation form the regional aquifer system. This system is separated from the deeper units by mostly impervious, dense, anhydrite layers, and shaly limestone zone of the Rus Formation (Al-Sulaimi et al., 2000).

9.4.1 Kuwait Group Aquifer

As mentioned before, the Kuwait Group consists of an alternation of sands, gravels, sandstones, clays, silts, limestones, and marls. It covers the entire surface of Kuwait capping the Dammam Limestone Formation (Al-Ruwaih et al., 1998). The thickness of the Kuwait Group increases from about 150 m to 400 m from southwest of Kuwait to the northeast but is reduced to only a few meters over the Ahmadi Ridge and other domal structures. Generally, the Kuwait Group is completely dry in the extreme southwest of Kuwait and almost completely saturated with water along the Arabian Gulf coast. The TDS of the Kuwait Group aquifer is increased from about 2000 mg/l to 12,000 mg/l from southwest to northeast for 150 km. Figure 9.8 shows west–east hydrological section.

Two aquifers separated by an aquitard have been identified in the Kuwait Group, in areas southwest of the 12,000 mg/l concentration contour of the groundwater in the

Fig. 9.6 Kuwait Group (Dibdibba, Lower Fars, and Ghar Formations): **a** Sheet of sands and gravel, (Dibdibba Formation), Khabary Al Awazem 2019. **b** Cavernous calcareous sandstone, water-bearing formation, Al shaqiah 2019. **c** Dibdibba Formation over Fars Formation, Jal Az Zour, 2016. **d** Dibdibba Formation over Fars Formation, note the fractures and cavities at the base of the section, Jal Az Zour, 2016. **e** Dibdibba Formation (top) followed by calcareous sandstone and limestone (Fars Formation) Jal Az Zour, 2017. **f** White chalky limestone at the base of geologic section, Jal Az Zour, 2017

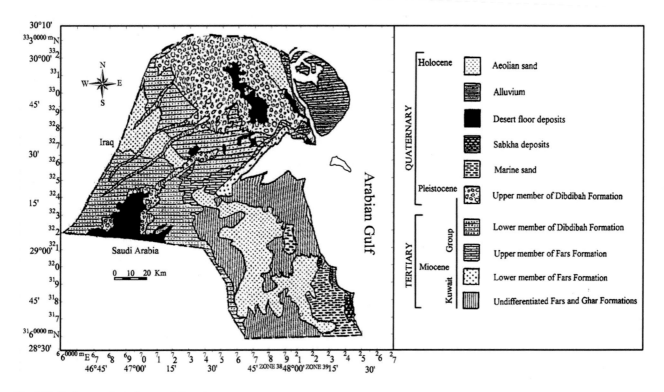

Fig. 9.7 Surface geological map of Kuwait

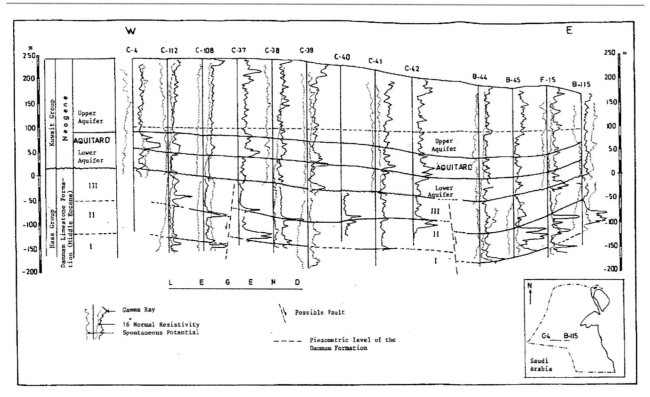

Fig. 9.8 Hydrological cross section (west–east) showing the hydrostratigraphic units of the Kuwait Group and Dammam Formation

underlying Dammam Limestone aquifer. These two aquifers of the Kuwait Group are believed to extend to the east and northeast of this contour line. The upper aquifer appears to be either a leaky water table aquifer or a leaky aquifer that is overlain by an aquitard containing a water table (Al-Sulaimi & El-Rabaa, 1994). The lower aquifer overlies the basal clays and/or the cherts capping the Dammam aquifer. Locally, however, the middle aquitard may grade into and dominate the lower aquifer forming together a relatively pervious aquitard. The two aquifers (and the underlying Dammam Formation, where semiconfined) form a system termed "couple leaky aquifers" or "mutually leaky aquifers" (Hantush, 1970). The water level in the Kuwait Group below the mean sea level (a.m.s.l) varies from zero along the coast to about 90 m in the southwest flowing generally towards the northeast (Al-Nasser, 1978).

9.4.2 Dammam Limestone Aquifer

The Dammam Formation underlies the entire state of Kuwait forming the main water-bearing formation in the country. It consists of chalky limestone, dolomitic limestone, limestone with zones of clay sand, and fossiliferous layers (Al-sharhan & Naim, 1997). The thickness of the Dammam Formation ranges from about 150 m to 280 m from southwest to northeast. The TDS content of the Dammam Formation

groundwater ranges from 2500 mg/l up to 200,000 mg/l from extreme southwest to northeast (Fig. 9.9).

The initial piezometric heads of the Dammam aquifer (Fig. 9.10) indicate that water levels are about 140 m from mean sea level southwest of Kuwait (Abusada, 1988), and sloping in northeast direction. In comparison to Kuwait Group aquifer, the head in the Dammam aquifer is 3 to 20 m higher (Al-Ruwaih & Shehata, 1998). This resulted in expectation of upward vertical leakage between aquifers.

9.5 Al-Raudhtain Freshwater Field

As mentioned in literature, Al-Raudhtain water field is the oldest and biggest fresh groundwater field in Kuwait. Worth mentioning that this freshwater field was discovered accidently in the early 1960s (Parsons Corporation, 1963–1964). The TDS of the freshwater lenses range from less than 350 mg/l to 1000 mg/l. Al-Raudhtain topography, hydrogeology, and hydrochemistry were explored and studied by many researchers (Parson Corporation, 1964; Senay, 1973; Khalafalla, 1977; Al-Nasser, 1978; Al-Ruwaih, 1985, 1987; Al-Ruwaih & Ali, 1986; Al-Sulaimi, 1988; Al-Sulaimi et al., 1993; Hadi, 1993; Mukhopadhyay et al., 1996; Al-Sulaimi & Pitty, 1995; Al-Ruwaih & Shehata, 1998; Al-Sulaimi & Mukhopadhyay, 2000).

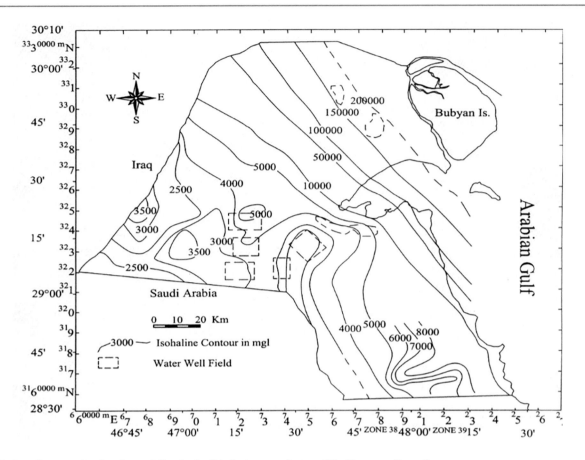

Fig. 9.9 Isosaline map showing the total dissolved solids in the groundwater of the Dammam Formation

Al-Raudhtain surface hydrologic unit is one of the major drainage basins in Kuwait. It exists in the northern part of the country, between latitudes 29° 40′N and 29° 59′N and longitudes 47° 20′E and 47° 44′E (Fig. 9.11). Generally Al-Raudhtain basin is a flat to slightly undulated area covered by sands and gravel. It is bounded by several dry wadis, which are flooded after heavy rain storms. Al-Raudhtain shallow hydrologic unit contains 12 wadis that drain into Al-Raudhtain playa from all directions (Fig. 9.12).

The total surface area of Al-Raudhtain drainage basin is about 670 km². The highest point in Al-Raudhtain drainage basin attains about 136 m above the sea level west of Kuwait, while the lowest point is close to 38 m above the sea level (Al-Sulaimi & Pitty, 1995). The lithology and stratigraphy of Al-Raudhtain hydrologic unit were studied by Senay (1973). The sediments of this hydrologic unit belong to the Holocene-Pleistocene period. They include drift sand, gravel, recent flood deposits silt, and clay besides the coarse sands and gravels of the Dibdibba Formation. The thickness of the Dibdibba Formation clastic attains approximately 107 m thick et al.-Raudhtain. The Dibdibba Formation is dominant with silts, cemented sands, gravels and with amount of minor

clay. Shallow excavation in or near the middle of Al-Raudhtain basin has shown that below a depth of 1 to 2 m, the sand and gravel are firmly cemented by lime and gypsum lime which is known as gatch locally. The gatch zone extends up to a depth of 3.6 m, after which the cementation weakens and the sands and gravel become loose and friable.

According to Senay (1977), the system consists of only two main aquifers. The first one combines the upper and middle aquifers. The saturated thickness of this first aquifer ranges from 12 to 36 m and contains fresh groundwater occurring within depression. The second aquifer consists of the lower aquifer and underlying beds. The effective thickness may vary from 10 to 18 m and it contains saline water in the lower part and while in the upper part it contains brackish water. In general, its salinity varies from 205 to 975 mg/l, although the salinity of some wells may exceed 1300 mg/l. The mixture of these waters gives an average salinity of 625 mg/l. The best water quality appears at the upper limits of saturation (205–800 mg/l). The water salinity increases rapidly with depth, and there is the central part of the field represents the lowest water salinity; water quality tends to deteriorate towards the east and west.

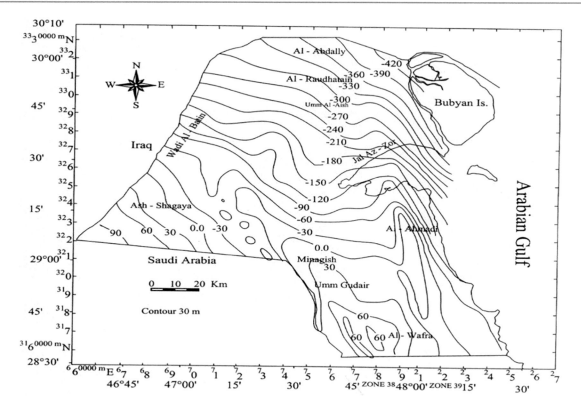

Fig. 9.10 Elevation contour map of the top of the Dammam Formation in Kuwait

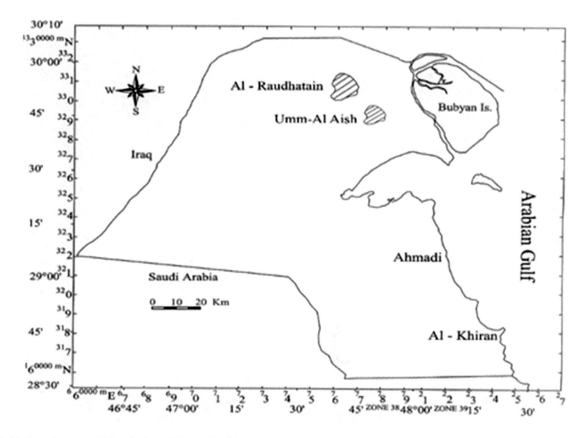

Fig. 9.11 Location map. Al-Raudhtain and Umm-Al-Aish

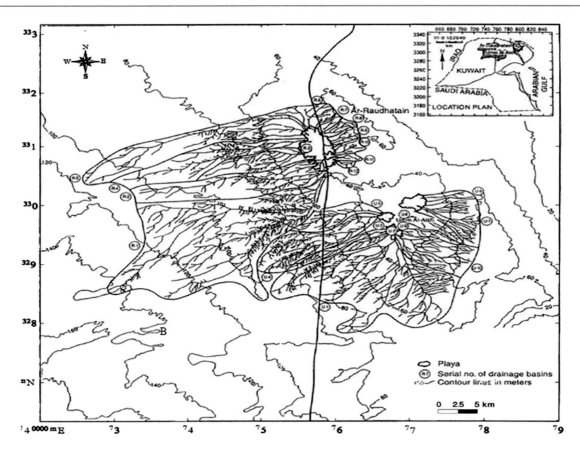

Fig. 9.12 Al-Raudhtain and Umm-Al-Aish drainage system

9.6 Groundwater Misuse and Consequences (the Case Al-Wafra Agricultural Area)

Misuse of irrigation water in Al-Wafra Agricultural Area is the main cause of soil salinization problem. During the period between 1995 and 2006, farmers were unofficially exploiting groundwater from Al Dammam Fractured Limestone aquifer (Salinity from 2500 mg/l to 10,000 mg/l). Some 600 flowing wells (Artesian) were developed. For at least ten years, there was no control over the flowing groundwater. Due to the lack of drainage systems at any level (farm or area levels), the irrigation water started to accumulate in low areas within the farms to form different depths, shapes, and sizes of water ponds. Due to high temperature (close to 45 °C in summer) and high evaporation in Kuwait (about 3000 mm/year), the soils started to be highly salinized. Consequently, the cultivations were deteriorated, and many farms were abandoned. While other farms were partially covered with water (depending on the farm

elevation relative to the level of flowing wells in surrounding farms). Figures 9.13 and 9.14 show abandoned Al-Wafra Farms.

In general, the local geomorphology was one of the controlling factors for the salinity problem in Al-Wafra Agricultural Area (Hussain, 2016). Alternation of hollows (playas called locally khabrat) and ridges/cliffs accelerates the flow of water towards the farms in the low-lying areas (natural hollows). Soil salinization has negative economic, social, hydrologic and environmental impacts. In general, land degradation and depletion of crop yields are the most significant socioeconomic consequences of misuse of irrigation water. Figure 9.15 shows soil salinity, shallow groundwater level, abandoned farm, and deterioration of date palms in Al-Wafra Farms.

In 2009/2010, 434 flowing wells representing 72% of the total flowing wells were controlled by Ministry of Water and Electricity (MWE). As a result, water ponds disappeared and soils dried up. However, symptoms of soil salinization are remarkably observed.

Fig. 9.13 Three abandoned Al-Wafra Farms, 2010 (before controlling flowing wells), **a** completely flooded **b**, **c** partly flooded

Fig. 9.14 Three abandoned Al-Wafra Farms, 2006 note the stagnant irrigation water in **a** and **c**

The level of salinity in some salt-affected farms in Al-Wafra Agricultural Area have reached a stage where the long-term productivity is deteriorated (EC more than 15 mS/cm), (Misak et al., 2014). Farmers invest in land improvement by covering the saline soils with a layer of fresh fine to medium sands of about 100 cm. thick they add manure to enhance soil fertility. Drip irrigation is applied. The cost of soil improvement is KD 0.65 ($ 2)/m². The cost includes materials and manpower. Figure 9.16 shows saline and cultivated soils.

Fig. 9.15 Soil salinization and water logging, Al-Wafra Farms (Misak, 2021). **a** Saline water pond (irrigation water drain). **b** Wet unproductive soil. **c** Salts (mainly sodium chloride) resulted from evaporation of irrigation water. **d** Shallow saline groundwater (about 100 cm from ground). **e** Killed date palm trees (saline soil). **f** Farm invaded by irrigation water (abandoned)

Fig. 9.16 Reclamation of saline soils after treatment (from saline nonproductive soil to non-saline and productive one)

9.7 Monitoring Water Ponds and Saline Soils, Al Wafra Agricultural Area (2008–2011)

In 2009/2010, some 434 flowing wells in Al Wafra Agricultural Area were controlled by the MEW in Kuwait. Consequently, the sizes of water ponds and saline wetlands were remarkably reduced. To monitor the water ponds and saline wetlands in Al Wafra Agricultural Area, five sites were selected. These sites were mapped using high-resolution satellite images of 2008 (before controlling flowing wells) and 2011 (after controlling flowing wells). Arc GIS 9.3 was used to accurately delineate the saline wetlands and water ponds (Table 9.2 and Figs. 9.17, 9.18 and 9.19).

9.8 Saline Wetlands

In 2008 (prior controlling the flowing wells which were the source of excessive irrigation water to Al Wafra Farms) the size of saline wetlands was 7,564,323 m^2. In 2011 (after controlling the flowing wells) the size was reduced to 151,292.2 m^2 (98%).

Site 1: In 2008 the size of the saline wetlands in site 1 was 601,983.2 m^2, while in 2011 it was decreased to 136,597.8 m^2 (77%)

Site 2: In 2008 the size of the saline wetlands in site 2 was 40,346 m^2, while in 2011 it was decreased to 3995.4 m^2 (90%)

Table 9.2 Comparing between 2008 and 2011 water ponds and saline soils

Site	Water ponds (m^2) 2008	Saline soils (wetlands), m^2 2008	Water ponds (m^2) 2011	Saline soils (wetlands) (m^2) 2011
1	67,368.6	601,983.2	18,120	136,597.8
2	8108.6	40,346	2941	3995.4
3	9464	908.2	3259.3	760.4
4	13,605.9	60,412.6	316.6	3921.9
5	12,947.9	52,782.3	613.5	6016.7
Total	111,495	756,4323	25,250.4	151,292.2
Average	**22,299**	**151,286.46**	**5050.08**	**30,258.44**

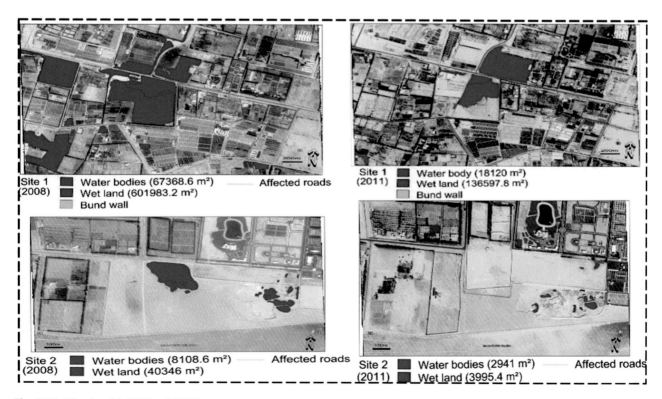

Fig. 9.17 Sites 1 and 2 (2008 and 2011)

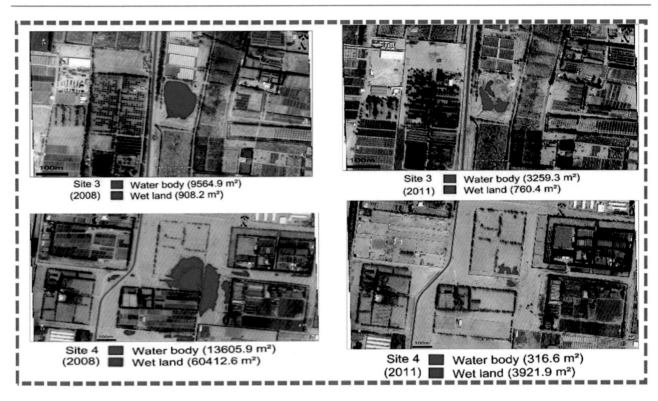

Fig. 9.18 Sites 3 and 4 (2008 and 2011)

Fig. 9.19 Site 5 (2008 and 2011)

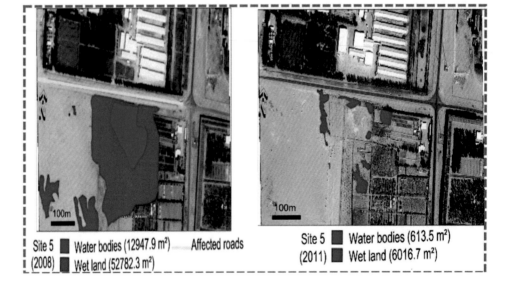

Site 3: In 2008 the size of the saline wetlands in site 3 was 908.2 m^2 while in 2011 it was decreased to 760.4 m^2 (16%)
Site 4: In 2008 the size of the saline wetlands in site 4 was 60,412.6 m^2 while in 2011 it was decreased to 3921.9 m^2 (93%)
Site 5: In 2008 the size of the saline wetlands in site 5 was 52,782.3 m^2 while in 2011 it was decreased to 6016.7 m^2 (88.6%).

9.9 Water Ponds

In 2008 (prior controlling the flowing wells which were the source of excessive irrigation water to the farms) the size of the total water ponds was 111,495 m^2. In 2011 (after controlling the flowing wells) the size of the water ponds remarkably reduced to 25,250.4 m^2 (77%)

Site 1: In 2008 the size of the water ponds in site 1 was 67,368.6 m^2 while in 2011 was decreased to18120 (73%)

Site 2: In 2008 the size of the water ponds in site 2 was 8108.6 m^2 while in 2011 the size was decreased to 2941 m^2 (64%)

Site 3: In 2008 the size of the water ponds in site 3 was 9464 m^2 while in 2011 the size was decreased to 3259.3 m^2 (65.5%)

Site 4: In 2008 the size of the water ponds in site 4 was 13,605.9 m^2 while in 2011 the size was decreased to 316.6 m^2 (97%)

Site 5: In 2008 the size of the water ponds in site 5 was 12,947.9 m^2 while in 2011 the size was decreased to 613.6 m^2 (95%). Figures 9.17, 9.18 and 9.19.

9.10 Summary and Conclusion

In Kuwait, fresh and brackish groundwater resources form only two main aquifers. These are Dammam Formation and Kuwait Group. The Kuwait Group consists of an alternation of sands, gravels, sandstones, clays, silts, limestones, and marls. In Kuwait Group water, TDS increases in general from about 2500 mg/l to 100,000 mg/l from southwest to northeast. Fresh groundwater occurs in the depressions of Al-Raudhatain, Umm Al-Aish, and Umm Nigga north of Kuwait. The Dammam Formation is considered the main aquifer. It consists of chalky limestone, dolomitic limestone, and limestone with some clayey, sandy, and fossiliferous zones. TDS of the groundwater of the Dammam Formation ranges from 2500 mg/l to 200,000 mg/l in the direction from southwest to the northeast. The misuse of groundwater within Dammam Formation causes the soil salinization problems. These high salinity levels of soil cause deterioration of farm productivity. Farmers invest in land improvement by covering the saline soils with a layer of fresh sands. They add manure to enhance soil fertility. Drip irrigation is applied. The cost of soil improvement is KD 0.65 ($ 2)/m^2.

References

Abusada, S. M. (1988). *The essential of groundwater resources of Kuwait, Technical Report.* Report No. KISR 2665, Kuwait Institute of Scientific Research, Kuwait.

Al-Awadi, E., Mukhopadhaya, A., & Al-Senafy, M. (1998). Geology and hydrogeology of the Dammam formation in Kuwait. *Hydrogeology Journal, 6,* 302–314.

Al-Fahad, K., & Al-Senafy, M. (2000). Impact of oil lakes and oil fires on groundwater contamination in Northern Kuwait. In L. Bjerg, P. Engesgaard, & T. Krom (Eds.), *Groundwater 2000,* Balkema/Rotterdam/Brookfield.

Al-Hamad, A. (1964). *Groundwater resources of Kuwait.* Report to Ministry of Electricity and Water. Kuwait, V. I. The Ralph. M. Parsons Company, New York, U.SA, pp. 1–22.

Al-Murad, M. (1994). *Evaluation of the Kuwait Aquifer system and assessment of future well fields abstraction using a numerical 3D flow model.* Thesis, Arabian Gulf University, Bahrain.

Al-Nasser, S. (1978). *Water.* Kuwait National Symposium on Science and Technology for Development, Kuwait Institute of Scientific Research, Kuwait.

Al-Rashed, M., Al-Senafy, M., Viswanathan, M., & Sumait, A. (1998). Groundwater utilization in Kuwait: Some problems and solutions. *Water Resource Development, 14*(1), 91–105.

Al-Ruwaih, F. (1984). Groundwater chemistry of Dibdiba formation. *North Kuwait. Groundwater, 22*(4), 412–417.

Al-Ruwaih, F. (1985). Hydroehemical classification of the groundwater of Umm Al-Aish, Kuwait. *Kuwait Journal of the University of Kuwait (Science), 12*(2), 288–296.

Al-Ruwaih, F. (1987). Groundwater classifications and quality trends of Al-Rawdhatain field, Kuwait. *Journal of the University of Kuwait (Science), 14,* 395–414.

Al-Ruwaih, F. (1999). Hydrogeology and hydrochemical facies evaluation of the Kuwait group aquifer. *Al-Atraf, Kuwait, Kuwait Journal of Science and Engineering, 26,* 337–354.

Al-Ruwaih, F. (2001). Hydrochemical investigation on the clastic and carbonate aquifers of Kuwait. *Bulletin of Engineering Geology and the Environment, 60,* 301–314.

Al-Ruwaih, F., & Ali, H. (1986). Resistivity Measurements for Groundwater Investigation in the Umm Al-Aish Area of Northern Kuwait. *Journal of Hydrology, 88,* 185–198.

Al-Ruwaih, F., & Hadi, K. (2005). Water quality trends and management of fresh groundwater at Rawdhatain. *Kuwait. European Journal of Scientific Research, 9*(1), 40–62.

Al-Ruwaih, F., & Shehata, M. (1998). The chemical evolution and hydrogeology of Al-Shagaya field B, Kuwait. *Water International, 23,* 75–83.

Al-Ruwaih, F., Sayed, S., & Al-Rashed, M. (1998). Geological controls on water quality in Arid Kuwait. *Journal of Arid Environments, 38,* 187–204.

Al-Ruwaih, F., Shehata, M., & Al-Awadi, E. (2000). Groundwater utilization and management in the state of Kuwait. *Water International Journal, 25,* 378–389.

Al-Sharhan, A., & Naim, A. (1997). *Sedimentary basins and petroleum geology of the middle east,* Elsevier Scince B.V, Amsterdam, Netherlands.

Al-Sulaimi, J. (1988). Calcrete and near surface geology of Kuwait city and suburb, Kuwait, Arabian Gulf. *Sedimentary Geology, 54,* 331–345.

Al-Sulaimi, J., Al-Rabaa, S., Muhanna, A., Amer, A., & Lenindre, Y. (1992). Assessment of groundwater resources in Kuwait using remote sensing technology (WH-002), Geology, Kuwait institute for scientist research, report No. 4038, vol. 3.

Al-Sulaimi, J., & Al-Ruwaih, F. (2004). Geological, structural and geochemical aspects of the main aquifer systems in Kuwait. *Kuwait Journal of Science and Engineering, 31*(1), 149–174.

Al-Sulaimi, J., & El-Rabaa, M. (1994). Morphostructural features of Kuwait. *Geomorphology, 11,* 151–167.

Al-Sulaimi, J., & Mukhopadhyay, A. (2000). An overview of the surface and near surface geology, geomorphology and natural resource of Kuwait. *Earth Science Reviews, 50,* 227–267.

Al-Sulaimi, J., & Pitty, A. (1995). Origin and depositional model of Wadi Al-Batin and its associated alluvial fan, Saudi Arabia and Kuwait. *Sedimentary Geology, 97,* 203–229.

Al-Sulaimi, J., Viswanathan, M., & Szekely, F. (1993). Effect of oil pollution on fresh groundwater in Kuwait. *Environmental Geology, 22,* 246–256.

Aleisa, E., & Al Shayji, K. (2019). Analysis on reclamation and reuse of wastewater in Kuwait. *Journal of Engineering Research, 7*(1).

Bergstrom, R., & Aten, R. (1964). Natural recharge and localization of fresh groundwater in Kuwait. *Journal of Hydrology, 2*, 213–231.

Bhandary, H., Sabarathinam, C., & Al-Khalid, A. (2018). Occurrence of hypersaline groundwater along the coastal aquifers of Kuwait. *Desalination, 436*, 15–27.

Hadi, K., Kumar, U., Al-Senafy, M., & Mukhopadhyay, A. (2018). Historical evaluation of hydrological and water quality changes of southern Kuwait groundwater system. *Arabian Journal of Geoscience, 11*, 413.

Hantush, M. (1970). *Memorandum on the Al-Shagaya Groundwater Project. Summary, Conclusions and Recommendations.* Kuwait Institute for Scientific Research, Kuwait. pp. 1–19.

Hussain, W. (2004). *Analysis of spatial variation of groundwater chemistry in Al-Raudhatain and Umm Al-Aish Fields Using GIS and Statistical Methods.* M.S. Thesis, Kuwait University, Kuwait.

Hussain, W. (2016). *Assessment, monitoring and treatment of water logging and soil salinity in Wafra Agricultural Area, Southern Part of Kuwait.* PhD. Thesis, Mansoura University, Egypt.

Khalafalla, M. (1977). *Applicability of the electrical resistivity method for groundwater research and prospection in Kuwait.* Thesis, Kuwait University, Kuwait.

Ministry of Electricity and Water (MEW). (2021). Statistical Year Book.

Misak, R., El Gamily, H., & Hussain, W. (2014). Threats to Agriculture Lands at Al-Wafra, Southern Part of Kuwait. *Journal of Arid Land Studies*, 24–1.

Misak, R. (2021). *Trends and targets of land degradation neutrality (LDN), the Case of Kuwait.* Book Chapter, (in press)

Mukhopadhyay, A., Al-Sulaimi, J., Al-Awadi, E., & Al-Ruwaih, F. (1996). An overview of the tertiary geology and hydrogeology of the northern part of Arabian gulf region with special reference to Kuwait. *Earth-Science Reviews, 40*, 259–295.

Mukhopadhyay, A., Al-Sulaimi, J., & Barat, J. (1994). Numerical modeling of groundwater resource management options in Kuwait. *Ground Water, 32*(6), 917–928.

Omar, S., Al-Yaqubi, A., & Senay, Y. (1981). Geology and groundwater hydrology of the state of Kuwait. *Journal of the Gulf and Arabian Peninsula Studies, l*, 5–67.

Owen, R. M., & Naser, S. N. (1958). Stratigraphy of the Kuwait-Basrah area. *Habitat of Oil, American Association of Petroleum Geologists, 42*, 1252–1278.

Parsons Corporation, 1963–1964. Groundwater resources of Kuwait, Vols. I, II, and III. RIGW. (2000). Assessment and solutions of water logging problems at the new agricultural development at the Nubarya. Report presented to the International Fund for Agricultural Development (IFAD), Rome, Italy.

Senay, Y. (1973). *Geohydrology of Al-Raudhatain Field* (pp. 24–43). Unpublished report to the Ministry of Electricity and Water.

Senay, Y. (1977). *Groundwater resource and artificial recharge in Raudhatain Water Field.* Ministry of Electricity and Water, Kuwait.

Viswanathan, M., Akber, A., & Rashed, T. (2002). Contamination of fresh groundwater lenses in Northern Kuwait. In M. M. Sheriff, V. P. Singh, & M. Al-Rashed (Eds.), *Environmental and Groundwater Pollution* V. 3.

Hala Al Jassar, Peter Petrov, Ali Al Hemoud, Abdullah Al-Enezi,
and Abeer Alsaleh

Abstract

Satellite Remote Sensing can provide a valuable source of information in different applications and is considered an important tool for disaster management and support for decision-making in the state of Kuwait. This is especially valid for the cloudless atmosphere of Arabia. In this chapter, some examples related to geology ideas are presented. The application of satellite remote sensing techniques using both passive and active sensors is presented over Kuwait desert and marine. This includes academic studies of soil moisture, land subsidence, and operational monitoring of recent flash floods, dust storms, and oil spills detected by optical and SAR instruments from various space satellite platforms over the territory of the state of Kuwait. Operational monitoring of dust storms, especially sever jets from Iraq was performed with MODIS NASA/Terra and Aqua instruments, the oil pollution in the Northern Gulf oil fields was very effectively detected by Sentinel SAR-C instrument of Copernicus/ESA, and some rain floods in urban areas in winter were analyzed by high-resolution instruments of Pleiades.

Keywords

MODIS • Sentinel • SAR • SMAP • Soil moisture • Land subsidence • Flash flood • Dust storm • Oil spills

10.1 Introduction

Satellites are unique sources of information about the surface of the Earth and its atmosphere and they are used by many scientists to improve the knowledge of our planet. The science of satellite remote sensing started back in the 1950s and is mainly based on our understanding of the physics of the interaction of electromagnetic radiation with different Earth's surface geophysical parameters or atmospheric gases and particles. Such understanding of the interaction of electromagnetic radiation including absorption, emission, scattering, and reflection is important for the interpretation of data and images collected by different satellite sensors.

Historically, the capability of the satellites to provide information about the Earth goes back to 1957, when Sputnik 1 was launched by the former Soviet Union. It was the first satellite that orbits around the Earth. Then came the USA TIROS series of satellites which were mainly used for metrological applications. The first satellite to monitor the Earth's surface has been recognized in 1972 when Landsat-1 was launched. Now we have more than 3000 satellites orbiting the Earth and serving many vital applications. By satellites images, the geologist can observe many geological features (i.e., faults, folds, stratigraphy, and landform), distribution of groundwater, and location of natural resources (i.e., oil, gas, and mineral deposits).

The majority of Earth's observational satellites carry passive sensors that measure either the reflected solar radiation in the visible or near-infrared regions or measure the natural emitted radiation from the Earth in the infrared or the microwave regions. Some satellites carry active sensors such as Altimeters, Radars, and Lidars that send signals and then record the reflected or backscattered radiation from different targets on Earth.

Kuwait is potentially facing both natural and anthropogenic hazards. This can range from dust storms, floods, fires, oil spills, and other hazards. Remote sensing is an important tool to support the decision-makers in the

H. Al Jassar (✉)
Department of Physics, Kuwait University, Kuwait City, Kuwait
e-mail: hala.aljassar@ku.edu.kw

P. Petrov · A. Al Hemoud · A. Al-Enezi · A. Alsaleh
Kuwait Institute for Scientific Research, Safat, Kuwait

government with vital technical information to manage these hazards through all disaster management cycles, i.e., disaster preparedness, mitigation, response, and recovery (Eguchi et al., 2008; Thomas & Kemec, 2007). Kuwait Institute for Scientific Research (KISR) established a Decision Support System for crises management in Kuwait using the necessary decision support tools including remote sensing.

A remote sensing facility was developed to provide near real-time remote sensing monitoring on a 24/7 basis. The sensing operational archive is integrated based on several satellite missions covering Kuwait including middle-resolution optical scanners MODIS, VIIRS; high-resolution constellations Planet, Pleiades, LANDSAT; and Sentinel family: Sentinel-2, Sentinel-3, and Sentinel-1 C-band SAR and TerraX X-band SAR. This allows appropriate timing for data collection during cloudy periods, day and night as well as space and spectral resolution for observed disaster phenomena. Operational remote sensing monitoring identifies the land and marine environmental high-risk and accident areas, which can be a threat to the Kuwaiti environment hazards (Misak and Al-Dousari, 2012).

Kuwait is currently facing a possible potentially extensive and harmful scope of both natural and anthropogenic hazards. Hazards in Kuwait vary in nature, magnitude, and consequences. They are differentiated into man-related hazards (anthropogenic/technologic) and natural hazards. Man-related hazards are reported including fires, nuclear accidents, oil spills, and notable industrial activities associated with environmental crises such as oil refineries, oil gathering centers (GCs), petrochemical industries as well as thermal power plants (Girin & Carpenter, 2017; Li et al., 2016). Natural hazards are classified into geophysical (earthquakes of low magnitude), meteorological (dust and sand storms and drought), and surface hydrological (flash floods) (Bou-Rabee, 1994; Misak et al., 2013; Hassan et al., 2021). Unfortunately, these natural and anthropogenic crises are expected to continue and maybe increased in frequency and severity with the advance of climate changes. Meteorological hazards including dust and sandstorms and flash floods and oil spills are major hazards afflicting Kuwait with a focus on Kuwait City in the last few years. In 2018, Kuwait experienced the most severe dust storms in the summer, as well as flash floods which resulted in economic losses and death. Remote Sensing was an important tool for the observation of these hazards and providing an early warning (Manche, 2014).

10.2 Field Measurements of Soil Moisture in Kuwait by Remote Sensing

Soil moisture is an important geophysical parameter for hydrological, climate, and weather model predictions. In arid regions around the world, soil moisture is the main environmental factor that restricts plant growth and vegetation restoration, and it is an important index reflecting soil characteristics (D'Odorico et al., 2007). Many factors such as climate, topography, elevation, slope, vegetation, and soil properties affect soil moisture.

Remote sensing and field measurement of soil moisture in the desert of Kuwait has a long history since the year 2000. About 100 soil samples were collected from several locations in the north area of Kuwait during the Shuttle Radar Topography Mission (SRTM) mission in February–March 2000 (Al Jassar et al., 2006). In the laboratory, the collected samples were immediately weighed and dried in an oven at a temperature of around 105 °C for 24 h. Then, the gravimetric soil moisture was estimated. The result of soil moisture contents for these samples was estimated from 1 to 11% with an average of 3.88%. In addition, the salinity, soil texture, specific density bulk density, and surface roughness were also measured. The analysis indicates that sand is the main content of all samples and the bulk density with an average of 1.2 g cm^{-3}, while the specific density average about 2.65 g cm^{-3}. In the same study, a simple statistical inversion model was developed to retrieve soil moisture from microwave satellite brightness temperatures, based on two frequencies (Al Jassar et al., 2006). The model was applied to retrieve soil moisture from Nimbus-7 SMMR brightness temperature (Tb) data at two bands (Tb(H) of 6.6 GHz and Tb(V) of 37 GHz) for six years (1979–1985). The volumetric soil moisture was found to range from 0.01 m^3 m^{-3} (in dry summer) to 0.13 m^3 m^{-3} (during the rainy season), with an average of 4%, Fig. 10.1. The simple inversion model is applicable to the desert of Kuwait, due to the scanty vegetation cover and moderate roughness.

The second major campaign was conducted from Dec-2005 to March-2006 nearby Al-Abdaly, North of Kuwait (Al-Jassar & Rao, 2010). During this campaign about Forty-five soil samples were collected in synchronous with the Advanced Microwave b Scanning Radiometer-Earth Observing System (AMSR-E) covering an area of one pixel of 25-km circular diameter Fig. 10.2a. A comparison was done between Field-estimated soil moisture values up to

Fig. 10.1 Graph representing the time series of retrieved soil moisture during the day (11 am) and night (12 midnight) passes (Al Jassar et al., 2006)

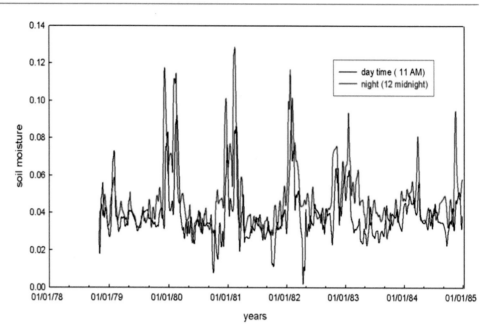

5 cm depth, AMSR-E soil moisture values, and our model results. The result indicates that the field soil moisture values are consistently lower than AMSR-E and model values. This difference can be considered within the error value Also, the result indicates the similarity value of AMSR-E soil moisture and our model values. To study the temporal and spatial variation of soil moisture, monthly average soil moisture maps for the State of Kuwait were produced from AMSR-E data Fig. 10.2b (Al-Jassar & Rao, 2010).

The third campaign experiment to measure soil moisture was from April 2011 to September 2013 within KISR experimental research station in Kabd area (Al Jassar & Rao, 2015). The samples were collected from 16 sites within a 50 km^2 area (Fig. 10.3). The field soil moisture measurements were compared in synchronous with soil moisture data from AMSR-E, which passes over the Kuwait desert, (Fig. 10.4) (Al Jassar & Rao, 2015). The lowest soil moisture was measured in September with a value of 0.01 m^3 m^{-3}, while the highest was in December during the wet season with a value of 0.11 m^3 m^{-3}. In a similar way, AMSR-E data from 2003 to 2011 shows a seasonal variation in Monthly average Volumetric Soil Moisture values (VSM) in Kuwait. The data shows the moisture is higher during the wet season and moisture is higher, in particular, during January with a value of 0.08 m^3 m^{-3} and lowest in the dry season in August with a value of 0.06 m^3 m^{-3}.

In the fourth field campaign, a total of 322 soil samples were collected in 2016 from Al-Salmi west of Kuwait within

an area of 36 km by 36 km (Fig. 10.5) (AlJassar et al., 2019). The samples were collected from7:30 a.m. to 5:30 p.m. on 20 February 2016 and 19 March 2016. The in-situ field soil moisture measurements were then compared with data obtained from Special Sensor Microwave/Imager (SSM/I), European Space Agency Soil Moisture and Ocean Salinity (SMOS), NASA Soil Moisture Active Passive (SMAP), and Advanced Microwave Scanning Radiometer 2 (AMSR2). During this campaign, a large range of soil moisture values was observed due to precedent rain events and subsequent dry down. The statistical analysis of the VSM of collected samples shows a low variability of Mean Relative Difference (M) RD = −0.005 m^{3}m^{-3}. This indicates the stability of volumetric water content spatially and temporally over the selected site. This variability of the MRD values indicates the presence of differences in soil moisture values within the study site, which is possibly related to the soil heterogeneity (Fig. 10.6) (AlJassar et al., 2019). In relation to the topography, The study found that there is no clear correlation between soil moisture and elevation. This could be related to the nature of drainage patterns in the desert environment.

Kuwait University team was selected by the NASA SMAP science team as an international partner in the calibration and validation of soil moisture data from the Active and Passive Satellite (SMAP) mission which was launched on 31st January 2015. Kuwait's site is located in the desert on the west side of Kuwait, and it is the only test site in the Middle East for the pre-launch and post-launch calibration

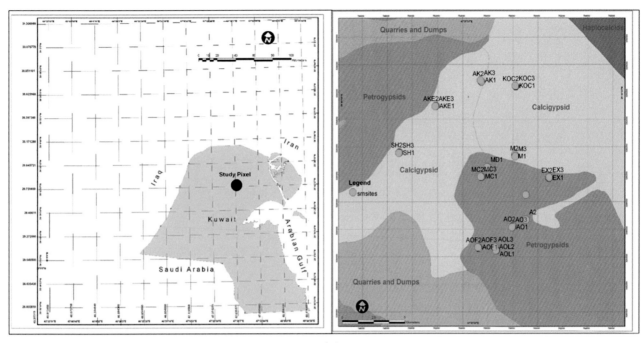

(a)

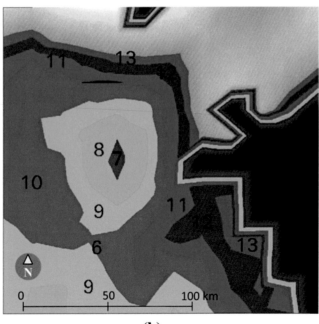

(b)

Fig. 10.2 **a** Location of AMSR-E pixel with 25 km in diameter where 45 samples were collected (Al-Jassar & Rao, 2010). **b** Monthly Average Soil Moisture Map of Kuwait

and validation activities of NASA SMAP satellite data (Fig. 10.7) (Colliander et al., 2017). The test site has exceptionally homogeneity of land surface conditions and possesses six permanent stations to measure soil moisture at different depths. SMAP soil moisture data are compared with the automated station's measurements of soil moisture in Fig. 10.8. The Kuwait University team conducted different gravimetric soil moisture measurements next to each station to validate the station's measurement at 5 cm depth. The ground soil moisture measurements were compared with SMAP 36 km, 9 km, and 3 km resolution of soil moisture data from 2015 to 2020.

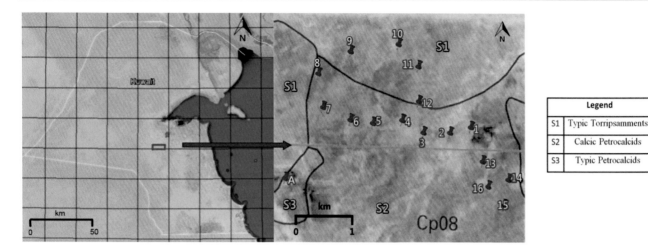

Fig. 10.3 Soil moisture field point at KISR Site. The AMSR-E 25-km grid is superimposed on the left image. The soil classification map is overlaid on the right image (Al Jassar & Rao, 2015)

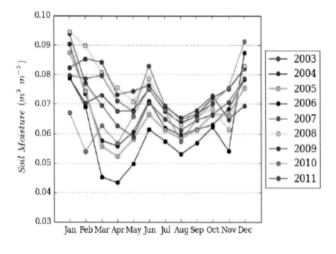

Fig. 10.4 Monthly average Soil moisture variations (2003–2011) (Al Jassar & Rao, 2015)

10.3 Land Subsidence in Burgan Oil Field in Kuwait

Land subsidence is a very common phenomenon wherever underground activities are in progress such as oil and water extraction, coal mining, and underground rail network. Land subsidence is a major problem in many countries and comes under disaster management. Though land subsidence is a problem in Kuwait oil fields due to oil extraction and other geological factors, no reliable data is available on this subject. Differential Synthetic Aperture Radar Interferometry technique is applied over Burgan oil field of Kuwait to assess the land subsidence. Thirty-five subsidence maps are generated with the temporal resolution varying from 35 to 630 days (Fig. 10.10) shows the subsidence map of the Burgan oil field which is the net result of processing 35 subsidence maps. It can be seen from Fig. 10.24 that as high as 4.1 mm/100 days subsidence is noticed in the southern portion of the Greater Burgan Oil Field. The subsidence slowly decreases as we move north of the oil field (Rao et al., 2011).

Another study addresses the spatial variability of land subsidence over the Minagish and Umm Gudair oil fields of Kuwait. Synthetic Aperture Radar Interferometry technique coupled with Interferometric Point Target Analysis (IPTA) approach is used in this study. 29 scenes of ENVISAT ASAR data (for the period January 2005–August 2009) (Rao & Al-Jassar, 2010) were used to make 20 pairs of interferograms (with high coherence and low noise) for IPTA analysis (Fig. 10.11). The output of this study is the land subsidence map of Minagish and Umm Gudair oil fields with a spatial resolution of 40 m. The results indicate that there is land subsidence of 8 mm/100 days on the southern part of the oil field (Umm Gudair) (Fig. 10.12).

10.4 Flash Flood

Kuwait like most parts of the Arabian Peninsula is a desert-type environment with scanty rainfall (Al-Awadhi et al., 2005). As a result, when heavy rains occur, they often cause flash floods that can be very destructive. In Kuwait, flash floods, which are generally associated with heavy rainfall events over short durations, have caused millions of dollars worth of damage in the country, wreaking havoc on roads, bridges, and homes (Misak and Al-Dousari, 2012). In 2018, Kuwait experienced a great flash flood during the period from 4 to 14 November, and many urban communities of the country were exposed to devastating floods, in particular, southern urban areas. It was an extreme event with an average rainfall value of 151 mm. This event

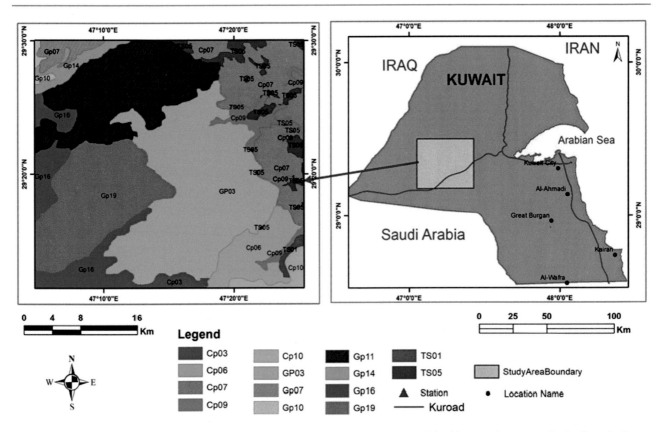

Fig. 10.5 The test site 36 × 36 km with 14 soil types showing six in-situ stations. The work in this paper focuses on the dominant landscape (Cp06, Cp07, GP03, Gp11, Gp 16, and Gp 19) (AlJassar et al., 2019)

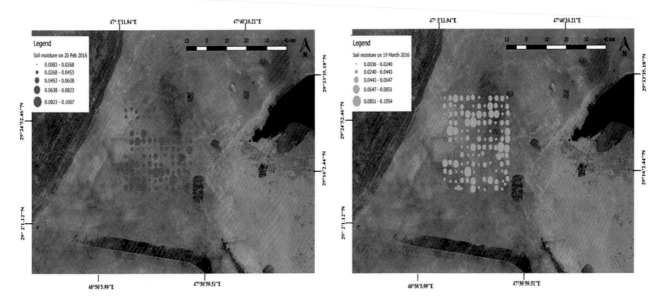

Fig. 10.6 Spatial distribution of the measured VSM on February 20th, 2016 (left). Spatial distribution of the measured VSM on March 19th, 2016 (right) (AlJassar et al., 2019)

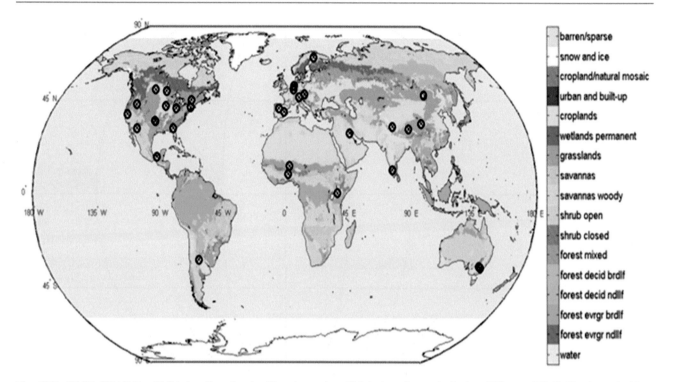

Fig. 10.7 NASA SMAP Core Validation Sites showing Kuwait test site which is the only test site in the middle east. [TJ (Colliander et al., 2017)

Fig. 10.8 Automatic Weather Station and field validation sampling during the NASA/SMAP project

considered the third-largest destructive flash floods occurred in Kuwait causing damages more than 900 million dollars, with roads and tunnels paralyzed by an accumulation of clay following heavy rains. The flooded areas have been evacuated and local municipalities have begun a huge cleanup operation to clear roads and highways. The rainfall was accompanied by thunderstorms and gusts reaching up to 115 km/h. In this study, we will present flash flooded urban areas as seen by satellite images with flood hazard analysis, extent, and depth of flood over the flood-prone area during this event.

10.4.1 Flood Event Analysis (November 2018)

In 2018, from the 4th to 14th of November, a rainfall event occurred; it was an extreme event with an average rainfall value of 150 mm. Among the 16 meteorological stations, the highest total rainfall during that period was at Kuwait Airport station which reached 190 mm, while the daily rainfall was the highest recorded at Al-Taweel station on the 14th of November which was about 101 mm. The daily rainfall recorded among the 16 stations is tabulated and illustrated in Table 10.1 and Fig. 10.13 (Misak & Raafat, 2015; KMD,

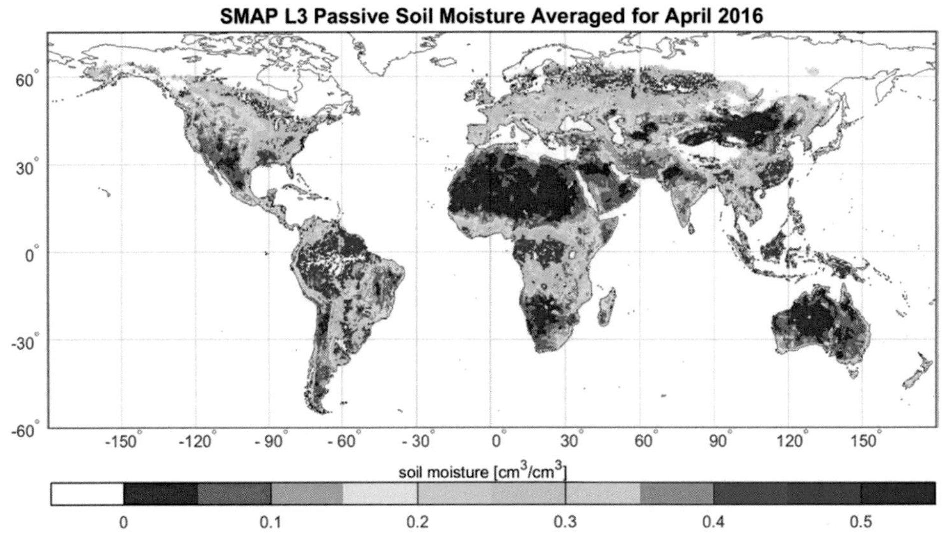

Fig. 10.9 SMAP Soil Moisture Over the Globe (JPL/NASA/Photojournal)

Fig. 10.10 Subsidence image of Greater Burgan Oil Field generated through least square technique. Most parts of the study area are free from subsidence. The southern portion shows subsidence of 4 mm/100 days and the northern portion shows upliftment of 1 mm/100 days (Rao et al., 2011)

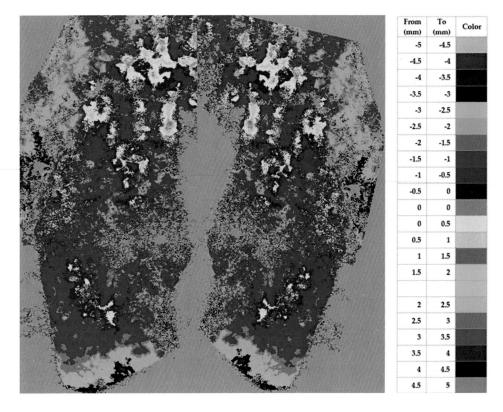

2020). Figure 10.13 clarifies the spatial variation of rainfall, total rainfall of the entire event, recorded in each station across Kuwait. The map shows that the southern sector of the state of Kuwait is highly subjected to intensive rainfall. The remote sensing was used to delineate the flooded area within the urban area, in addition, using to uphold the rainfall event, a MODIS satellite image of cloud cover over the Kuwait region was extracted from 4th November to 14th November 2018. The data was downloaded from the USGS website for the respective days and the enhanced image along with the wind direction in the region is illustrated from Figs. 10.15, 10.16 and 10.17. The images depict that the direction of the wind during rainy days is mostly from the south to north or south to north-east and the cloud cover is the highest on the 14th of November (Fig. 10.14).

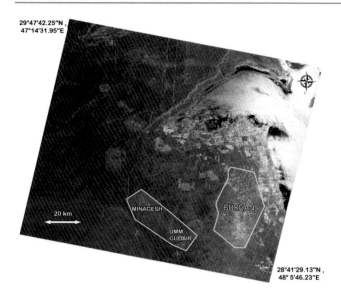

Fig. 10.11 ENVISAT scene acquired on 20-JAN-2007 showing Burgan, Minagish, and Umm Gudair Oil fields (Rao & Al-Jassar, 2010)

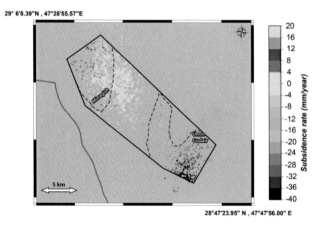

Fig. 10.12 The final average velocity map (Geo-coded) of Minagish and Umm Gudair oil fields. The subsidence was estimated with reference to 28.97 N, 47.59E. The color coding is given on the right side of figure. The boundary of the oil fields is superimposed (Rao & Al-Jassar, 2010)

10.4.2 Flash Hazard Assessment

An estimation of probabilistic flood extent and corresponding flood water depths for the entire State of Kuwait during the rainfall of Nov 2018 (Hassan et al., 2021) was done using hydrological and hydraulic modeling. For that purpose, the State of Kuwait was divided into 70 basins with a catchment area from 1 km^2 to 10,300 km^2. Only 10 large basins were having a catchment area between 500 km^2 to 10,300 km^2 (Fig. 10.18). Others were small basins having an area under 500 km^2 and mostly these small basins were located along the coast. The flood extent and the depth were

also simulated for one of the very heavy rain events that occurred from 4th–14th November 2018 in Kuwait. Flood extent for the entire Kuwait is governed by rainfall, drainage profile, and topography of the study area. Flood modeling for Kuwait shows that the southern area of Kuwait was highly susceptible to higher floodwater depth during heavy rainfall events. The flood depths were computed at most up to and to less extent between 1 to 2 m (Fig. 10.19). In southern Kuwait near Al-Taweel station, where there is the highest rainfall, the flooding depth reached more than 3 m.

10.4.3 Urban Areas Flash Floods in 2018

In Nov 2018 flash flood events, the urban area in Kuwait witnessed intensive flooding because the quantity of rainfall is more than sewage and drainage systems are capable to drain within the urban areas. The southern urban area is mostly affected by flash floods in particular Al-Ahmadi Governorate. Figure 10.19 shows many water bodies accumulated in low terrain areas and mostly in sabkha. This governorate is located about 33 km south of Kuwait City with a total area of about 5,000 km^2 and a population of 679,527 inhabitants. Most of the communities are located in the eastern of the Ridge called AL-Ahmadi ridge. It is a ridge that runs parallels to the southern coastline and rises to heights of about 137 m above sea level and is dissected a 116 shallow wadi and tributaries of the length of approximately 69 km located on the eastern and the western side of Kuwait (Hassan et al., 2021) (Fig. 10.20). The eastern wadis are pouring into the Arabian Gulf crossing some urban areas. In case of heavy rains for a few hours, similar to the Nov 2018 flash flood, the floods start from the top of the ridge and the rainfall water continues to flow east through drainage channels directed to the urban areas (Fig. 10.21). The western wadis flow toward the Burgan oil field and nearby areas. Figure 10.22 shows satellite images of some oil facilities including pipelines and oil gathering centers that are damaged by flash flooding.

Sabah Al-Ahmad is a recently established residential area located in the State of Kuwait about 76 km away from Kuwait city. Satellite images with GIS maps show several drainages trending from the Northwest to Southeast direction crossing the city (Fig. 10.23). During the 2018 rainfall, the city was flooded and many houses were drawn due to the crossing of this drainage within the urban area (Fig. 10.24). Most of the drainage drain into the inland sabkha South of Sabah Al-Ahmad. These sabkhas caused a body of water that accumulated due to limited infiltration of rainfall and hence flooded back to the city.

Table 10.1 Daily Rainfall recorded at various stations in Kuwait. *(KMD 2020)*

S. No	Station	Rainfall (mm), November 2018											Total event rainfall (mm)
		4th	5th	6th	7th	8th	9th	10th	11th	12th	13th	14th	
1	Abdaly	17.03	0	0.54	0	0.48	1.73	0.24	7.41	0	0	9.93	37.36
2	Mitribah	13.93	0	0.23	0	10.93	0	2.88	11.37	0	0	18.46	57.8
3	Huwaimliyah	3.7	0.2	0.3	0	8.3	10.1	0.3	17	0.1	0	20.5	60.5
4	Ras Az-Subiya	22	11	39	0	1	35.5	1.5	0	0	0	41.5	151.5
5	Al-Mutla	20	1.5	8.5	0	2	4	2	0	0	0	44	82
6	Kuwait City	0	0	0	0	0	0	0	0	0	0	0	0
7	KISR	10	4.5	26	0	2.5	11.5	5	0	0	0	48.5	108
8	Jahra	*17.5*	3.3	6.3	0	3.9	18.5	3.3	0	0	0	48	100.8
9	Kuwait Airport	10.13	1.67	*40.5*	0	3.76	*49.41*	6.1	0	0	0	79	*190.57*
10	Um Omara	12	2.5	19	0.5	6.5	2	40	0.5	0	0	44.5	127.5
11	Salmy	9.65	0	6.21	0	5.57	3.47	5.72	0	0	0.76	57.57	88.95
12	Al-Taweel	10.5	2	27.5	0	4	29	6	2.5	0.5	0	*101*	183
13	Um Al-Haiman	16	3.5	13.5	0	2	9.5	42.5	12.5	0	0	66	165.5
14	Ras Az-Zoor	27	2.5	7	2.5	0.5	3	12	23.5	0	0	72	150
15	Al-Wafra	13.5	3.5	14	0.5	7.5	0	*49.5*	2	0	0	77	167.5
16	Wafra	2.09	5.5	6.14	0.29	2.99	0.61	0	24.92	0	0	29.82	72.36

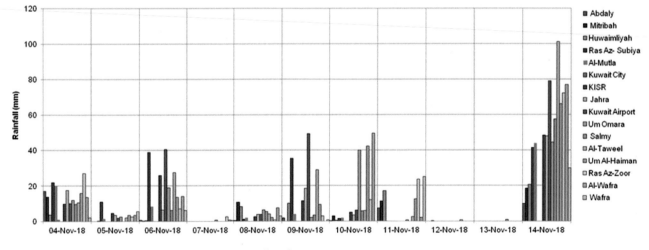

Fig. 10.13 Daily Rainfall plot for 2018 Flood event at all stations

10.5 Transboundary Dust Storm Jets from Southern Iraq to Kuwait

Kuwait is in the northwestern part of the Arabian Gulf and shares borders with Saudi Arabia and Iraq. (Al-Awadhi et al., 2014; Al-Dabbas et al., 2012; Al-Dousari et al., Ahmed, 2017) defined the major source areas of dust for Kuwait. They are from the western desert of Iraq, the Mesopotamian flood plain in Iraq, the northern desert of Saudi Arabia, drained marshes in southern Iraq, and the dry marshes and abandoned farms in Iran. Figure 10.25 depicts the regional dust sources around Kuwait and the seasonal wind patterns in and around Kuwait. Since the winds dominantly blow from north and northwesterly directions during the summer months (mostly dry conditions) starting from April to October, this leads to frequent dust storms over Kuwait during the summer months as compared to the rest of

Fig. 10.14 Spatial variation of total rainfall from 4 to 14th November 2018

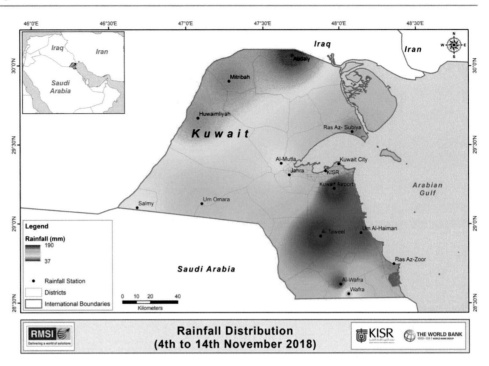

Fig. 10.15 Cloud cover extent and wind movement on 04th November 2018

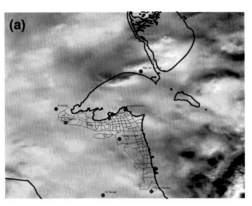

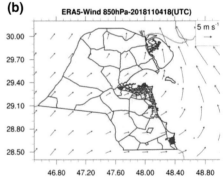

Fig. 10.16 Cloud cover extent and wind movement on 10th November 2018

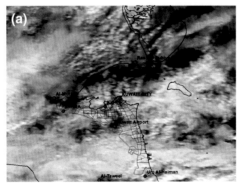

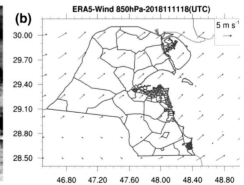

Fig. 10.17 Cloud cover extent and wind movement on 14th November 2018

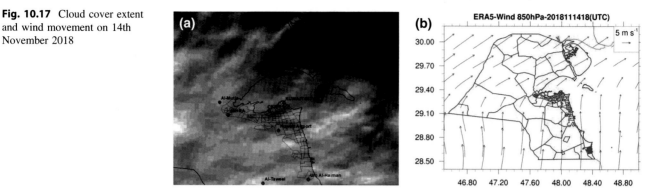

Fig. 10.18 Map showing the ten Major basin areas in Kuwait

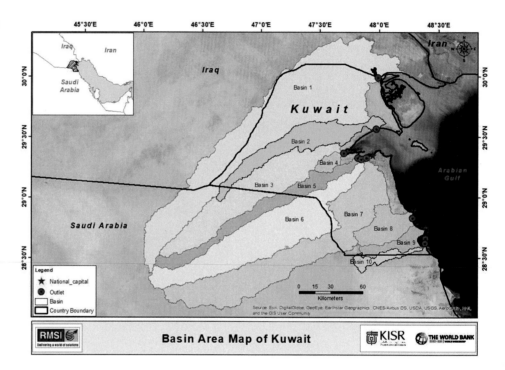

Fig. 10.19 Flood inundation map for November 2018 rainfall event

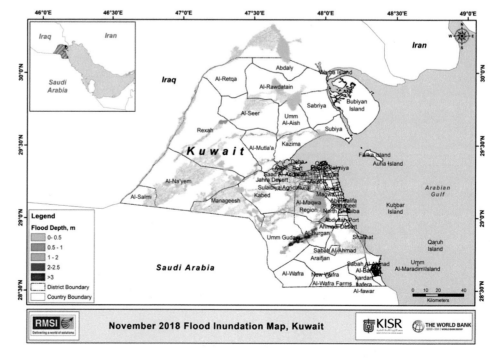

Fig. 10.20 Satellite images in southern Kuwait showing water bodies in low land areas

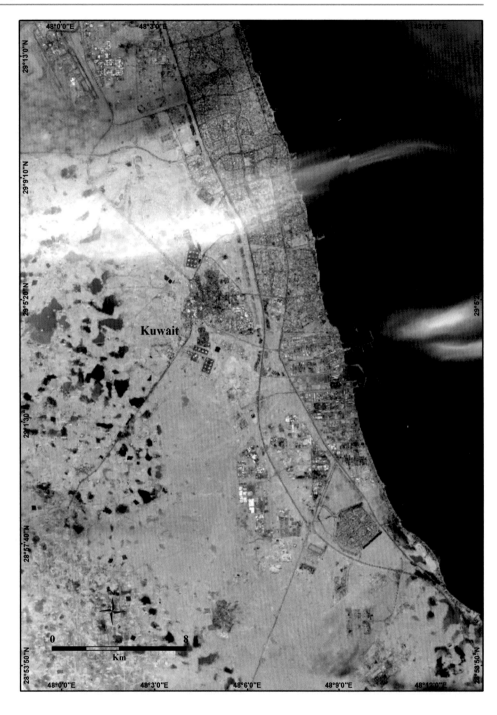

the years called Shamal winds (Yassin et al., 2018). This causes frequently transboundary dust storm events in the state of Kuwait.

Kuwait is heavily influenced by dust storms from southern Iraq, particularly from the area located 250 km from its northern border which is considered a "hot spot" (Fig. 10.26). The "hot spot" recorded the largest monthly fallen dust weights in Iraq (200-250 g/m^2/month) (Al-Dabbas et al., 2012). Remote sensing satellites have identified that intensive dust jets originate from this "hot spot" area, and it is most active in the summer months between May and July. To a lesser extent, other arid sources in the vast desert area in western Iraq also contribute to dust events; however, these dust storms do not constitute a major concern to Kuwait (Al-Awadhi et al., 2014; Al-Dousari et al., 2017). The particular "hot spot" area is located within the Mesopotamian flood plain, Samawah and Abu Jir lineaments; located between three major southern Iraqi cities (Al-Samawah, Al-Diwaniya, and Al-Nasriya); and stretches along with two provinces (Al-Muthana and Thi-Qar).

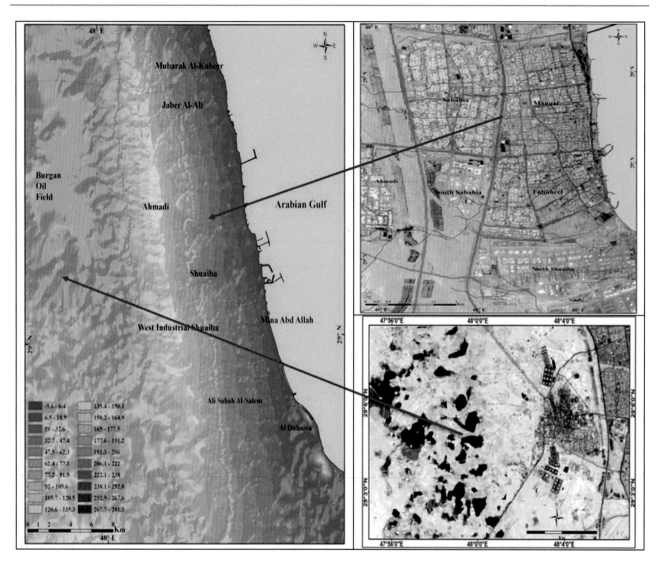

Fig. 10.21 Wadi System over Kuwait and in the eastern and western side of Al-Ahmadi Ridge south of Kuwait

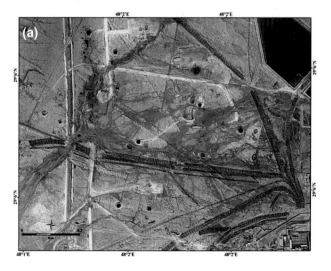
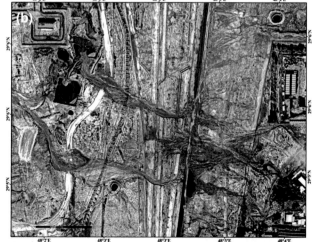

Fig. 10.22 Flooded area within Al-Ahmadi and Burgan oil field

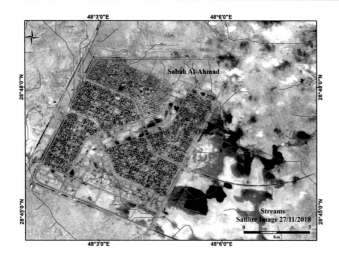

Fig. 10.23 Satellite image and wadis within Sabah Al-Ahmad Urban area on 12 November

Moderate Resolution Imaging Spectroradiometer (MODIS) images from NASA's Terra and Aqua satellites have identified that the size of the "hot spot" area has been shrinking during the past 35 years (1986–2020) (Table 10.2 and Fig. 10.27). MODIS Terra and Aqua satellite images were collected from 2008 to 2021 (14 years). Remote sensing satellite images from NASA GSFC OBPG Ocean color Level 0 from Local Area Coverage were processed and generated. Image processing included the true color based on RGB composition of Bands 1, 4, 3 and false color based on RGB compositions of Bands 2, 1, 1 and Bands 6, 5, 4. The B211 composition was mainly used for best MODIS resolution (250 m/pxl) and shows water as black color due to the use of red and NIR bands, vegetation as red color, shallow water as cyan color, and bared desert as white color. The real color composition B143 was used to obtain dust jet delineation of moderate resolution (500 m/pxl) and shows water as dark blue, dust jets as light yellow/bluish, and local sand jets as dark brownish from dry farmlands or overgrazing areas.

Local dust storm jets within Iraq and Kuwait were originally discovered by (Peter & Al-Awadi, 2010) and confirmed by (Al-Hemoud et al., 2020). Historical MODIS images showed that thick clouds of dust just originate from the "hot spot" area in southern Iraq and cross over Kuwait (Fig. 10.28). Intensive dust storm jets from the "hot spot" area can travel long distances crossing Kuwait into the northern part of the Arabian and eastern Saudi Arabia (Fig. 10.29). Oftentimes, dust jets are carried by the northwesterly winds directly into Kuwait urban areas (Fig. 10.30), and the increased frontogenesis between the hot and dry atmosphere over the desert and the moist intrusions from the Arabian can create a cyclone over the area (Francis et al., 2021).

Recently, on June 11, 2021, an intense dust storm blew from the "hot spot" area in southern Iraq and had a detrimental impact on Kuwait (Figs. 10.31 and 10.32). MODIS Terra acquired a true-color image of long, thick plumes of dust jets stretching from southern Iraq into Kuwait and the Arabian Gulf. Dust jet emissions started to rise from the "hot spot" area as strong winds whipped over the region on June 10, 2021. By June 11, 2021, the original gray-colored plume traveled over 500 km and stretched over Kuwait and the western coast of the Arabian Gulf. Two other dust jets were originated from two arid areas; the first dust jet originated from the drylands south of Basra, Iraq, while the second dust jet originated at the borders between the southwest of Iran and southeast of Iraq (east of Al-Faw). The dust jet continued the following day on June 12, 2021 (Fig. 10.33). A zoomed resolution (80 m) showed the high-density dust jet as whitish and the agriculture as reddish (Fig. 10.33a). The "hot spot" area is properly delineated during a non-dust day on May 16, 2021 (Fig. 10.33b).

Fig. 10.24 Flooding of Sabah Al-Ahmad Urban area in November 2018

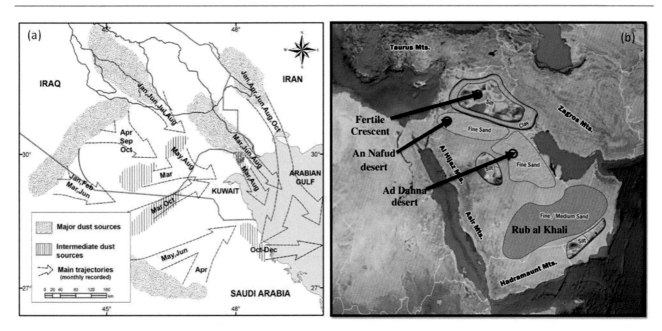

Fig. 10.25 Schematic diagram of the (**a**) Seasonal wind patterns in and around Kuwait (**b**) Sources of dust emission in and around Kuwait. *Source* Al-Dousari and Al-Awadhi (2012)

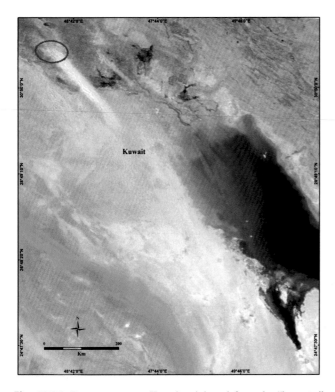

Fig. 10.26 Dust storm over Kuwait originated from the "hot spot" area, MODIS Aqua (500 m/pixel), June 05, 2008

Table 10.2 The size of the "hot spot" area in southern Iraq (1986–2020)

Year	Area (km^2)
1986	5,851
2006	3,185
2016	2,204
2020	1,769

10.6 Oil Spill Mapping

Oil is considered the most important energy source in the world. The oil spill is described as a form of pollution caused by the release of liquid petroleum hydrocarbons into the environment, especially marine areas, due to human activities (Li et al., 2016). These may be instigated by various activities, such as the release of crude oil from oil tankers, pipelines, railcars, offshore platforms, drilling rigs, and wells, as well as spills of refined petroleum products and their byproducts.

The impacts of accidental oil spills can contaminate the coastal zones and impact the high ecological quality. Spills can cause severe damage to the flora and fauna, aquaculture,

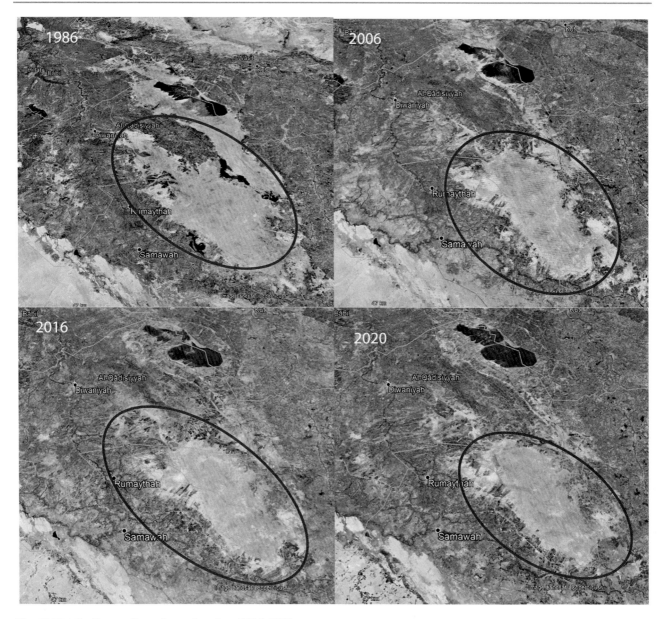

Fig. 10.27 The "hot spot" area in southern Iraq (1986–2020)

and fisheries along the coast and the marine environment. The impact of an oil spill may eventually continue for several years, or maybe decades. The human activities in the Arabian Gulf cause a combination of intentional and accidental oil spills and pollution in the marine environment. The oil spills in the Arabian Gulf can be related to war-related activities, incidents release from oil tankers, offshore platforms, offshore drilling, and spills and pipeline leakages. Due to the high volume of shipping traffic in the Arabian Gulf, oil pollution is considered the main environmental concern in the Regional Organization for Protection of the Marine Environment (ROPME) Sea Area.

Kuwait is a part of the Arabian Gulf, a geographic region formed by countries having extensive reserves of crude oil. Like other countries in the Gulf region, the Kuwaiti economy relies heavily on petroleum exports, making it more

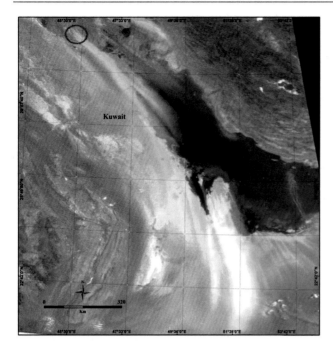

Fig. 10.28 Dust storm over Kuwait originated from the "hot spot" area, MODIS Aqua (500 m/pixel), July 05, 2012

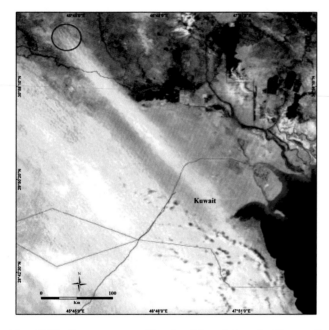

Fig. 10.29 Dust storm over Kuwait originated from the "hot spot" area, MODIS Aqua (250 m/pixel), July 28, 2018

susceptible to frequent oil spills and consequent marine pollution. The impact seems to vary a great deal from spill to spill, depending on the kind of oil, physical conditions, and the ecosystem affected. Based on available data, attempts were made to identify the location in the Kuwait marine environment where maximum events occurred along with the areal extents of oil spills. These oil spills also affect the economy of Kuwait. The oil spills during the Gulf war are estimated to have depleted about 2% of Kuwait's oil reserves. At $20 a barrel, Kuwait lost about $60 million a day in revenues, which translates to $22 billion a year. On August 10, 2017, a large oil spill spot near the Kuwaiti coast, south of the State of Kuwait, leaked into the water approximately 3 km from the coast, and it is considered the largest leak in Kuwait after the gulf war. In this part, we will present the role of the remote sensing monitoring of oil spill in Kuwait and in the particular oil spill in 2017.

This study focuses on remote sensing and GIS-based mapping of oil spill events to estimate the extent of oil spills and identify hotspots in the State of Kuwait where the frequency of oil spill events is higher. Based on available data, satellite imagery is processed for estimating the extent of an oil spill. A total of 34 images were available for offshore oil spill mapping—26 images from 2018 and 6 images from 2017. However, out of these, 29 images were analyzed, due to the limitation of images available for validation.

The details of offshore oil spill data are summarized in Tables 10.3 and 10.4. Spaceborne Synthetic Aperture Radar (SAR) has proven to be very useful in oil spill detection and monitoring. Wide coverage provided by SAR-equipped satellites such as European Envisat gives a good opportunity for the development of operational oil spill applications. SAR system relies on the detection variation of the sea surface roughness. Oil films can be detected as dark patches relative to the surrounding water. In this study, we have used Sentinel-1 imagery for the detection of deepwater horizon oil spills.

Attempts were made to categorize these oil spill events considering the distance of events from Kuwait City. Oil spill events were classified into four categories, namely those between 0–50 km, 50–100 km, 100–150 km, and those more than 150 km. The extent of each oil spill event was determined by processing satellite data available from different sources and presented in Fig. 10.34. The areal extents for these events vary in range from 3 km^2 to 750 km^2. The

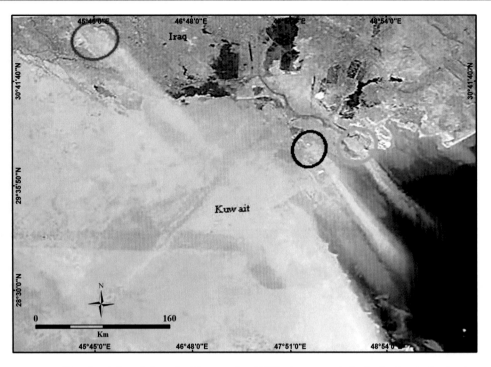

Fig. 10.30 Dust storm over Kuwait originated from the "hot spot" area, MODIS Aqua (500 m/pixel June 11, 2021. **a** red circle: "hot spot" area in southern Iraq, **b** black circle: drylands south of Basra, Iraq (Al-Faw), **c** green circle: dry land at the borders between the southwest of Iran and southeast of Iraq

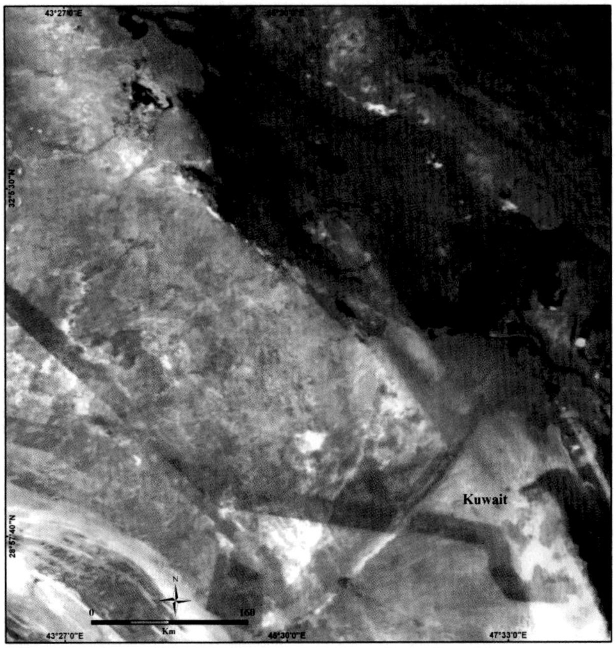

Fig. 10.31 Dust storm from the "hot spot" area, MODIS Terra (250 m/pixel), June 11, 2021

majority of these events occurred between 100–150 km from Kuwait City and represent 44% of all selected offshore oil spill events (Fig. 10.35).

Figure 10.36 shows the oil spill rose diagram of the offshore oil spill events. Most of these events occurred in the southeastern direction with reference to Kuwait City. This is due to the influence of transport ships in this direction, southern ports such as Port of Mina Az Zawr, Mina Saud, Mina Al-Ahmadi, Mina Abdallah, and Ash Shuaybah. In the last few years, the number of supertankers increased in the number of these ports. Oil patches near these waiting areas are often seen in satellite images (Fig. 10.37).

On the 10th of August, a large oil spill was reported south of Kuwait near Az-zour port and about 30 km East of Qaru island (Fig. 10.38). It is estimated that about more than 35, 000 barrels of crude oil have leaked into the marine. This spill is considered the worst for the state of Kuwait in the last decades. The oil spill polluted the coastal and the marine environment south of Kuwait and causes an economic loss by shutting down some industrial ports (Fig. 10.39). Remote sensing, in this oil spill monitoring, utilizes satellites Sentinel-1, image data. These images indicated that the oil

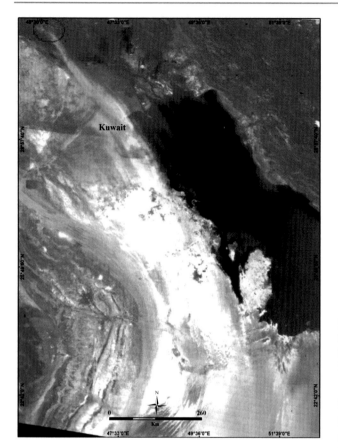

Fig. 10.32 Dust storm over Kuwait originated from the "hot spot" area, MODIS Terra (500 m/pixel), June 12, 2021

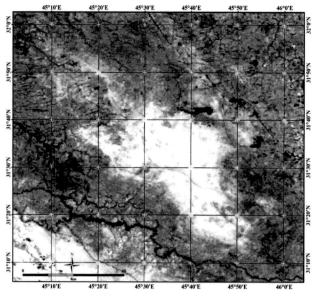

Fig. 10.33 A zoomed resolution (80 m) of the "hot spot" area: **a** during a dust jet MODIS Aqua (80 m/pxl), June 12, 2021 (Fig. 10.9a); **b** during a clear day, MODIS Aqua (250 m/pxl), May 16, 2021

spill started in the south of Kuwait near Az-zour and then spread to the North nearby Funaits area and Nord of Ahmadi on 14th of August and finally by 16 Aug 2017 Kuwait crude oil spill patches reached Kuwait Bay and started to scatter (Fig. 10.40).

Part of these oil spills is from natural causes. The southern marine region of the State of Kuwait is distinguished by its islands that contain coral reefs and are surrounded by sand sediments arising from the erosion of coral reefs, which are important ecosystems for the rest of the marine organisms that interact with them in terms of habitat and food resources (Al-Mohanna et al., 2014). They may be subject to potential pollution from the natural oil spill from

the seabed in the marine area between the islands of Qaruh and Umm Al-Maradim, which was documented in the early nineties in that area (Literathy, 1993, 2001). Near Qaru Island, natural oil seepage is more frequent and appears at the surface as a sheen. The reason for this rise is due to the hydraulic transfer of petroleum hydrocarbons in the water column as soon as they leak from the bottom. This creates background contamination level, which affects marine life in the area. Figure 10.41 shows SAR satellite images (SENTINEL) of some large natural oil seepage around Qaru Island.

10.7 Summary

This chapter comprises the application of remote sensing in various fields such as soil moisture variation study, land subsidence across oil fields, flash floods, dust storms, and oil spills across the State of Kuwait using various satellite images.

Table 10.3 Data availability for offshore oil spill events from KISR

S. no	Event date	Offshore location	Satellite image
1	05-Aug-17	Qaruh Island	Sentinel-1
2	08-Aug-17	Qaruh Island	Sentinel-1
3	10-Aug-17	Al Basra Terminal	Sentinel-1
4	14-Aug-17	Bubian	Sentinel-1
5	22-Aug-17	Bubian	Sentinel-1
6	29-Aug-17	Qaruh Island	Sentinel-1
7	22-Sep-17	Bubian	Sentinel-1
8	03-Mar-18	Qaruh Island	Sentinel-1
9	14-Mar-18	Kafji	Sentinel-1
10	14-Mar-18	Failaka	Sentinel-1
11	22-Mar-18	Kafji	Not Available
12	18-Apr-18	Eastern Boarders	Sentinel-1
13	20-Apr-18	Qaruh Island	Sentinel-1 (23-April-2018)
14	05-May-18	Kafji	Sentinel-1
15	13-May-18	Failaka	Sentinel-1
16	12-Jun-18	Eastern Boarders	Sentinel-1
17	30-Jun-18	Ahmadi	Sentinel-1
18	02-Jul-18	Qaruh Island	Sentinel-1 (03-July-2018)
19	03-Jul-18	Qaruh Island	Sentinel-1
20	04-Jul-18	Failaka	Sentinel-1
21	07-Jul-18	Bubian	Sentinel-1
22	14-Jul-18	Qaruh Island	Sentinel-1 (16-July-2018)
23	23-Jul-18	Qaruh Island	Sentinel-1(24-July-2018)
24	24-Jul-18	Qaruh Island	Sentinel-1
25	01-Aug-18	Qaruh Island	Not Available
26	26-Aug-18	Al Basra Terminal	Not Available
27	28-Aug-18	Qaruh Island	Sentinel-1
28	02-Sep-18	Al Basra Terminal	Sentinel-1
29	04-Sep-18	Qaruh Island	Sentinel-1
30	06-Sep-18	Al Basra Eastern Border	Not Available
31	11-Sep-18	Qaruh Island	Sentinel-1 (14-Sept-2018)
32	07-Dec-18	Qaruh Island	Sentinel-1
33	15-Dec-18	Bubian	Sentinel-1

Table 10.4 Offshore oil spill extent area

S. No	Event date	Location	Estimated offshore Oil Spill Area (km²)				
			Distance from Kuwait City as a reference point				
			0–50 km	50–100 km	100–150 km	>150 km	Total area (km²)
Oil Spill Events provided by KISR/EPA Offshore							
1	05-Aug-17	Qaruh Island	0	18.01	248.56	109.95	376.52
2	08-Aug-17	Qaruh Island	NA	NA	NA	NA	NA
3	10-Aug-17	Al Basra Terminal	0	0	150.71	0	150.71
4	14-Aug-17	Bubian	64.73	51.57	112.45	9.5	238.25
5	22-Aug-17	Bubian	0	0	29.45	0	29.45
6	29-Aug-17	Qaruh Island	0	0	119.71	0	119.71
7	22-Sep-17	Bubian	0	0	379.86	0	379.86
8	03-Mar-18	Qaruh Island	0	0	0	25.23	25.23
9	14-Mar-18	Kafji	0	0	159.86	10.37	170.23
10	14-Mar-18	Failaka	0	8.37	84.53	45.86	138.76
11	22-Mar-18	Kafji	NA	NA	NA	NA	NA
12	18-Apr-18	Eastern Borders	0	0	0	745.09	745.09
13	20-Apr-18	Qaruh Island	0	0	0	0	0
14	05-May-18	Kafji	0	0	127.89	378.08	505.97
15	13-May-18	Failaka	110.72	0	989.28	0	1100
16	12-Jun-18	Eastern Borders	0	0	10.96	45	55.96
17	30-Jun-18	Ahmadi	0	10.45	30.56	2.95	43.96
18	02-Jul-18	Qaruh Island	NA	NA	NA	NA	NA
19	03-Jul-18	Qaruh Island	NA	NA	NA	NA	NA
20	04-Jul-18	Failaka	0	53.3	107	0	160.3
21	07-Jul-18	Bubian	0	0	299.83	11.86	311.69
22	14-Jul-18	Qaruh Island	3.86	0	156.56	8.58	169
23	23-Jul-18	Qaruh Island	218.78	0	198.78	0	417.56
24	24-Jul-18	Qaruh Island	218.78	0	198.78	0	417.56
25	01-Aug-18	Qaruh Island	NA	NA	NA	NA	NA
26	26-Aug-18	Al Basra Terminal	NA	NA	NA	NA	NA
27	28-Aug-18	Qaruh Island	0	0	226.44	108.56	335
28	02-Sep-18	Al Basra Terminal	0	0	72.89	12.01	84.9
29	04-Sep-18	Qaruh Island	NA	NA	NA	NA	NA
30	06-Sep-18	Al Basra Eastern Border	0	0	224.21	208.35	432.56
31	11-Sep-18	Qaruh Island	0	0	44.17	22.29	66.46
32	07-Dec-18	Qaruh Island	NA	NA	NA	NA	NA
33	15-Dec-18	Bubian	0	3.01	0	0	3.01

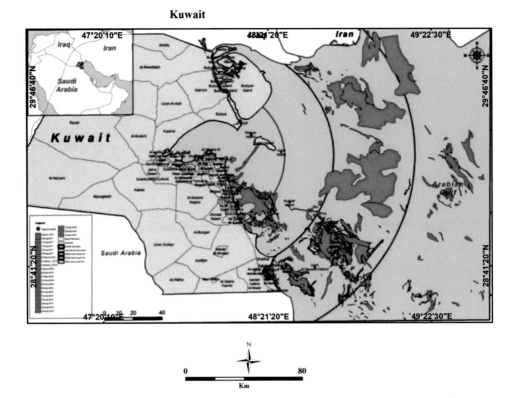

Fig. 10.34 Map showing offshore oil spills in the Arabian Gulf nearby the state of Kuwait

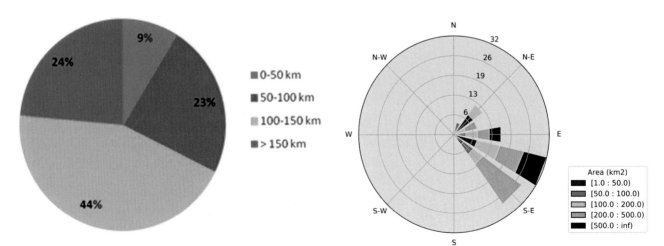

Fig. 10.35 Distribution of Offshore Oil Spill Events in percentage with distance from Kuwait City

Fig. 10.36 Oil spill rose diagram of offshore oil spills in Kuwait

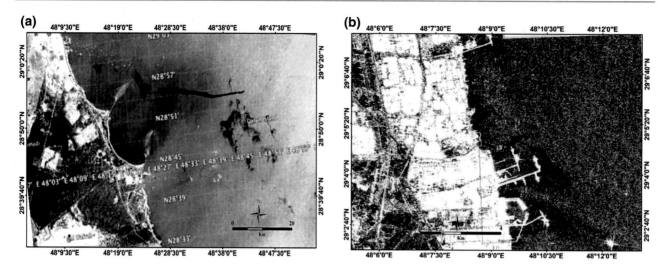

Fig. 10.37 Oil patches near Ahmadi port south of Kuwait, 21 November 2018. SAR—C, S1

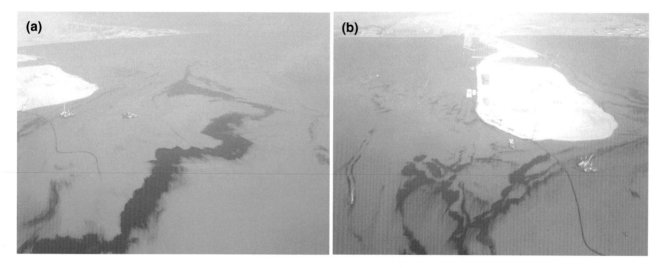

Fig. 10.38 Aerial view of the oil spill nearby Az-zour port south of Kuwait in August 2017

Fig. 10.39 Impact of oil spill south of Kuwait causes pollution to the coastal and marine environment

Fig. 10.40 SAR-C Sentinel-1 ESA images present the movement of the Oil spill, on 10th (**a**), 14th (**b**), and 16th (**c**) Aug 2017

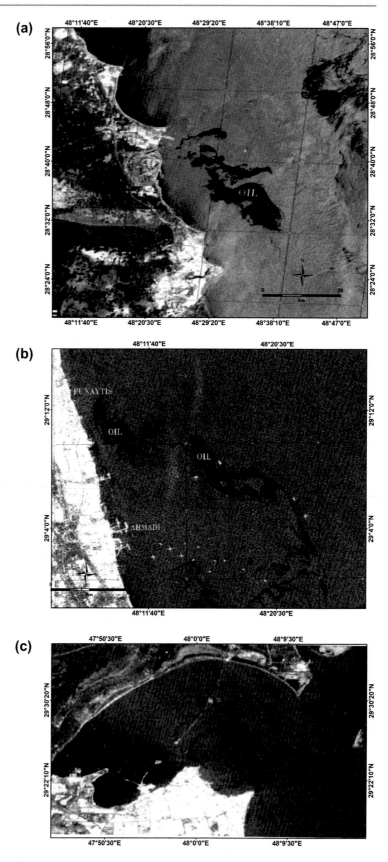

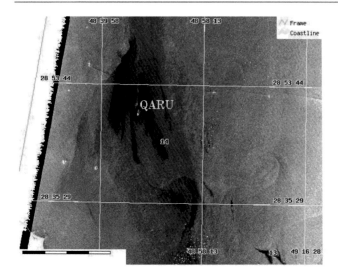

Fig. 10.41 Kuwait Qaru Island, large natural seepage continue, 17 Aug 2017, length 60 km with 24 km, Area 330 km^2

References

Al-Awadhi, J. M., Omar, S. A., & Misak, R. F. (2005). Land degradation indicators in Kuwait. *Land Degradation & Development, 16*(2), 163–176.

Al-Awadhi, J. M., Al-Dousari, A. M., & Khalaf, F. I. (2014). Influence of land degradation on the local rate of dust fallout in Kuwait. *Atmospheric and Climate Sciences.*

Al-Dabbas, M. A., Abbas, M. A., & Al-Khafaji, R. M. (2012). Dust storms loads analyses—Iraq. *Arabian Journal of Geosciences, 5*(1), 121–131.

Al-Dousari, A., Doronzo, D., & Ahmed, M. (2017). Types, indications and impact evaluation of sand and dust storms trajectories in the Arabian Gulf. *Sustainability, 9*(9), 1526.

Al-Hemoud, A., Al-Dousari, A., Al-Dashti, H., Petrov, P., Al-Saleh, A., Al-Khafaji, S., ... & Koutrakis, P. (2020). Sand and dust storm trajectories from Iraq Mesopotamian flood plain to Kuwait. *Science of The Total Environment, 710*, 136291.

Al Jassar, H. K., Rao, K. S., & Sabbah, I. (2006). A model for the retrieval and monitoring of soil moisture over desert area of Kuwait. *International Journal of Remote Sensing, 27*(2), 329–348.

Al-Jassar, H. K., & Rao, K. S. (2010). Monitoring of soil moisture over the Kuwait desert using remote sensing techniques. *International Journal of Remote Sensing, 31*(16), 4373–4385.

Al Jassar, H. K., & Rao, K. S. (2015). Assessment of soil moisture through field measurements and AMSR-E remote sensing data analysis over Kuwait desert. *Kuwait Journal of Science, 42*(2).

AlJassar, H. K., Temimi, M., Entekhabi, D., Petrov, P., AlSarraf, H., Kokkalis, P., & Roshni, N. (2019). Forward simulation of multi-frequency microwave brightness temperature over desert soils in Kuwait and comparison with satellite observations. *Remote Sensing, 11*(14), 1647.

Al-Mohanna, S. Y., Al-Zaidan, A. S., & George, P. (2014). Green turtles (Chelonia mydas) of the north-western Arabian Gulf, Kuwait: The need for conservation. *Aquatic Conservation: Marine and Freshwater Ecosystems, 24*(2), 166–178.

Bou-Rabee, F. I. R. Y. A. L. (1994). Earthquake hazard in Kuwait. *Journal of the University of Kuwait(science). Kuwait, 21*(2), 253–262.

Colliander, A., Jackson, T. J., Bindlish, R., Chan, S., Das, N., Kim, S. B., ... & Yueh, S. (2017). Validation of SMAP surface soil moisture products with core validation sites. *Remote Sensing of Environment, 191*, 215–231.

D'Odorico, P., Caylor, K., Okin, G. S., & Scanlon, T. M. (2007). On soil moisture–vegetation feedbacks and their possible effects on the dynamics of dryland ecosystems. *Journal of Geophysical Research: Biogeosciences, 112*(G4).

Eguchi, R. T., Huyck, C. K., Ghosh, S., & Adams, B. J. (2008). The application of remote sensing technologies for disaster management. In *The 14th World Conference on Earthquake Engineering*, Vol. 17.

Francis, D., Chaboureau, J. P., Nelli, N., Cuesta, J., Alshamsi, N., Temimi, M., ... & Xue, L. (2021). Summertime dust storms over the Arabian Peninsula and impacts on radiation, circulation, cloud development and rain. *Atmospheric Research, 250*, 105364.

Girin, M., & Carpenter, A. (2017). Shipping and oil transportation in the Mediterranean Sea. In *Oil Pollution in the Mediterranean Sea: Part I* (pp. 33–51). Springer, Cham.

Hassan, A., Albanai, J. A., & Goudie, A. (2021). *Modeling and managing flash flood Hazards in the State of Kuwait.*

KMD, Kuwait Metrological Department (2020). https://www.met.gov.kw/Climate/climate_hist.php?lang=arb. Accessed November 2018.

Literathy, P. (1993). Considerations for the assessment of environmental consequences of the 1991 Gulf War. *Marine Pollution Bulletin, 27*, 349–356.

Literathy, P. (2001). Polar and non-polar aromatic micropollutants in water (drinking-water) resources. *Water Science and Technology: Water Supply, 1*(4), 149–157.

Li, P., Cai, Q., Lin, W., Chen, B., & Zhang, B. (2016). Offshore oil spill response practices and emerging challenges. *Marine Pollution Bulletin, 110*(1), 6–27.

Manche, C. J. (2014). *A Remote Sensing-based Early Warning System for Algal Blooms in Kuwait Bay and Coastal Waters* (Doctoral dissertation, Western Michigan University).

Misak, Hamdy, & Al-Dousari, A. M. (2012). Managing natural hazards in HIMA the case of Kuwait, International Workshop: Towards an Implementation Strategy for the HIMA Governance Systems, KISR, Kuwait.

Misak, R. F., Khalaf, F. I., & Omar, S. A. (2013). Managing the hazards of drought and shifting sands in dry lands: The case study of Kuwait. In *Developments in Soil Classification, Land Use Planning and Policy Implications* (pp. 703–729). Springer, Dordrecht.

Misak, & Raafat. (2015). GIS Representation of Flash Floods and GIS Representation of Sea Level Rise, Crisis Decision Support Program, Environment and Life Sciences Research Center, KISR, Kuwait.

Peter & Al-Awadi. (2010). *Early warning system for environmental monitoring of ROPME Region, 61st International Astronautical Congress 2010.* Curran Associates Inc.

Rao, K. S., & Al-Jassar, H. K. (2010). Error analysis in the digital elevation model of Kuwait desert derived from repeat pass synthetic aperture radar interferometry. *Journal of Applied Remote Sensing, 4* (1), 043546.

Rao, K. S., Al-Jassar, H. K., Abdullah, F. H., Al-Kanderi, J., Al-Saeed, M., & Al-Kandari, A. (2011). Least square approach for time series analysis of land subsidence over Greater Burgan Oil Field, Kuwait, using Synthetic Aperture Radar Interferometry. *Kuwait J. Sci. Eng, 38*(1A), 141–171.

Thomas, D. S., & Kemec, S. (2007). The role of geographic information systems/remote sensing in disaster management. In *Handbook of disaster research* (pp. 83–96). Springer, New York, NY.

Yassin, M. F., Almutairi, S. K., & Al-Hemoud, A. (2018). Dust storms backward Trajectories' and source identification over Kuwait. *Atmospheric Research, 212*, 158–171.